CLYMER®
MANUALS

KAWASAKI
KZ, Z & ZX750 • 1980-1985

WHAT'S IN YOUR TOOLBOX?

More information available at Clymer.com
Phone: 805-498-6703

Haynes Publishing Group
Sparkford Nr Yeovil
Somerset BA22 7JJ England

Haynes North America, Inc
859 Lawrence Drive
Newbury Park
California 91320 USA

ISBN-10: 0-89287-356-6
ISBN-13: 978-0-89287-356-2

M450, 8S1, 14-352

ABCDEFGHIJKLMNOPQR

Common spark plug conditions

NORMAL

Symptoms: Brown to grayish-tan color and slight electrode wear. Correct heat range for engine and operating conditions.

Recommendation: When new spark plugs are installed, replace with plugs of the same heat range.

WORN

Symptoms: Rounded electrodes with a small amount of deposits on the firing end. Normal color. Causes hard starting in damp or cold weather and poor fuel economy.

Recommendation: Plugs have been left in the engine too long. Replace with new plugs of the same heat range. Follow the recommended maintenance schedule.

CARBON DEPOSITS

Symptoms: Dry sooty deposits indicate a rich mixture or weak ignition. Causes misfiring, hard starting and hesitation.

Recommendation: Make sure the plug has the correct heat range. Check for a clogged air filter or problem in the fuel system or engine management system. Also check for ignition system problems.

ASH DEPOSITS

Symptoms: Light brown deposits encrusted on the side or center electrodes or both. Derived from oil and/or fuel additives. Excessive amounts may mask the spark, causing misfiring and hesitation during acceleration.

Recommendation: If excessive deposits accumulate over a short time or low mileage, install new valve guide seals to prevent seepage of oil into the combustion chambers. Also try changing gasoline brands.

OIL DEPOSITS

Symptoms: Oily coating caused by poor oil control. Oil is leaking past worn valve guides or piston rings into the combustion chamber. Causes hard starting, misfiring and hesitation.

Recommendation: Correct the mechanical condition with necessary repairs and install new plugs.

GAP BRIDGING

Symptoms: Combustion deposits lodge between the electrodes. Heavy deposits accumulate and bridge the electrode gap. The plug ceases to fire, resulting in a dead cylinder.

Recommendation: Locate the faulty plug and remove the deposits from between the electrodes.

TOO HOT

Symptoms: Blistered, white insulator, eroded electrode and absence of deposits. Results in shortened plug life.

Recommendation: Check for the correct plug heat range, over-advanced ignition timing, lean fuel mixture, intake manifold vacuum leaks, sticking valves and insufficient engine cooling.

PREIGNITION

Symptoms: Melted electrodes. Insulators are white, but may be dirty due to misfiring or flying debris in the combustion chamber. Can lead to engine damage.

Recommendation: Check for the correct plug heat range, over-advanced ignition timing, lean fuel mixture, insufficient engine cooling and lack of lubrication.

HIGH SPEED GLAZING

Symptoms: Insulator has yellowish, glazed appearance. Indicates that combustion chamber temperatures have risen suddenly during hard acceleration. Normal deposits melt to form a conductive coating. Causes misfiring at high speeds.

Recommendation: Install new plugs. Consider using a colder plug if driving habits warrant.

DETONATION

Symptoms: Insulators may be cracked or chipped. Improper gap setting techniques can also result in a fractured insulator tip. Can lead to piston damage.

Recommendation: Make sure the fuel anti-knock values meet engine requirements. Use care when setting the gaps on new plugs. Avoid lugging the engine.

MECHANICAL DAMAGE

Symptoms: May be caused by a foreign object in the combustion chamber or the piston striking an incorrect reach (too long) plug. Causes a dead cylinder and could result in piston damage.

Recommendation: Repair the mechanical damage. Remove the foreign object from the engine and/or install the correct reach plug.

CONTENTS

QUICK REFERENCE DATA

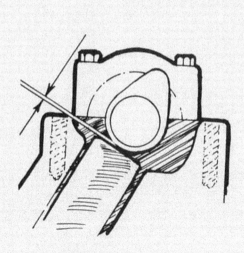

Non-turbo:
0.028-0.032 in (0.7-0.8 mm)
Turbo:
0.019-0.023 in. (0.5-0.6 mm)

VALVE CLEARANCE

SPARK PLUG GAP

FRONT FORK SPECIFICATIONS*

1982 KZ/750H		
Dry capacity	304-312 cc	10.3-10.5.5 oz.
Wet capacity	290 cc	9.8 oz.
All other 1982 models		
Dry capacity	251-259 cc	8.5-8.8 oz.
Wet capacity	240 cc	8.2 oz.
1983-on		
KZ750H	308-316 cc	10.4-10.7 oz.
KZ/Z750L	293-301 cc	9.9-10.2 oz.
ZX750A and L	245-253 cc	8.3-8.6 oz.
ZX750E	267-275 cc	9.0-9.3 oz.
Oil level (fork tube extended without spring)		
1982 KZ/Z750H	434-438 mm	17.1-17.2 in.
All other 1982 models	380-384 mm	15.0-15.1 in.
1983 KZ750H	436-440 mm	17.2-17.3 in
Oil level (fork tube compressed without spring)		
1983-on KZ/750L	101-105 mm	4.0-4.1 in.
1983-on ZX750A	183-187 mm	7.2-7.4 in.
1984-on ZX750E	174-178 mm	6.8-7.0 in.
Air pressure		
1982 KZ750R		
Standard	0.7 kg/cm²	10 psi
Range	0.6-0.9 kg/cm²	8.5-13 psi
1983 KZ750H	0.5-1.0 kg/cm²	7-14 psi
1983-on KZ/750L	0.6-0.9 kg/cm²	8.5-13 psi
1983-on ZX750A	0.4-0.6 kg/cm²	5.7-8.5 psi
1984-on ZX750E	0.4-0.6 kg/cm²	5.7-8.5 psi

* Fork oil capacities will vary slightly from model to model. Always measure fork oil level to obtain the most accurate oil capacity in each fork tube.

TUNE-UP SPECIFICATIONS

Spark plug gap	
Non-turbo	0.028-0.032 in. (0.7-0.8 mm)
Turbo	0.019-0.023 in. (0.5-0.6 mm)
Spark plug type	
Normal conditions	
U.S. models	NGK B8ES, ND WZ4ES-U
Non-U.S. models	NGK BR8ES, ND WZ4ESR-U
Turbo models	NGK BR9EV
Cold weather[1]	
U.S. models	NGK B7ES, ND WZZES-U
Non-U.S. models	NGKBR7ES
Turbo models	NGK BR9EV
Valve clearance (cold)	
Non-turbo models	
Intake and exhaust	
1984-on ZX750A	0.005-0.009 in. (0.13-0.23 mm)
All other models	0.003-0.007 in. (0.08-0.18 mm)
Turbo models	
Intake	0.005-0.009 in. (0.13-0.23 mm)
Exhaust	0.003-0.007 in. (0.08-0.18 mm)
Idle speed	
Non-turbo models	1,000-1,100 rpm
Turbo models	
U.S. models	1,150-1,250 rpm
Non-U.S. models	1,000-1,100 rpm
Idle mixture (non-U.S. models)[2]	2 turns out from seated, then adjust for highest idle speed

1. Below 50° F (10° C) or low-speed riding.
2. Does not apply to turbocharged models.

CHASSIS ADJUSTMENTS

Clutch cable play	About 1 1/8 in. (2-3 mm)
Drive chain play (on centerstand)	
1983-on ZX750A; 1984-on ZX750E	1 3/8-1 5/8 in. (35-40 mm)
All other models	1 in. (25 mm)
Throttle cable play	About 1/8 in. (2-3 mm)

LUBRICANTS AND FUEL

Engine oil	SAE 10W40, 10W50, 20W40, 20W50, rated ''SE''
Front fork oil	SAE 10W
Fuel	87 pump octane
	91 research octane

CLYMER®

KAWASAKI
KZ, Z & ZX750 • 1980-1985

INTRODUCTION

This detailed, comprehensive manual covers Kawasaki KZ750 series motorcycles. The expert text gives complete information on maintenance, repair and overhaul. Hundreds of photos and drawings guide you through every step. The book includes all you need to know to keep your Kawasaki running right.

General information on all models and specific information on 1980-1981 models appears in Chapters One-Ten. The Supplement at the end of the book contains information on 1982 and later models that differs from earlier models.

Where repairs are practical for the owner/mechanic, complete procedures are given. Equally important, difficult jobs are pointed out. Such operations are usually more economically performed by a dealer or independent garage.

A shop manual is a reference. You want to be able to find information fast. As in all Clymer books, this one is designed with this in mind. All chapters are thumb tabbed. Important items are indexed at the rear of the book. Finally, all the most frequently used specifications and capacities are summarized on the *Quick Reference* pages at the front of the book.

Keep the book handy. Carry it in your glove box. It will help you to better understand your Kawasaki, lower repair and maintenance costs, and generally improve your satisfaction with your bike.

CHAPTER ONE

GENERAL INFORMATION

The troubleshooting, maintenance, tune-up, and step-by-step repair procedures in this book are written specifically for the owner and home mechanic. The text is accompanied by helpful photos and diagrams to make the job as clear and correct as possible.

Troubleshooting, maintenance, tune-up, and repair are not difficult if you know what to do and what tools and equipment to use. Anyone of average intelligence, with some mechanical ability, and not afraid to get their hands dirty can perform most of the procedures in this book.

In some cases, a repair job may require tools or skills not reasonably expected of the home mechanic. These procedures are noted in each chapter and it is recommended that you take the job to your dealer, a competent mechanic, or a machine shop.

MANUAL ORGANIZATION

This chapter provides general information, safety and service hints. Also included are lists of recommended shop and emergency tools as well as a brief description of troubleshooting and tune-up equipment.

Chapter Two provides methods and suggestions for quick and accurate diagnosis and repair of problems. Troubleshooting procedures discuss typical symptoms and logical methods to pinpoint the trouble.

Chapter Three explains all periodic lubrication and routine maintenance necessary to keep your motorcycle running well. Chapter Three also includes recommended tune-up procedures, eliminating the need to constantly consult chapters on the various subassemblies.

Subsequent chapters cover specific systems such as the engine, transmission, and electrical system. Each of these chapters provides disassembly, inspection, repair, and assembly procedures in a simple step-by-step format. If a repair is impractical for the home mechanic it is indicated. In these cases it is usually faster and less expensive to have the repairs made by a dealer or competent repair shop. Essential specifications are included in the appropriate chapters.

When special tools are required to perform a task included in this manual, the tools are illustrated. It may be possible to borrow or rent these tools. The inventive mechanic may also be able to find a suitable substitute in his tool box, or to fabricate one.

The terms NOTE, CAUTION, and WARNING have specific meanings in this manual. A NOTE provides additional or explanatory information. A

CAUTION is used to emphasize areas where equipment damage could result if proper precautions are not taken. A WARNING is used to stress those areas where personal injury or death could result from negligence, in addition to possible mechanical damage.

SERVICE HINTS

Time, effort, and frustration will be saved and possible injury will be prevented if you observe the following practices.

Most of the service procedures covered are straightforward and can be performed by anyone reasonably handy with tools. It is suggested, however, that you consider your own capabilities carefully before attempting any operation involving major disassembly of the engine.

Some operations, for example, require the use of a press. It would be wiser to have these performed by a shop equipped for such work, rather than to try to do the job yourself with makeshift equipment. Other procedures require precision measurements. Unless you have the skills and equipment required, it would be better to have a qualified repair shop make the measurements for you.

Repairs go much faster and easier if the parts that will be worked on are clean before you begin. There are special cleaners for washing the engine and related parts. Brush or spray on the cleaning solution, let stand, then rinse it away with a garden hose. Clean all oily or greasy parts with cleaning solvent as you remove them.

WARNING
Never use gasoline as a cleaning agent. It presents an extreme fire hazard. Be sure to work in a well-ventilated area when using cleaning solvent. Keep a fire extinguisher, rated for gasoline fires, handy in any case.

Much of the labor charge for repairs made by dealers is for the removal and disassembly of other parts to reach the defective unit. It is frequently possible to perform the preliminary operations yourself and then take the defective unit in to the dealer for repair, at considerable savings.

Once you have decided to tackle the job yourself, make sure you locate the appropriate section in this manual, and read it entirely. Study the illustrations and text until you have a good idea of what is involved in completing the job satisfactorily. If special tools are required, make arrangements to get them before you start. Also, purchase any known defective parts prior to starting on the procedure. It is frustrating and time-consuming to get partially into a job and then be unable to complete it.

Simple wiring checks can be easily made at home, but knowledge of electronics is almost a necessity for performing tests with complicated electronic testing gear.

During disassembly of parts keep a few general cautions in mind. Force is rarely needed to get things apart. If parts are a tight fit, like a bearing in a case, there is usually a tool designed to separate them. Never use a screwdriver to pry apart parts with machined surfaces such as cylinder head or crankcase halves. You will mar the surfaces and end up with leaks.

Make diagrams wherever similar-appearing parts are found. You may think you can remember where everything came from — but mistakes are costly. There is also the possibility you may get sidetracked and not return to work for days or even weeks — in which interval, carefully laid out parts may have become disturbed.

Tag all similar internal parts for location, and mark all mating parts for position. Record number and thickness of any shims as they are removed. Small parts such as bolts can be identified by placing them in plastic sandwich bags that are sealed and labeled with masking tape.

Wiring should be tagged with masking tape and marked as each wire is removed. Again, do not rely on memory alone.

Disconnect battery ground cable before working near electrical connections and before disconnecting wires. Never run the engine with the battery disconnected; the alternator could be seriously damaged.

Protect finished surfaces from physical damage or corrosion. Keep gasoline and brake fluid off painted surfaces.

Frozen or very tight bolts and screws can often be loosened by soaking with penetrating oil like Liquid Wrench or WD-40, then sharply striking the bolt head a few times with a hammer and punch (or screwdriver for screws). Avoid heat unless absolutely necessary, since it may melt, warp, or remove the temper from many parts.

Avoid flames or sparks when working near a charging battery or flammable liquids, such as gasoline.

No parts, except those assembled with a press fit, require unusual force during assembly. If a part is hard to remove or install, find out why before proceeding.

Cover all openings after removing parts to keep dirt, small tools, etc., from falling in.

When assembling two parts, start all fasteners, then tighten evenly.

Wiring connections and brake shoes, drums, pads, and discs and contact surfaces in dry clutches should be kept clean and free of grease and oil.

When assembling parts, be sure all shims and washers are replaced exactly as they came out.

Whenever a rotating part butts against a stationary part, look for a shim or washer. Use new gaskets if there is any doubt about the condition of old ones. Generally, you should apply gasket cement to one mating surface only, so the parts may be easily disassembled in the future. A thin coat of oil on gaskets helps them seal effectively.

Heavy grease can be used to hold small parts in place if they tend to fall out during assembly. However, keep grease and oil away from electrical, clutch, and brake components.

High spots may be sanded off a piston with sandpaper, but emery cloth and oil do a much more professional job.

Carburetors are best cleaned by disassembling them and soaking the parts in a commercial carburetor cleaner. Never soak gaskets and rubber parts in these cleaners. Never use wire to clean out jets and air passages; they are easily damaged. Use compressed air to blow out the carburetor, but only if the float has been removed first.

Take your time and do the job right. Do not forget that a newly rebuilt engine must be broken in the same as a new one. Refer to your owner's manual for the proper break-in procedures.

SAFETY FIRST

Professional mechanics can work for years and never sustain a serious injury. If you observe a few rules of common sense and safety, you can enjoy many safe hours servicing your motorcycle. You could hurt yourself or damage the motorcycle if you ignore these rules.

1. Never use gasoline as a cleaning solvent.

2. Never smoke or use a torch in the vicinity of flammable liquids such as cleaning solvent in open containers.

3. Never smoke or use a torch in an area where batteries are being charged. Highly explosive hydrogen gas is formed during the charging process.

4. Use the proper sized wrenches to avoid damage to nuts and injury to yourself.

5. When loosening a tight or stuck nut, be guided by what would happen if the wrench should slip. Protect yourself accordingly.

6. Keep your work area clean and uncluttered.

7. Wear safety goggles during all operations involving drilling, grinding, or use of a cold chisel.

8. Never use worn tools.

9. Keep a fire extinguisher handy and be sure it is rated for gasoline (Class B) and electrical (Class C) fires.

EXPENDABLE SUPPLIES

Certain expendable supplies are necessary. These include grease, oil, gasket cement, wiping rags, cleaning solvent, and distilled water. Also, special locking compounds, silicone lubricants, and engine and carburetor cleaners may be useful. Cleaning solvent is available at most service stations and distilled water for the battery is available at supermarkets.

SHOP TOOLS

For complete servicing and repair you will need an assortment of ordinary hand tools (**Figure 1**).

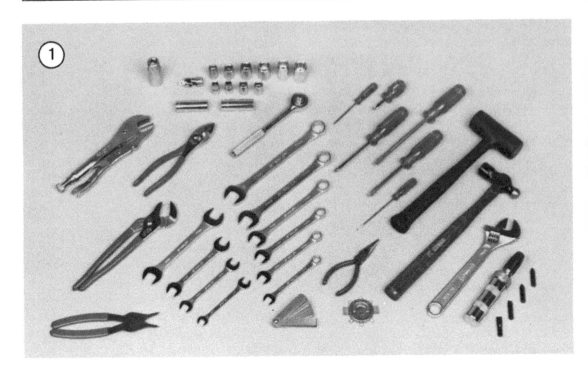

As a minimum, these include:

a. Combination wrenches
b. Sockets
c. Plastic mallet
d. Small hammer
e. Impact driver
f. Snap ring pliers
g. Gas pliers
h. Phillips screwdrivers
i. Slot (common) screwdrivers
j. Feeler gauges
k. Spark plug gauge
l. Spark plug wrench

Special tools required are shown in the chapters covering the particular repair in which they are used.

Engine tune-up and troubleshooting procedures require other special tools and equipment. These are described in detail in the following sections.

EMERGENCY TOOL KITS

Highway

A small emergency tool kit kept on the bike is handy for road emergencies which otherwise could leave you stranded. The tools and spares listed below and shown in **Figure 2** will let you handle most roadside repairs.

a. Motorcycle tool kit (original equipment)
b. Impact driver
c. Silver waterproof sealing tape (duct tape)
d. Hose-clamps (3 sizes)
e. Silicone sealer
f. Lock 'N' Seal
g. Flashlight
h. Tire patch kit
i. Tire irons
j. Plastic pint bottle (for oil)
k. Waterless hand cleaner
l. Rags for clean up

Off-Road

A few simple tools and aids carried on the motorcycle can mean the difference between walking or riding back to camp or to where repairs can be made. See **Figure 3**.

A few essential spare parts carried in your truck or van can prevent a day or weekend of trail riding from being spoiled. See **Figure 4**.

On the Motorcycle

a. Motorcycle tool kit (original equipment)
b. Drive chain master link
c. Tow line
d. Spark plug
e. Spark plug wrench
f. Shifter lever
g. Clutch/brake lever
h. Silver waterproof sealing tape (duct tape)
i. Loctite Lock 'N' Seal

In the Truck

a. Control cables (throttle, clutch, brake)
b. Silicone sealer
c. Tire patch kit
d. Tire irons
e. Tire pump
f. Impact driver
g. Oil

WARNING
*Tools and spares should be carried on
the motorcycle — not in clothing where
a simple fall could result in serious injury from a sharp tool.*

TROUBLESHOOTING AND TUNE-UP EQUIPMENT

Voltmeter, Ohmmeter, and Ammeter

For testing the ignition or electrical system, a good voltmeter is required. For motorcycle use, an instrument covering 0-20 volts is satisfactory. One which also has a 0-2 volt scale is necessary for testing relays, points, or individual contacts where voltage drops are much smaller. Accuracy should be ± ½ volt.

An ohmmeter measures electrical resistance. This instrument is useful for checking continuity (open and short circuits), and testing fuses and lights.

The ammeter measures electrical current. Ammeters for motorcycle use should cover 0-50 amperes and 0-250 amperes. These are useful for checking battery charging and starting current.

Several inexpensive voM's (volt-ohm-milli-ammeter) combine all three instruments into one which fits easily in any tool box. See **Figure 5**. However, the ammeter ranges are usually too small for motorcycle work.

Hydrometer

The hydrometer gives a useful indication of battery condition and charge by measuring the

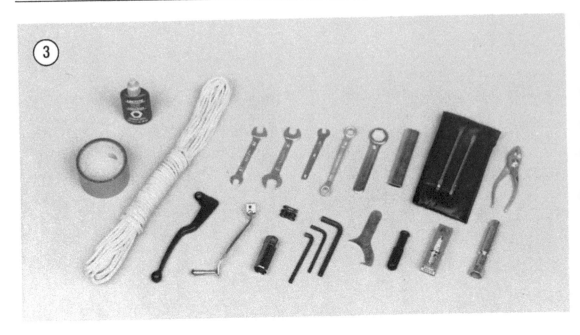

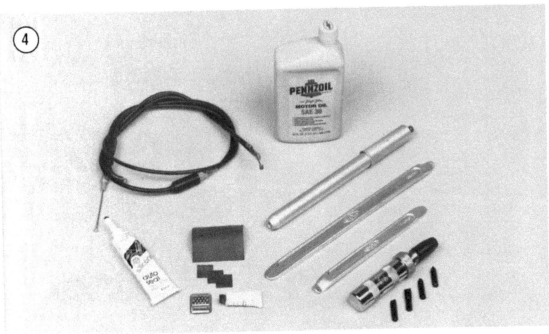

specific gravity of the electrolyte in each cell. See **Figure 6**. Complete details on use and interpretation of readings are provided in the electrical chapter.

Compression Tester

The compression tester measures the compression pressure built up in each cylinder. The results, when properly interpreted, can indicate general cylinder, ring, and valve condition. See **Figure 7**. Extension lines are available for hard-to-reach cylinders.

Dwell Meter (Contact Breaker Point Ignition Only)

A dwell meter measures the distance in degrees of cam rotation that the breaker points remain closed while the engine is running. Since

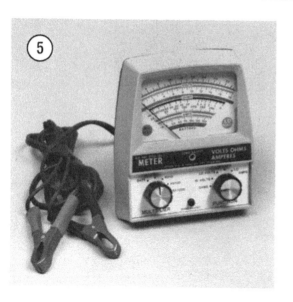

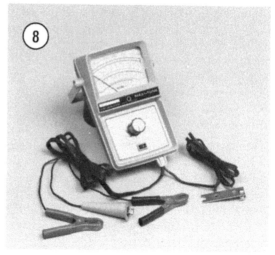

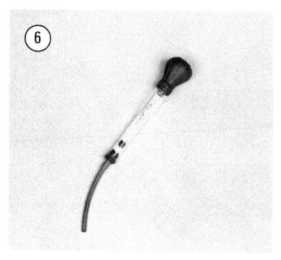

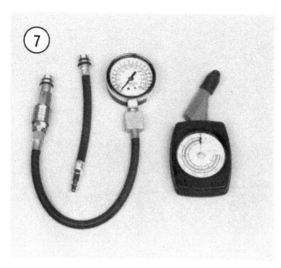

this angle is determined by breaker point gap, dwell angle is an accurate indication of breaker point gap.

Many tachometers intended for tuning and testing incorporate a dwell meter as well. See **Figure 8**. Follow the manufacturer's instructions to measure dwell.

Tachometer

A tachometer is necessary for tuning. See **Figure 8**. Ignition timing and carburetor adjustments must be performed at the specified idle speed. The best instrument for this purpose is one with a low range of 0-1,000 or 0-2,000 rpm for setting idle, and a high range of 0-4,000 or more for setting ignition timing at 3,000 rpm. Extended range (0-6,000 or 0-8,000 rpm) instruments lack accuracy at lower speeds. The instrument should be capable of detecting changes of 25 rpm on the low range.

> NOTE: *The motorcycle's tachometer is not accurate enough for correct idle adjustment.*

Strobe Timing Light

This instrument is necessary for tuning, as it permits very accurate ignition timing. The light flashes at precisely the same instant that No. 1 cylinder fires, at which time the timing marks on the engine should align. Refer to Chapter Three for exact location of the timing marks for your engine.

Suitable lights range from inexpensive neon bulb types ($2-3) to powerful xenon strobe lights ($20-40). See **Figure 9**. Neon timing lights are difficult to see and must be used in dimly lit areas. Xenon strobe timing lights can be used outside in bright sunlight.

Tune-up Kits

Many manufacturers offer kits that combine several useful instruments. Some come in a convenient carry case and are usually less expensive than purchasing one instrument at a time. **Figure 10** shows one of the kits that is available. The prices vary with the number of instruments included in the kit.

Manometer (Carburetor Synchronizer)

A manometer is essential for accurately synchronizing carburetors on multi-cylinder engines. The instrument detects intake pressure differences between carburetors and permits them to be adjusted equally. A suitable manometer costs about $25 and comes with detailed instructions for use. See **Figure 11**.

Fire Extinguisher

A fire extinguisher is a necessity when working on a vehicle. It should be rated for both *Class B* (flammable liquids — gasoline, oil, paint, etc.) and *Class C* (electrical — wiring, etc.) type fires. It should always be kept within reach. See **Figure 12**.

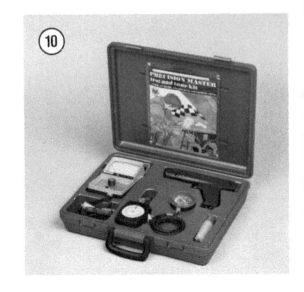

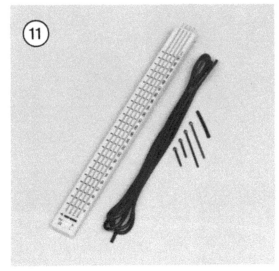

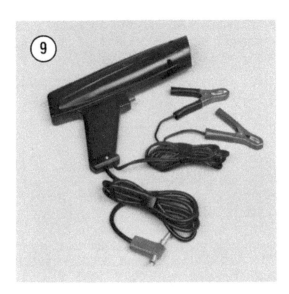

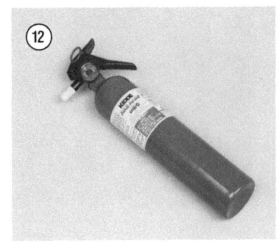

CHAPTER TWO

TROUBLESHOOTING

Troubleshooting motorcycle problems is relatively simple. To be effective and efficient, however, it must be done in a logical step-by-step manner. If it is not, a great deal of time may be wasted, good parts may be replaced unnecessarily, and the true problem may never be uncovered.

Always begin by defining the symptoms as closely as possible. Then, analyze the symptoms carefully so that you can make an intelligent guess at the probable cause. Next, test the probable cause and attempt to verify it; if it's not at fault, analyze the symptoms once again, this time eliminating the first probable cause. Continue on in this manner, a step at a time, until the problem is solved.

At first, this approach may seem to be time consuming, but you will soon discover that it's not nearly so wasteful as a hit-or-miss method that may never solve the problem. And just as important, the methodical approach to troubleshooting ensures that only those parts that are defective will be replaced.

The troubleshooting procedures in this chapter analyze typical symptoms and show logical methods for isolating and correcting trouble. They are not, however, the only methods; there may be several approaches to a given problem, but all good troubleshooting methods have one thing in common — a logical, systematic approach.

ENGINE

The entire engine must be considered when trouble arises that is experienced as poor performance or failure to start. The engine is more than a combustion chamber, piston, and crankshaft; it also includes a fuel delivery system, an ignition system, and an exhaust system.

Before beginning to troubleshoot any engine problems, it's important to understand an engine's operating requirements. First, it must have a correctly metered mixture of gasoline and air (**Figure 1**). Second, it must have an air-tight combustion chamber in which the mixture can be compressed. And finally, it requires a precisely timed spark to ignite the compressed mixture. If one or more is missing, the engine won't run, and if just one is deficient, the engine will run poorly at best.

Of the three requirements, the precisely timed spark — provided by the ignition system — is most likely to be the culprit, with gas/air mixture (carburetion) second, and poor compression the least likely.

STARTING DIFFICULTIES

Hard starting is probably the most common motorcycle ailment, with a wide range of problems likely. Before delving into a reluctant or non-starter, first determine what has changed

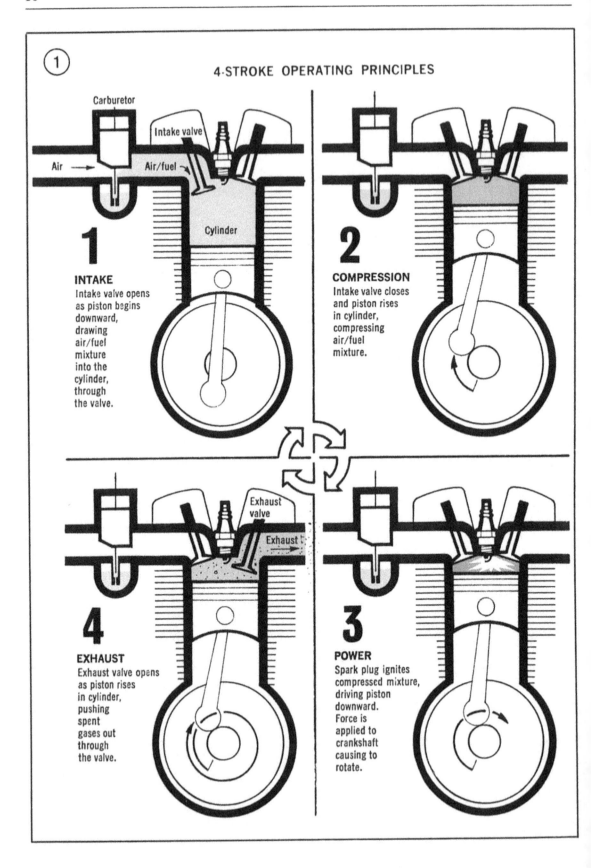

① 4-STROKE OPERATING PRINCIPLES

1 INTAKE
Intake valve opens as piston begins downward, drawing air/fuel mixture into the cylinder, through the valve.

2 COMPRESSION
Intake valve closes and piston rises in cylinder, compressing air/fuel mixture.

4 EXHAUST
Exhaust valve opens as piston rises in cylinder, pushing spent gases out through the valve.

3 POWER
Spark plug ignites compressed mixture, driving piston downward. Force is applied to crankshaft causing to rotate.

Carburetor

Intake valve

Air

Air/fuel

Cylinder

Exhaust valve

Exhaust

since the motorcycle last started easily. For instance, was the weather dry then and is it wet now? Has the motorcycle been sitting in the garage for a long time? Has it been ridden many miles since it was last fueled?

Has starting become increasingly more difficult? This alone could indicate a number of things that may be wrong but is usually associated with normal wear of ignition and engine components.

While it's not always possible to diagnose trouble simply from a change of conditions, this information can be helpful and at some future time may uncover a recurring problem.

Fuel Delivery

Although it is the second most likely cause of trouble, fuel delivery should be checked first simply because it is the easiest.

First, check the tank to make sure there is fuel in it. Then, disconnect the fuel hose at the carburetor, open the valve and check for flow (**Figure 2**). If fuel does not flow freely make sure the tank vent is clear. Next, check for blockage in the line or valve. Remove the valve and clean it as described in the fuel system chapter.

If fuel flows from the hose, reconnect it and remove the float bowl from the carburetor, open the valve and check for flow through the float needle valve. If it does not flow freely when the float is extended and then shut off when the flow is gently raised, clean the carburetor as described in the fuel system chapter.

When fuel delivery is satisfactory, go on to the ignition system.

Ignition

Remove the spark plug from the cylinder and check its condition. The appearance of the plug is a good indication of what's happening in the combustion chamber; for instance, if the plug is wet with gas, it's likely that engine is flooded. Compare the spark plug to **Figure 3**. Make certain the spark plug heat range is correct. A "cold" plug makes starting difficult.

After checking the spark plug, reconnect it to the high-tension lead and lay it on the cylinder head so it makes good contact (**Figure 4**). Then,

with the ignition switched on, crank the engine several times and watch for a spark across the plug electrodes. A fat, blue spark should be visible. If there is no spark, or if the spark is weak, substitute a good plug for the old one and check again. If the spark has improved, the old plug is faulty. If there was no change, keep looking.

Make sure the ignition switch is not shorted to ground. Remove the spark plug cap from the end of the high-tension lead and hold the exposed end of the lead about ⅛ inch from the cylinder head. Crank the engine and watch for a spark arcing from the lead to the head. If it's satisfactory, the connection between the lead and the cap was faulty. If the spark hasn't improved, check the coil wire connections.

If the spark is still weak, remove the ignition cover and remove any dirt or moisture from the points or sensor. Check the point or air gap against the specifications in the *Quick Reference Data* at the beginning of the book.

If spark is still not satisfactory, a more serious problem exists than can be corrected with simple adjustments. Refer to the electrical system chapter for detailed information for correcting major ignition problems.

Compression

Compression — or the lack of it — is the least likely cause of starting trouble. However, if compression is unsatisfactory, more than a simple adjustment is required to correct it (see the engine chapter).

An accurate compression check reveals a lot about the condition of the engine. To perform this test you need a compression gauge (see Chapter One). The engine should be at operating temperature for a fully accurate test, but even a cold test will reveal if the starting problem is compression.

Remove the spark plug and screw in a compression gauge (**Figure 5**). With assistance, hold the throttle wide open and crank the engine several times, until the gauge ceases to rise. Normal compression should be 130-160 psi, but a reading as low as 100 psi is usually sufficient for the engine to start. If the reading is much lower than normal, remove the gauge and pour about a tablespoon of oil into the cylinder.

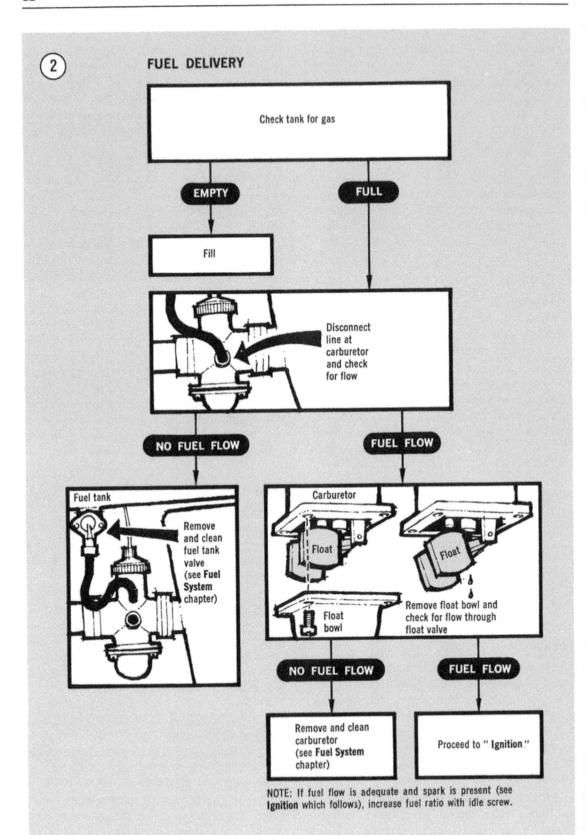

FUEL DELIVERY

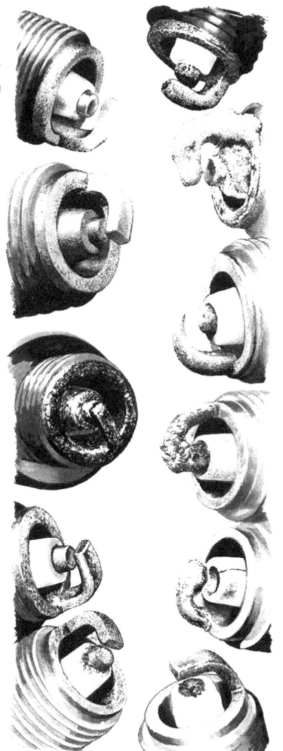

NORMAL
• Appearance—Firing tip has deposits of light gray to light tan.
• Can be cleaned, regapped and reused.

CARBON FOULED
• Appearance—Dull, dry black with fluffy carbon deposits on the insulator tip, electrode and exposed shell.
• Caused by—Fuel/air mixture too rich, plug heat range too cold, weak ignition system, dirty air cleaner, faulty automatic choke or excessive idling.
• Can be cleaned, regapped and reused.

OIL FOULED
• Appearance—Wet black deposits on insulator and exposed shell.
• Caused by—Excessive oil entering the combustion chamber through worn rings, pistons, valve guides or bearings.
• Replace with new plugs (use a hotter plug if engine is not repaired).

LEAD FOULED
• Appearance — Yellow insulator deposits (may sometimes be dark gray, black or tan in color) on the insulator tip.
• Caused by—Highly leaded gasoline.
• Replace with new plugs.

LEAD FOULED
• Appearance—Yellow glazed deposits indicating melted lead deposits due to hard acceleration.
• Caused by—Highly leaded gasoline.
• Replace with new plugs.

OIL AND LEAD FOULED
• Appearance—Glazed yellow deposits with a slight brownish tint on the insulator tip and ground electrode.
• Replace with new plugs.

FUEL ADDITIVE RESIDUE
• Appearance — Brown colored hardened ash deposits on the insulator tip and ground electrode.
• Caused by—Fuel and/or oil additives.
• Replace with new plugs.

WORN
• Appearance — Severely worn or eroded electrodes.
• Caused by—Normal wear or unusual oil and/or fuel additives.
• Replace with new plugs.

PREIGNITION
• Appearance — Melted ground electrode.
• Caused by—Overadvanced ignition timing, inoperative ignition advance mechanism, too low of a fuel octane rating, lean fuel/air mixture or carbon deposits in combustion chamber.

PREIGNITION
• Appearance—Melted center electrode.
• Caused by—Abnormal combustion due to overadvanced ignition timing or incorrect advance, too low of a fuel octane rating, lean fuel/air mixture, or carbon deposits in combustion chamber.
• Correct engine problem and replace with new plugs.

INCORRECT HEAT RANGE
• Appearance—Melted center electrode and white blistered insulator tip.
• Caused by—Incorrect plug heat range selection.
• Replace with new plugs

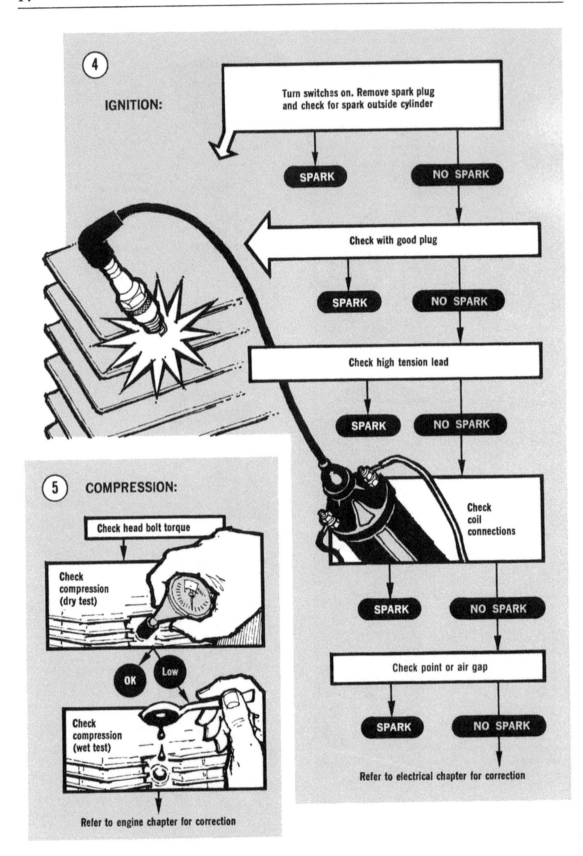

④ IGNITION:

Turn switches on. Remove spark plug and check for spark outside cylinder

SPARK   NO SPARK

Check with good plug

SPARK   NO SPARK

Check high tension lead

SPARK   NO SPARK

Check coil connections

SPARK   NO SPARK

Check point or air gap

SPARK   NO SPARK

Refer to electrical chapter for correction

⑤ COMPRESSION:

Check head bolt torque

Check compression (dry test)

OK   Low

Check compression (wet test)

Refer to engine chapter for correction

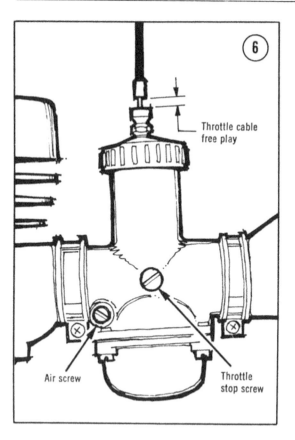

Throttle cable free play

Air screw

Throttle stop screw

Crank the engine several times to distribute the oil and test the compression once again. If it is now significantly higher, the rings and bore are worn. If the compression did not change, the valves are not seating correctly. Adjust the valves and check again. If the compression is still low, refer to the engine chapter.

> NOTE: *Low compression indicates a developing problem. The condition causing it should be corrected as soon as possible.*

POOR PERFORMANCE

Poor engine performance can be caused by any of a number of things related to carburetion, ignition, and the condition of the sliding and rotating components in the engine. In addition, components such as brakes, clutch, and transmission can cause problems that seem to be related to engine performance, even when the engine is in top running condition.

Poor Idling

Idling that is erratic, too high, or too low is most often caused by incorrect adjustment of the carburetor idle circuit. Also, a dirty air filter or an obstructed fuel tank vent can affect idle speed. Incorrect ignition timing or worn or faulty ignition components are also good possibilities.

First, make sure the air filter is clean and correctly installed. Then, adjust the throttle cable free play, the throttle stop screw, and the idle mixture air screw (**Figure 6**) as described in the routine maintenance chapter.

If idling is still poor, check the carburetor and manifold mounts for leaks; with the engine warmed up and running, spray WD-40 or a similar light lube around the flanges and joints of the carburetor and manifold (**Figure 7**). Listen for changes in engine speed. If a leak is present, the idle speed will drop as the lube "plugs" the leak and then pick up again as it is drawn into the engine. Tighten the nuts and clamps and test again. If a leak persists, check for a damaged gasket or a pinhole in the manifold. Minor leaks in manifold hoses can be repaired with silicone sealer, but if cracks or holes are extensive, the manifold should be replaced.

A worn throttle slide may cause erratic running and idling, but this is likely only after many thousands of miles of use. To check, remove the carburetor top and feel for back and forth movement of the slide in the bore; it should be barely perceptible. Inspect the slide for large worn areas and replace it if it is less than perfect (**Figure 8**).

If the fuel system is satisfactory, check ignition timing and breaker point gap (air gap in electronic ignition). Check the condition of the system components as well. Ignition-caused idling problems such as erratic running can be the fault of marginal components. See the electrical system chapter for appropriate tests.

Rough Running or Misfiring

Misfiring (see **Figure 9**) is usually caused by an ignition problem. First, check all ignition connections (**Figure 10**). They should be clean, dry, and tight. Don't forget the kill switch; a loose connection can create an intermittent short.

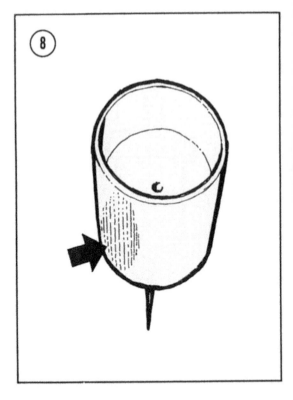

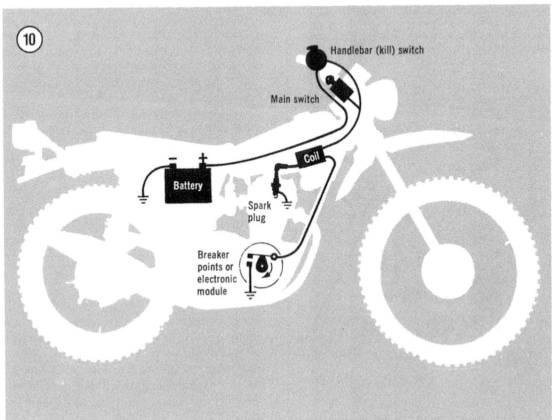

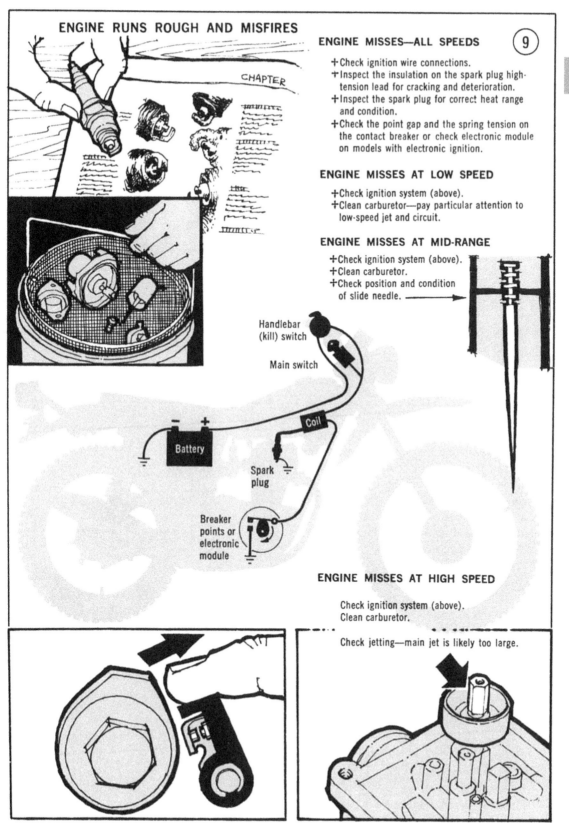

ENGINE RUNS ROUGH AND MISFIRES

ENGINE MISSES—ALL SPEEDS

+ Check ignition wire connections.
+ Inspect the insulation on the spark plug high-tension lead for cracking and deterioration.
+ Inspect the spark plug for correct heat range and condition.
+ Check the point gap and the spring tension on the contact breaker or check electronic module on models with electronic ignition.

ENGINE MISSES AT LOW SPEED

+ Check ignition system (above).
+ Clean carburetor—pay particular attention to low-speed jet and circuit.

ENGINE MISSES AT MID-RANGE

+ Check ignition system (above).
+ Clean carburetor.
+ Check position and condition of slide needle.

Handlebar (kill) switch

Main switch

Coil

Battery

Spark plug

Breaker points or electronic module

ENGINE MISSES AT HIGH SPEED

Check ignition system (above).
Clean carburetor.

Check jetting—main jet is likely too large.

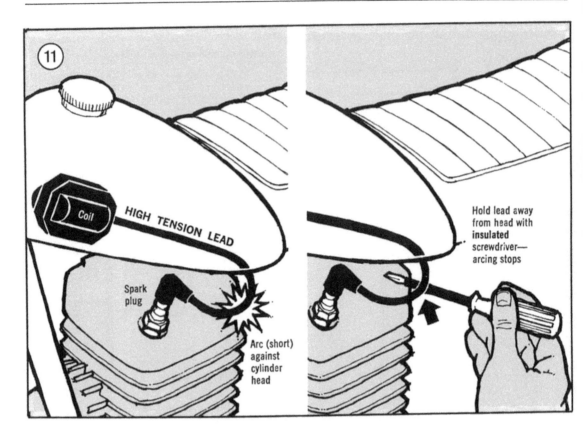

Check the insulation on the high-tension spark plug lead. If it is cracked or deteriorated it will allow the spark to short to ground when the engine is revved. This is easily seen at night. If arcing occurs, hold the affected area of the wire away from the metal to which it is arcing, using an insulated screwdriver (**Figure 11**), and see if the misfiring ceases. If it does, replace the high-tension lead. Also check the connection of the spark plug cap to the lead. If it is poor, the spark will break down at this point when the engine speed is increased.

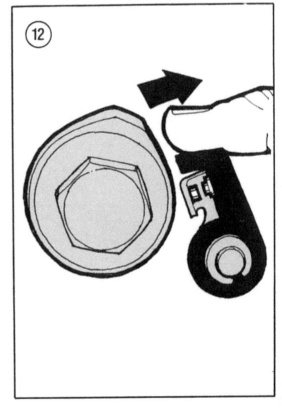

The spark plug could also be poor. Test the system with a new plug.

Incorrect point gap or a weak contact breaker spring can cause misfiring. Check the gap and the alignment of the points. Push the moveable arm back and check for spring tension (**Figure 12**). It should feel stiff.

On models with electronic ignition, have the electronic module tested by a dealer or substitute a known good unit for a suspected one.

If misfiring occurs only at a certain point in engine speed, the problem may very likely be

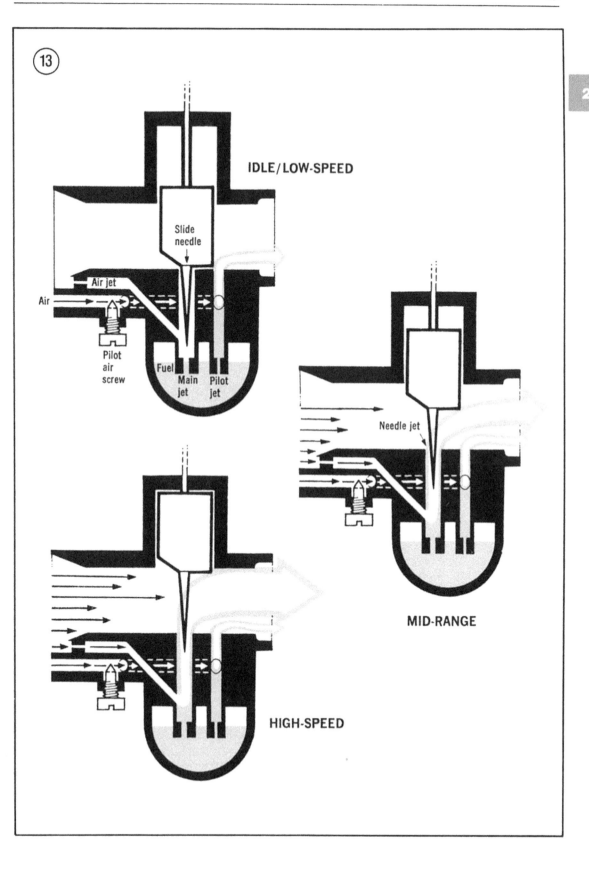

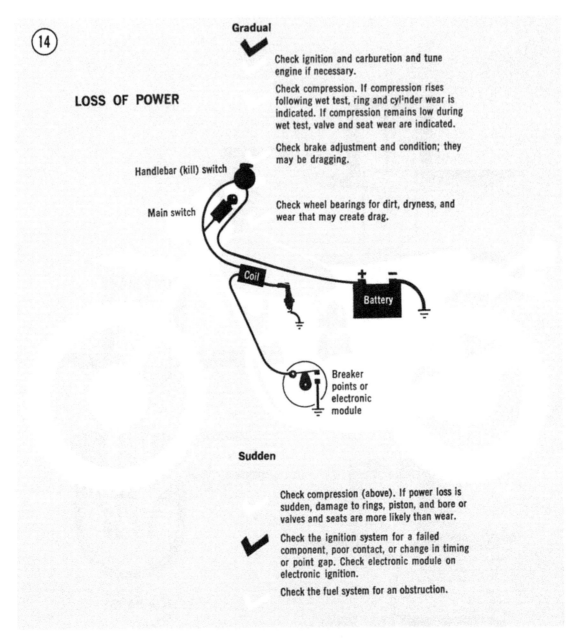

carburetion. Poor performance at idle is described earlier. Misfiring at low speed (just above idle) can be caused by a dirty low-speed circuit or jet (**Figure 13**). Poor midrange performance is attributable to a worn or incorrectly adjusted needle and needle jet. Misfiring at high speed (if not ignition related) is usually caused by a too-large main jet which causes the engine to run rich. Any of these carburetor-related conditions can be corrected by first cleaning the carburetor and then adjusting it as described in the tune-up and maintenance chapter.

Loss of Power

First determine how the power loss developed (**Figure 14**). Did it decline over a long period of time or did it drop abruptly? A gradual loss is normal, caused by deterioration of the engine's state of tune and the normal wear of the cylinder and piston rings and the valves and seats. In such case, check the condition of the

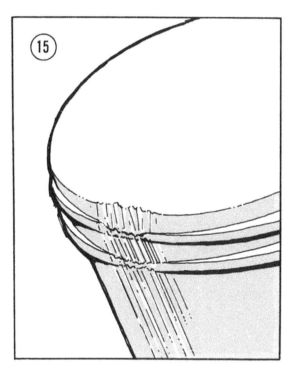

ignition and carburetion and measure the compression as described earlier.

A sudden power loss may be caused by a failed ignition component, obstruction in the fuel system, damaged valve or seat, or a broken piston ring or damaged piston (**Figure 15**).

If the engine is in good shape and tune, check the brake adjustment. If the brakes are dragging, they will consume considerable power. Also check the wheel bearings. If they are dry, extremely dirty, or badly worn they can create considerable drag.

Engine Runs Hot

A modern motorcycle engine, in good mechanical condition, correctly tuned, and operated as it was intended, will rarely experience overheating problems. However, out-of-spec conditions can create severe overheating that may result in serious engine damage. Refer to **Figure 16**.

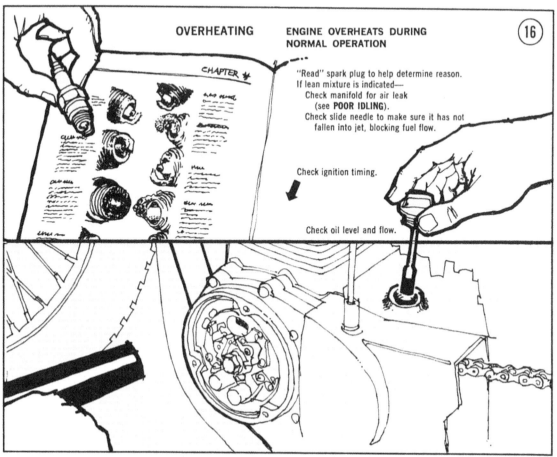

Overheating is difficult to detect unless it is extreme, in which case it will usually be apparent as excessive heat radiating from the engine, accompanied by the smell of hot oil and sharp, snapping noises when the engine is first shut off and begins to cool.

Unless the motorcycle is operated under sustained high load or is allowed to idle for long periods of time, overheating is usually the result of an internal problem. Most often it's caused by a too-lean fuel mixture.

Remove the spark plug and compare it to **Figure 3**. If a too-lean condition is indicated, check for leaks in the intake manifold (see *Poor Idling*). The carburetor jetting may be incorrect but this is unlikely if the overheating problem has just developed (unless, of course, the engine was jetted for high altitude and is now being run near sea level). Check the slide needle in the carburetor to make sure it hasn't come loose and is restricting the flow of gas through the main jet and needle jet (**Figure 17**).

Check the ignition timing; extremes of either advance or retard can cause overheating.

Piston Seizure and Damage

Piston seizure is a common result of overheating (see above) because an aluminum piston expands at a greater rate than a steel cylinder. Seizure can also be caused by piston-to-cylinder clearance that is too small; ring end gap that is too small; insufficient oil; spark plug heat range too hot; and broken piston ring or ring land.

A major piston seizure can cause severe engine damage. A minor seizure — which usually subsides after the engine has cooled a few minutes — rarely does more than scuff the piston skirt the first time it occurs. Fortunately, this condition can be corrected by dressing the piston with crocus cloth, refitting the piston and rings to the bore with recommended clearances, and checking the timing to ensure overheating does not occur. Regard that first seizure as a warning and correct the problem before continuing to run the engine.

CLUTCH AND TRANSMISSION

1. *Clutch slips*—Make sure lever free play is sufficient to allow the clutch to fully engage

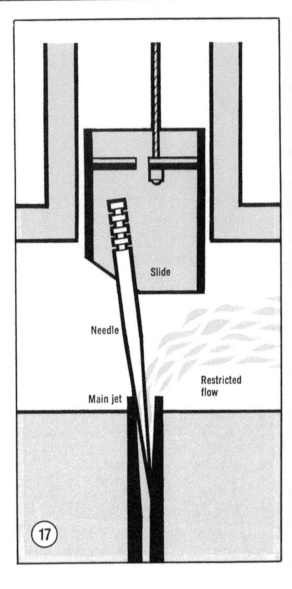

Slide

Needle

Main jet

Restricted flow

(17)

(**Figure 18**). Check the contact surfaces for wear and glazing. Transmission oil additives also can cause slippage in wet clutches. If slip occurs only under extreme load, check the condition of the springs or diaphragm and make sure the clutch bolts are snug and uniformly tightened.

2. *Clutch drags*—Make sure lever free play isn't so great that it fails to disengage the clutch. Check for warped plates or disc. If the transmission oil (in wet clutch systems) is extremely dirty or heavy, it may inhibit the clutch from releasing.

3. *Transmission shifts hard*—Extremely dirty oil can cause the transmission to shift hard.

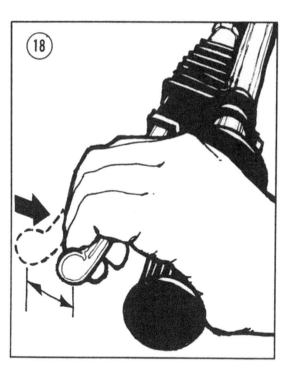

Check the selector shaft for bending (**Figure 19**). Inspect the shifter and gearsets for wear and damage.

4. *Transmission slips out of gear*—This can be caused by worn engagement dogs or a worn or damaged shifter (**Figure 20**). The overshift travel on the selector may be misadjusted.

5. *Transmission is noisy*—Noises usually indicate the absence of lubrication or wear and damage to gears, bearings, or shims. It's a good idea to disassemble the transmission and carefully inspect it when noise first occurs.

DRIVE TRAIN

Drive train problems (outlined in **Figure 21**) arise from normal wear and incorrect maintenance.

CHASSIS

Chassis problems are outlined in **Figure 22**.

1. *Motorcycle pulls to one side*—Check for loose suspension components, axles, steering

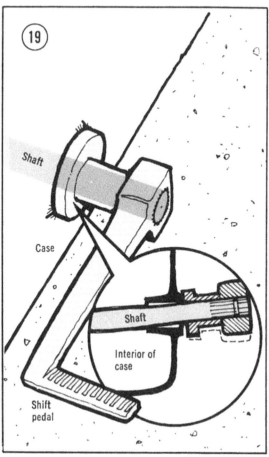

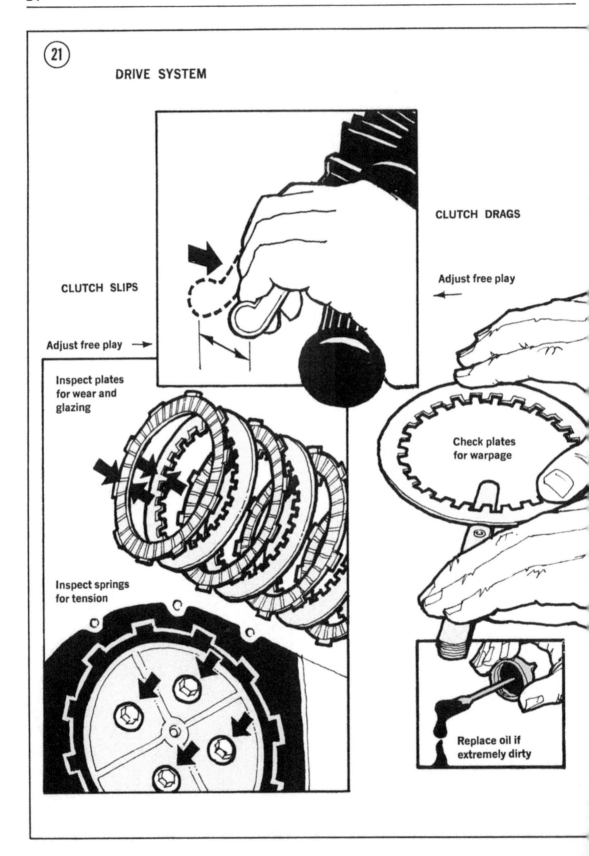

(21)

DRIVE SYSTEM

CLUTCH DRAGS

CLUTCH SLIPS

Adjust free play →

Adjust free play ←

Inspect plates
for wear and
glazing

Check plates
for warpage

Inspect springs
for tension

Replace oil if
extremely dirty

TRANSMISSION SLIPS OUT OF GEAR

TRANSMISSION SHIFTS HARD

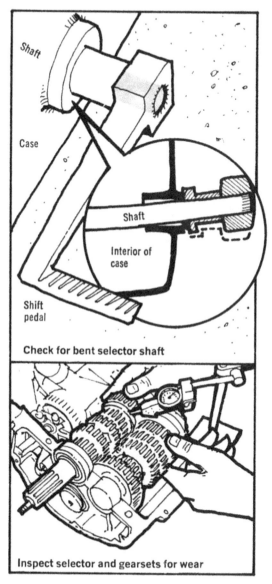

Check for bent selector shaft

Inspect selector and gearsets for wear

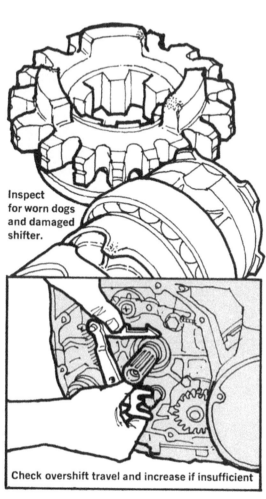

Inspect for worn dogs and damaged shifter.

Check overshift travel and increase if insufficient

TRANSMISSION IS NOISY

Check oil level

Disassemble and inspect (see Transmission chapter)

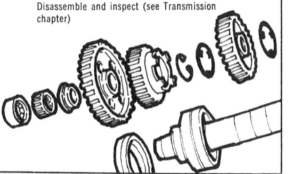

㉒

SUSPENSION AND HANDLING

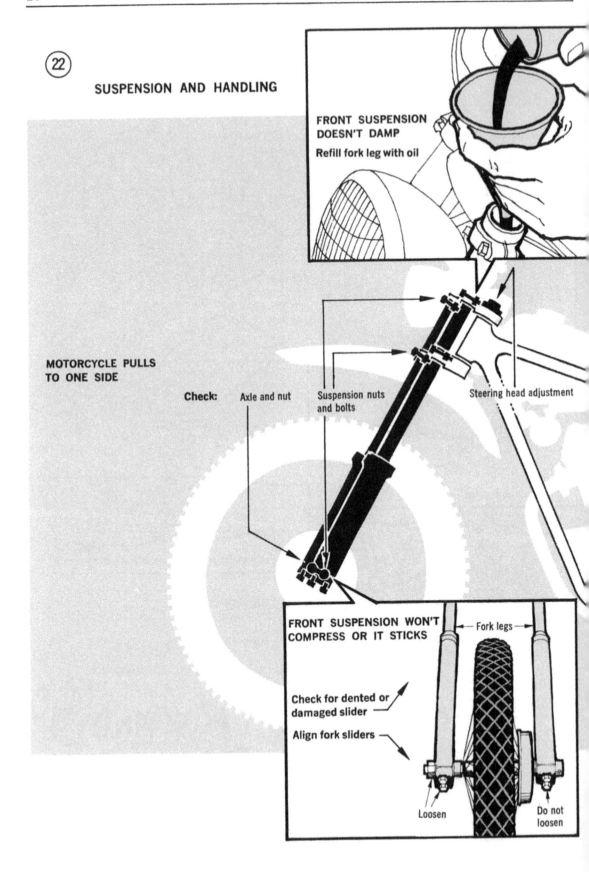

FRONT SUSPENSION
DOESN'T DAMP

Refill fork leg with oil

MOTORCYCLE PULLS
TO ONE SIDE

Check: Axle and nut   Suspension nuts   Steering head adjustment
                                  and bolts

FRONT SUSPENSION WON'T
COMPRESS OR IT STICKS

Fork legs

Check for dented or
damaged slider

Align fork sliders

Loosen   Do not
                    loosen

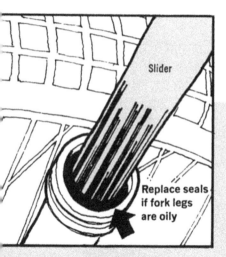

Slider

Replace seals if fork legs are oily

SUSPENSION AND HANDLING CONTINUED

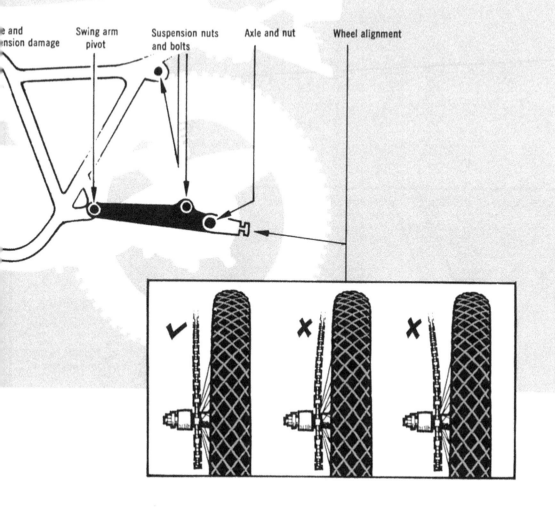

e and
nsion damage

Swing arm
pivot

Suspension nuts
and bolts

Axle and nut

Wheel alignment

SUSPENSION AND HANDLING CONTINUED

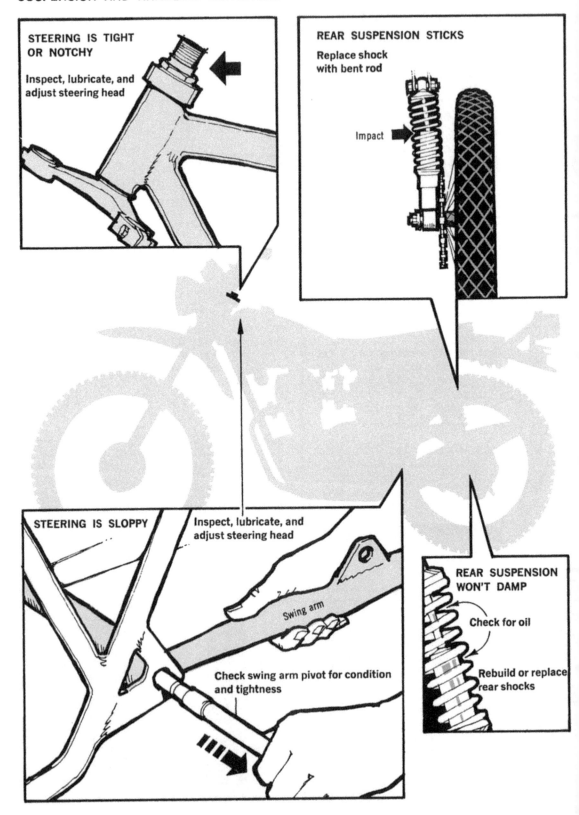

STEERING IS TIGHT OR NOTCHY

Inspect, lubricate, and adjust steering head

REAR SUSPENSION STICKS

Replace shock with bent rod

Impact

STEERING IS SLOPPY

Inspect, lubricate, and adjust steering head

Swing arm

Check swing arm pivot for condition and tightness

REAR SUSPENSION WON'T DAMP

Check for oil

Rebuild or replace rear shocks

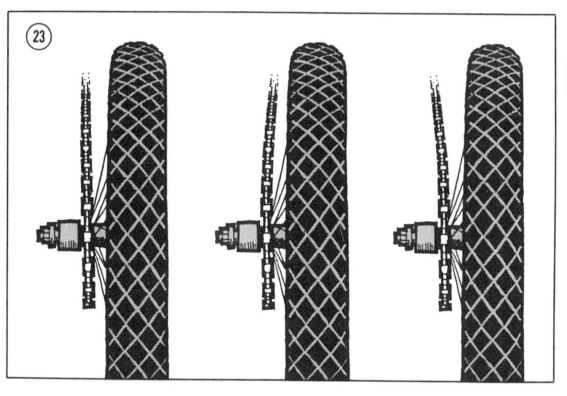

head, swing arm pivot. Check wheel alignment (**Figure 23**). Check for damage to the frame and suspension components.

2. *Front suspension doesn't damp*—This is most often caused by a lack of damping oil in the fork legs. If the upper fork tubes are exceptionally oily, it's likely that the seals are worn out and should be replaced.

3. *Front suspension sticks or won't fully compress*—Misalignment of the forks when the wheel is installed can cause this. Loosen the axle nut and the pinch bolt on the nut end of the axle (**Figure 24**). Lock the front wheel with the brake and compress the front suspension several times to align the fork legs. Then, tighten the pinch bolt and then the axle nut.

The trouble may also be caused by a bent or dented fork slider (**Figure 25**). The distortion required to lock up a fork tube is so slight that it is often impossible to visually detect. If this type of damage is suspected, remove the fork leg and remove the spring from it. Attempt to operate the fork leg. If it still binds, replace the slider; it's not practical to repair it.

4. *Rear suspension does not damp*—This is usually caused by damping oil leaking past

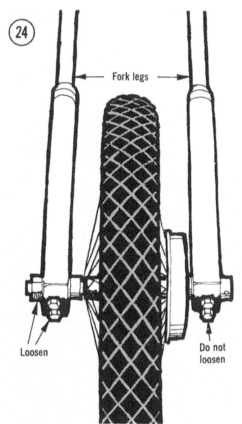

Fork legs

Loosen

Do not loosen

worn seals. Rebuildable shocks should be refitted with complete service kits and fresh oil. Non-rebuildable units should be replaced.

5. *Rear suspension sticks*—This is commonly caused by a bent shock absorber piston rod (**Figure 26**). Replace the shock; the rod can't be satisfactorily straightened.

6. *Steering is tight or "notchy"*—Steering head bearings may be dry, dirty, or worn. Adjustment of the steering head bearing pre-load may be too tight.

7. *Steering is sloppy*—Steering head adjustment may be too loose. Also check the swing arm pivot; looseness or extreme wear at this point translate to the steering.

BRAKES

Brake problems arise from wear, lack of maintenance, and from sustained or repeated exposure to dirt and water.

1. *Brakes are ineffective*—Ineffective brakes are most likely caused by incorrect adjustment. If adjustment will not correct the problem, remove the wheels and check for worn or glazed linings. If the linings are worn beyond the service limit, replace them. If they are simply glazed, rough them up with light sandpaper.

In hydraulic brake systems, low fluid levels can cause a loss of braking effectiveness, as can worn brake cylinder pistons and bores. Also check the pads to see if they are worn beyond the service limit.

2. *Brakes lock or drag*—This may be caused by incorrect adjustment. Check also for foreign matter embedded in the lining and for dirty and dry wheel bearings.

ELECTRICAL SYSTEM

Many electrical system problems can be easily solved by ensuring that the affected connections are clean, dry, and tight. In battery equipped motorcycles, a neglected battery is the source of a great number of difficulties that could be prevented by simple, regular service to the battery.

A multimeter, like the volt/ohm/milliammeter described in Chapter One, is invaluable for efficient electrical system troubleshooting.

See **Figures 27 and 28** for schematics showing

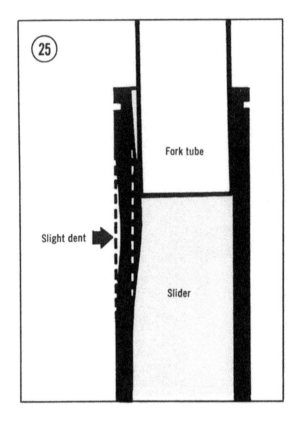

Fork tube

Slight dent

Slider

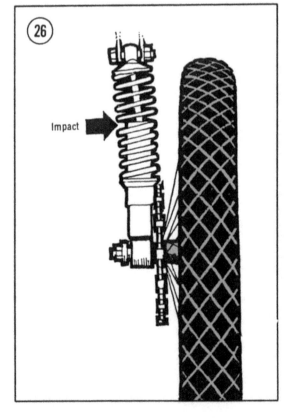

Impact

BASIC IGNITION CIRCUITS

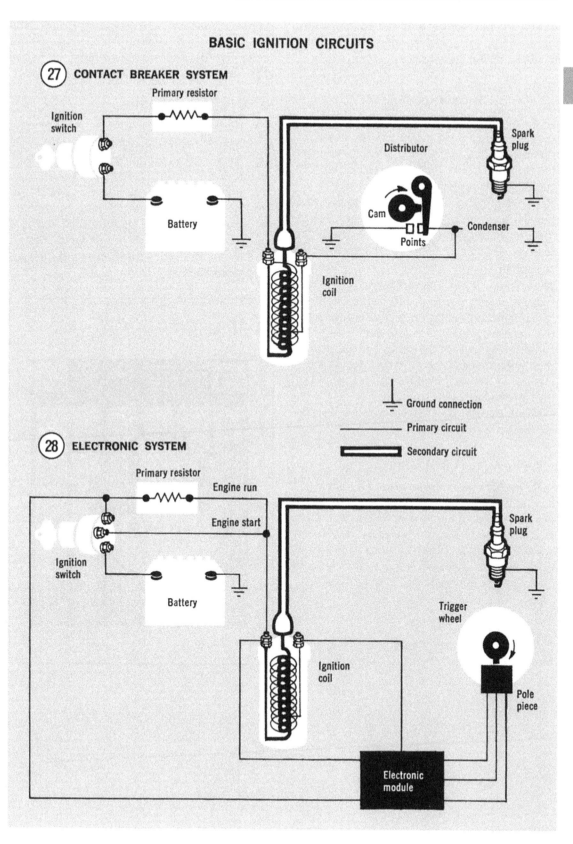

27 CONTACT BREAKER SYSTEM

Primary resistor

Ignition switch

Battery

Distributor

Spark plug

Cam

Points

Condenser

Ignition coil

Ground connection

Primary circuit

Secondary circuit

28 ELECTRONIC SYSTEM

Primary resistor

Engine run

Engine start

Spark plug

Ignition switch

Battery

Trigger wheel

Ignition coil

Pole piece

Electronic module

2

simplified conventional and electronic ignition systems. Typical and most common electrical troubles are also described.

CHARGING SYSTEM

1. *Battery will not accept a charge*—Make sure the electrolyte level in the battery is correct and that the terminal connections are tight and free of corrosion. Check for fuses in the battery circuit. If the battery is satisfactory, refer to the electrical system chapter for alternator tests. Finally, keep in mind that even a good alternator is not capable of restoring the charge to a severely discharged battery; it must first be charged by an external source.

2. *Battery will not hold a charge*—Check the battery for sulfate deposits in the bottom of the case (**Figure 29**). Sulfation occurs naturally and the deposits will accumulate and eventually come in contact with the plates and short them out. Sulfation can be greatly retarded by keeping the battery well charged at all times. Test the battery to assess its condition.

If the battery is satisfactory, look for excessive draw, such as a short.

LIGHTING

Bulbs burn out frequently—All bulbs will eventually burn out, but if the bulb in one particular light burns out frequently check the light assembly for looseness that may permit excessive vibration; check for loose connections that could cause current surges; check also to make sure the bulb is of the correct rating.

FUSES

Fuse blows—When a fuse blows, don't just replace it; try to find the cause. Consider a fuse a warning device as well as a safety device. And never replace a fuse with one of greater amperage rating. It probably won't melt before the insulation on the wiring does.

WIRING

Wiring problems should be corrected as soon as they arise — before a short can cause a fire that may seriously damage or destroy the motorcycle.

A circuit tester of some type is essential for locating shorts and opens. Use the appropriate wiring diagram at the end of the book for reference. If a wire must be replaced make a notation on the wiring diagram of any changes in color coding.

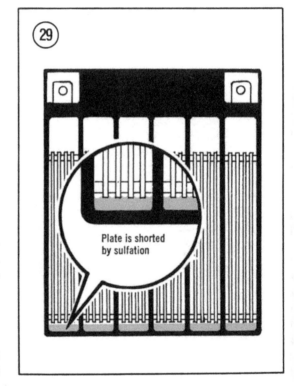

Plate is shorted by sulfation

NOTE: If you own a 1982 or later model, first check the Supplement at the back of the book for any new service information.

CHAPTER THREE

3

LUBRICATION, MAINTENANCE AND TUNE-UP

SCHEDULED MAINTENANCE

This chapter covers all the regular maintenance you have to perform to keep your machine in top shape.

Regular maintenance is the best guarantee of a trouble-free, long-lasting motorcycle. In addition, while performing the routine jobs, you will probably notice any other developing problems at an early stage when they are simple and inexpensive to correct.

Table 1 is a recommended minimum maintenance schedule (**Tables 1-9** are at the end of the chapter). However, you will have to determine your own maintenance requirements based on the type of riding you do and the place you ride. If you ride in dusty areas or at high speeds or if you make a lot of short 10 or 15 minute rides, service the items more often. Perform the maintenance at each *TIME* or *MILEAGE* interval, whichever comes first.

NOTE
If you have a brand new motorcycle, we recommend you take the bike to your dealer for the initial break-in maintenance at 500 miles (800 km).

Emission Controlled Motorcycles

This manual covers U.S. emission controlled motorcycles. We urge you to follow all procedures specifically designated for your bike. If you don't follow the maintenance schedule in this manual or if you alter engine parts or change their settings from the standard factory specifications (ignition timing, carburetor idle mixture, exhaust system, etc.), your bike may not comply with government emissions standards.

In addition, since most emission controlled bikes are carburetted on the lean side, any changes to emission-related parts (such as exhaust system modifications) could cause the engine to run so lean that engine damage would result.

BATTERY

The battery electrolyte level should be checked regularly, particularly during hot weather. Motorcycle batteries are marked with electrolyte level limit lines (**Figure 1**). Always maintain the fluid level between the lines, adding distilled water as required. Distilled water is available at most supermarkets and its use will prolong the life of the battery, especially in areas where tap water is hard (has a high mineral content).

Inspect the fluid level in all the cells. The battery is under the seat (**Figure 2**). Refer to *Battery* in Chapter Eight before removing the battery from the motorcycle.

Don't overfill the battery or you'll lose some electrolyte, weakening the battery and causing corrosion. Never allow the electrolyte level to drop below the top of the plates or the plates may be permanently damaged.

ENGINE OIL AND FILTER

Oil Level Inspection

1. Wait several minutes after shutting off the engine before making the check, to give all oil enough time to run down into the crankcase.
2. Put the bike on its centerstand (or hold it level).
3. Look at the oil inspection window near the bottom of the right engine cover (**Figure 3**). The oil level should lie between the upper and lower lines at the window.
4. If the oil level is below the lower line, remove the filler cap and add oil slowly, in small quantities, through the filler (**Figure 4**). Add enough to raise the oil level up to (but not above) the top line. Be sure to give the oil enough time to run down into the crankcase before rechecking the level in the inspection window. Use SAE 10W/40, 10W/50, 20W/40, or 20W/50 motor oil marked for service "SE" or "SF."
5. Install the filler cap.

Oil and Filter Change

Change the oil according to the maintenance schedule (**Table 1**). The filter should be changed every other oil change. If you ride hard or in dusty areas or if you take a lot of short trips, change the oil more frequently.

Try to stay with one brand of oil. The use of oil additives is not recommended; anything you add to the engine oil also gets on the clutch plates and could cause clutch slippage or damage.

1. Ride the bike to warm it up fully, then turn it off.
2. Put the bike up on its centerstand.
3. Put a drain pan under the crankcase and remove the drain plug (A, **Figure 5**). After the oil has drained, install the drain plug and torque it to 27 ft.-lb. (3.8 mkg).
4. The oil filter should be replaced every other engine oil change. If you are not changing the filter, skip to Step 10.
5. To remove the oil filter, unscrew the filter cover bolt (B, **Figure 5**).
6. Remove the cover and filter, discard the filter and clean the cover and the bolt. Inspect the O-rings on the cover and on the filter bolt (**Figure 6**). Replace them if damaged.
7. Insert the bolt into the cover and install the filter cup, spring, washer and cover plate (**Figure 7**).

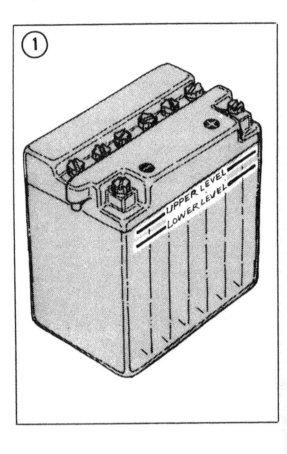

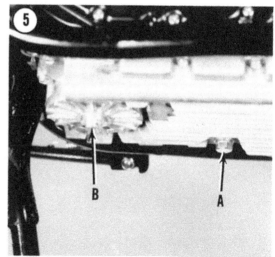

3

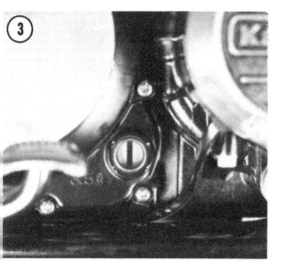

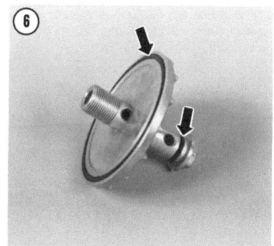

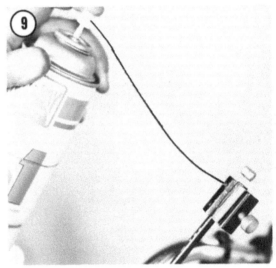

8. Check that the oil filter grommets are in place at both ends of the filter (**Figure 8**) and turn the filter onto the filter bolt.

NOTE
Before installing the cover, clean off the mating surface of the crankcase—do not allow any road dirt to enter the oil system.

9. Install the filter assembly in the crankcase and torque the filter bolt to 14.5 ft.-lb. (2.0 mkg).

10. Remove the oil filler cap and add the specified oil until it just reaches the upper line at the inspection window. Be sure to give the oil enough time to run down into the crankcase before checking the level in the inspection window.

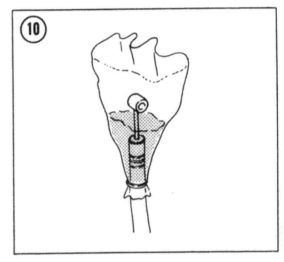

11. Screw in the filler cap and start the engine; let it idle and check for leaks.

12. Turn off the engine and recheck for the correct oil level.

GENERAL LUBRICATION

The following items should be lubricated according to the maintenance schedule (**Table 1**) and after cleaning the motorcycle. Special lubricants are available for control cables and other applications, but regular lubrication is more important than the type of lubricant you use: oil, grease, WD40, LPS3, etc.

Control Cables

The most positive method of control cable lubrication involves the use of a lubricator like the one shown in **Figure 9**. Disconnect the cable at the lever, attach the lubricator and inject lubricant into the cable sheath until it runs out of the other end.

If you do not have a lubricator, make a funnel from stiff paper or a plastic bag and tape it securely to one end of the cable (**Figure 10**). Hold the cable upright and add lubricant to the funnel. Work the cable in and out to help the lubricant work down the cable.

Control Pivots

Lubricate the brake pedal and linkage and centerstand pivots (**Figure 11**), the footpeg and sidestand pivots (**Figure 12**) and the control lever pivots and control cable ends (**Figure 13**).

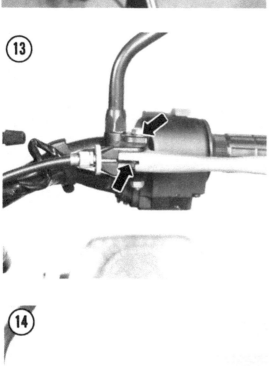

Throttle Grip Lubrication

1. Remove the screws that assemble the twist grip housing. Raise the top half of the housing.
2. Slide the grip back (disconnecting the throttle cable if necessary). Grease the handlebar under the grip and the cable end (**Figure 14**).
3. Reassemble the twist grip housing, fitting the upper housing peg into the hole in the handlebar. Check that the grip works smoothly.

Speedometer/Tachometer Cables

Disconnect the cables at the lower end. Pull the inner cable out, apply a light coat of grease and reinstall the cables. You may have to rotate the wheel to allow the speedometer cable to seat. If the tachometer cable won't seat, rotate the engine with the starter. Tighten the cable fasteners securely.

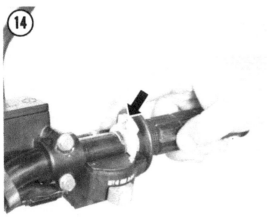

Ignition Advance Lubrication

To lubricate the ignition advance mechanism, refer to *Ignition Advance* in Chapter Eight.

CLUTCH ADJUSTMENT

Clutch Lever Play

The clutch cable should have about 1/8 in. (2-3 mm) play at the cable end of the lever before the clutch starts to disengage (**Figure 15**). Minor adjustments can be made at the hand lever; loosen the locknut, turn the adjuster as required and tighten the locknut.

According to the maintenance schedule (**Table 1**), and whenever the hand lever adjustment range is used up, adjust the clutch release as described here.

Clutch Release Adjustment

As the clutch cable stretches, cable play will exceed the range of the cable adjusters. As the clutch plates and discs inside the engine wear, the clutch release must be adjusted even when the cable play is within tolerance or the clutch can drag and cause rapid wear. Adjust the clutch as follows.

1. In front of the engine, loosen the clutch mid-cable adjuster locknut and shorten the adjuster all the way (**Figure 16**).
2. At the clutch lever, loosen the locknut and turn the adjuster until 3/16-1/4 in. (5-6 mm) of threads are showing between the locknut and the adjuster body (**Figure 17**).
3. Remove the 2 clutch adjuster cover screws and the cover above the shift pedal (**Figure 18**).
4. Loosen the locknut (**Figure 19**), then turn the screw out until it turns freely.
5. Turn the screw *in* until it becomes hard to turn. Then turn the screw *out* 1/2 turn. Hold the screw in position and tighten the locknut.
6. In front of the engine, lengthen the mid-cable adjuster until it has just taken all the slack out of the cable and the clutch lever has no free play. Tighten the locknut.
7. Check that the lower end of the clutch cable (below the engine) is fully seated in its socket.
8. At the clutch lever, turn the adjuster as required to get about 1/8 in. (2-3 mm) of cable play at the clutch lever.
9. Install the clutch adjuster cover.

DRIVE CHAIN

Clean, lubricate, adjust and check the drive chain for wear according to the maintenance

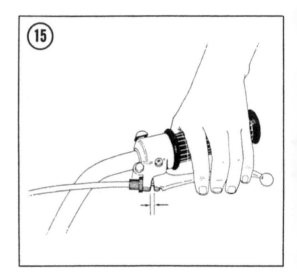

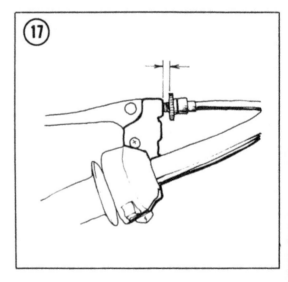

schedule (**Table 1**). The drive chain has no master link, for maximum strength.

Drive Chain Lubrication

Many lubricants are available that are specially formulated for drive chains. If a special lubricant is not available, Kawasaki recommends SAE 90 gear oil for chain lubrication; it is less likely to be thrown off the chain than lighter oils.

NOTE
*The drive chain has a permanent internal bushing lubricant sealed in by O-rings between the side plates (**Figure 20**). Do not use a solvent or aerosol lubricant not designed for use on O-rings.*

Chain Play Inspection

The drive chain must have adequate play so that the chain is not strung tight when the swing arm is horizontal (when the rider is seated). On the other hand, too much play may cause the chain to jump off the sprockets with potentially disasterous results.

1. Put the motorcycle on its centerstand.
2. Turn the rear wheel slowly until you locate the part of the chain that stretches tightest between the 2 sprockets on the bottom chain run (the chain wears unevenly).
3. With thumb and forefinger, lift up and press down the chain at that point, measuring the distance the chain moves vertically. The chain should have about 1 in. (25 mm) of vertical

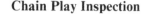

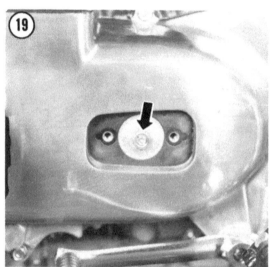

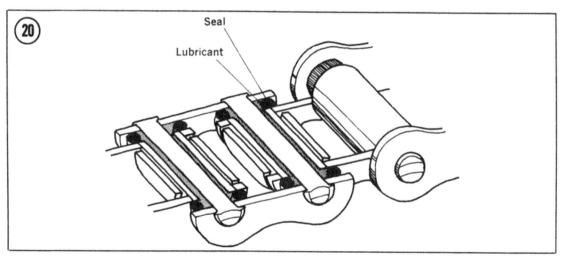

Seal

Lubricant

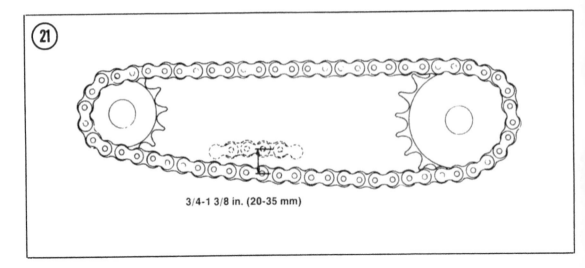

3/4-1 3/8 in. (20-35 mm)

travel at midpoint (**Figure 21**). If it has less than 3/4 in. (20 mm) or more than 1 3/8 in. (35 mm) of travel, adjust the chain play.

Drive Chain Adjustment

When adjusting the drive chain, you must also maintain rear wheel alignment. A mis-aligned rear wheel can cause poor handling and pulling to one side or the other, as well as increased chain, sprocket and tire wear. All models have wheel alignment marks on the swing arm and chain adjusters. If the alignment marks are kept at the same position left and right, the rear wheel should be aligned correctly.

1. Remove the rear axle nut cotter pin and loosen the axle nut (A, **Figure 22**).
2. Loosen the rear torque link nut (B, **Figure 22**).
3. Loosen the locknuts on both chain adjusters (A, **Figure 23**).
4. *If the chain was too tight*, back out both adjuster bolts an equal amount and kick the rear wheel forward until the chain is too loose.
5. Turn both adjuster bolts in an equal amount until the chain play is within specification. The notch in each chain adjuster should be positioned the same distance along the left and right side swing arm alignment marks (B, **Figure 23**).
6. When chain play is correct, check wheel alignment by sighting along the chain from the rear sprocket. It should leave the sprocket in a straight line (**Figure 24**). If it is cocked to one

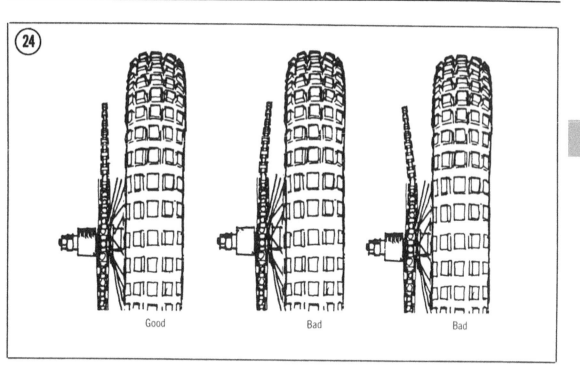

Good   Bad   Bad

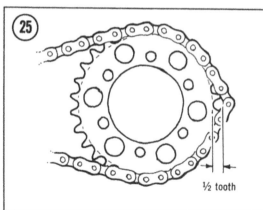

½ tooth

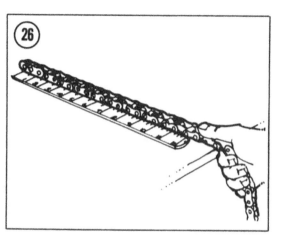

side or the other, adjust wheel alignment by turning one adjuster bolt or the other. Recheck chain play.

7. Torque the axle nut to 90 ft.-lb. (12 mkg), install a new cotter pin and spread its ends.

8. Tighten the chain adjuster locknuts and the rear torque link nut.

9. Recheck chain play.

Drive Chain Wear

Kawasaki recommends replacing the drive chain when it has worn longer than 2% of its original length. A quick check will give you an indication of when to measure chain wear. At the rear sprocket, pull one of the links away from the sprocket. If the link pulls away more than 1/2 the height of a sprocket tooth, the chain has probably worn out (**Figure 25**). To measure chain wear perform the following:

1. Remove the drive chain and stretch it out tight on a table top; see *Drive Chain Removal* in Chapter Ten.

2. Lay a scale along the top chain run and measure the length of any 20 links in the chain, from the center of the first pin you select to the 21st pin (**Figure 26**). If the 20 link length is more than 15.3 in (389 mm), install a new drive chain; see *Swing Arm Removal* in Chapter Ten.

3. If the drive chain is worn, inspect the rear wheel and engine sprockets for undercutting or sharp teeth (**Figure 27**). If wear is evident, replace the sprockets too or you'll soon wear out your new drive chain.

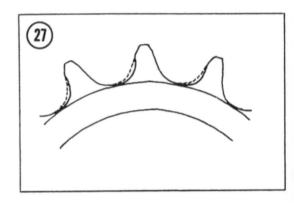

SWING ARM

The swing arm bearings must be lubricated with grease according to the maintenance schedule. On 1980 models, there is a grease fitting on the swing arm crossmember. Later models require removal of the swing arm to lubricate the bearings; see *Swing Arm Removal* in Chapter Ten.

1. Use a grease gun to force grease into the fitting on the swing arm, until the grease runs out both ends of the swing arm.

2. If grease will not run out of the ends of the swing arm, unscrew the grease fitting from the swing arm. Clean the fitting and make certain that its ball check valve is free. Reinstall the fitting.

3. Apply the grease gun again. If grease does not run out both ends of the swing arm, remove the swing arm, clean out the old grease, lubricate the bearings and install the swing arm; see *Swing Arm Removal* in Chapter Ten.

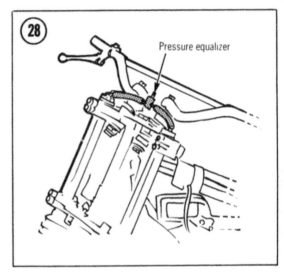

Pressure equalizer

STEERING

Steering Play Inspection

1. Prop up the motorcycle so that the front tire clears the ground.

2. Center the front wheel. Push lightly against the left handlebar grip to start the wheel turning to the right, then let go. The wheel should continue turning under its own momentum until the forks hit their stop. Try the same in the other direction.

NOTE
On some bikes, the wiring and control cables tend to stop the wheel movement. If the steering drags, make sure it's not because of wiring stiffness.

3. If, with a light push in either direction, the front wheel will not turn all the way to the stop, the steering adjustment is too tight.

4. Center the front wheel and kneel in front of it. Grasp the bottoms of the fork legs. Try to pull the forks toward you and then try to push them toward the engine. If you feel play, the steering adjustment is too loose.

5. If the steering is too tight or too loose, adjust it as described under *Steering Adjustment* in Chapter Ten.

Steering Head Lubrication

The steering head should be disassembled and the bearings cleaned, inspected for wear and lubricated with a waterproof grease according to the maintenance schedule (**Table 1**); see *Steering Adjustment* in Chapter Ten.

FRONT FORK

Fork Air Pressure

An air pressurized fork is standard equipment on this motorcycle. Both the fork springs and air pressure support the motorcycle and rider. The air pressure can be

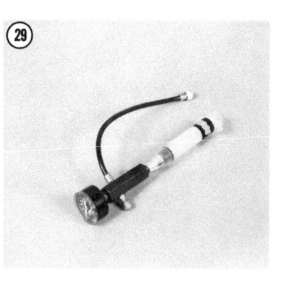

a. Increase air pressure for high-speed riding.
b. If the suspension is too hard, reduce air pressure.
c. If the suspension is too soft, increase air pressure.
d. Occasional bottoming of the forks shows that you are taking good advantage of all their travel. Severe or frequent bottoming should be avoided by increasing air pressure.

1. Support the bike with the front wheel off the ground.
2. Remove the air valve caps (**Figure 30**).
3. Connect a pump to the valve and pump the forks up to about 25 psi.

> *CAUTION*
> *Do not exceed 36 psi or the fork seals will be damaged.*

4. Slowly bleed off the pressure to reach the desired value. Refer to **Table 3** for standard air pressure and recommended range. Kawasaki recommends balancing light fork air pressure with light rear shock preload and damping; heavy fork air pressure with heavy rear shock preload and damping.

> *NOTE*
> *Each application of a pressure gauge bleeds off some air pressure merely in the process of applying and removing the gauge.*

5. Install the valve caps.

Fork Inspection

Apply the front brake and pump the fork up and down hard. You should hear the fork oil as it flows through its passages and there should be no binding. Inspect for fork oil leakage around the fork seals. If there is evidence of leakage, check the fork oil level; see *Fork Oil Change.*

Fork Oil Change

This procedure tells how to change the fork oil without removing the fork from the motorcycle. If the fork is removed and disassembled as described in Chapter Seven, more of the old oil will be drained.

varied to suit the load and your ride preference, but it is very important to have the same pressure in both forks to prevent an unbalanced suspension with poor handling. The maximum allowable air pressure difference between the forks is 1.5 psi, so be very careful when adding or bleeding air from the forks. The best way to guarantee equal fork air pressure is to install an accessory pressure equalization line between the forks (**Figure 28**).

Don't use a high-pressure hose or air bottle to pressurize the fork, a tire pump is a lot closer to the scale you need. S&W offers a combination hand pump/pressure gauge that is ideal (**Figure 29**).

Keep the following points in mind when adjusting the front forks.

*Release fork air pressure before remov-
ing the fork caps. Air pressure and spring
preload may eject the caps forcibly.*

1. Release any air pressure by removing the
valve cap and depressing the valve core (**Figure
30**).
2. Place a drain pan under the fork and remove the drain screw (**Figure 31**). Let the fork
drain for a few minutes, then pump the fork
(keeping your hand on the brake lever) to help
expel the oil. Install the drain screw and repeat
for the other fork leg.

WARNING
*Do not allow the fork oil to contact the
brake disc or pads. Stopping power will
be greatly reduced. If the brakes are
contaminated, clean the disc with a
non-oily solvent and install new brake
pads.*

3. Support the bike under the engine so the
front wheel clears the ground.
4. Loosen the upper triple clamp bolts, then
unscrew the fork cap plugs (**Figure 32**).
5. Remove the fork springs from inside the
fork tubes.
6. Fill the fork tubes with slightly less than the
specified quantity of oil; see **Table 4** at the end
of the chapter.

NOTE
*The amount of oil poured in is not as
accurate a measurement as the actual
level of the oil. You may have to add
more oil later in this procedure.*

7. After filling both tubes, slowly pump the
forks up and down by hand several times to
distribute the oil throughout the fork damper.
8. With the fork spring removed and the fork
fully extended, measure the distance from the
top of the fork tube to the surface of the oil
(**Figure 33**).
9. Add oil, if required, to bring the level up to
specification; see **Table 4**. Don't overfill the
fork legs.

CAUTION
*An excessive amount of oil can cause a
hydraulic locking of the forks during
compression, destroying the oil seals.*

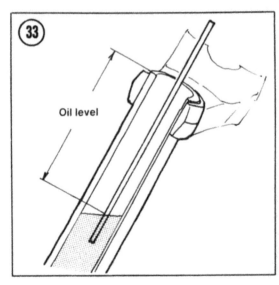

Oil level

10. Install the fork springs and the top spring seat or spacer.

11. Install the fork top caps and tighten the upper triple clamp bolts.

12. Pressurize the air forks.

REAR SHOCK ABSORBERS

The rear shock absorbers feature adjustable spring preload and damping. Damping adjustment affects mainly extension (rebound) damping.

To adjust spring preload, use the spanner or screwdriver in your motorcycle tool kit to turn both preload adjusters to one of the 5 settings (**Figure 34**). You will feel the adjuster become harder to turn as you set it for heavy preload.

To adjust damping, turn both damper wheels to one of the 4 click stops marked on the wheel (**Figure 35**). Position No. 1 is the lightest damping; position No. 4 is the heaviest damping.

Kawasaki recommends balancing light rear shock preload and damping with light fork air pressure; heavy rear shock preload and damping with heavy fork air pressure. Increase preload and damping for high speed riding.

> *WARNING*
> *Both rear shock absorbers must be set at the same preload and damping settings for safe handling.*

Rear Shock Inspection

Check that both rear shock absorber preload adjusters are set at the same notch (**Figure 34**). Check that both rear shock absorber damping adjusters are set at the same notch (**Figure 35**). Force the rear of the bike up and down. You should hear the fluid working in the shocks. Check for fluid leakage. If there is fluid leakage replace the shocks; they are not rebuildable. Check the shock mounting bolts for tightness and their rubber bushings for wear (**Figure 36**).

TIRES

Tire Pressure

Tire pressure must be checked with the tires cold. Correct tire pressure depends a lot on the load you are carrying and how fast you are going. A simple, accurate gauge (**Figure 37**) can be purchased for a few dollars and should be carried in your motorcycle tool kit. See **Table 5** or **Table 6** at the end of the chapter for tire inflation specifications.

Tire Wear

Check the tread for excessive wear, deep cuts and imbedded objects such as stones, nails or glass. If you find a nail in a tire, mark its location with a light crayon before pulling it out. See *Tire Changing* in Chapter Nine. Check local traffic regulations concerning minimum tread depth. Measure with a small ruler (**Figure 38**). Kawasaki recommends replacement when the front tread depth is 0.04 in. (1 mm) or less. For the rear tire, the recommended limits are 0.08 in. (2 mm) for speeds below 70 mph and 0.12 in. (3 mm) for higher speeds.

WHEELS

These motorcycles have 1-piece cast aluminum wheels. Check the cast wheels for cracks, bends or warping. If a wheel is damaged or cracked, replace it; repair is not possible. Periodic runout inspection is not necessary, but if the wheel has been subjected to a heavy impact or if you have any cause to suspect the wheel doesn't run "true" refer to Chapter Nine, *Cast Wheel Runout.*

WHEEL BEARING LUBRICATION

The ball bearings in the wheel hubs and the speedometer gear housing at the front wheel should be lubricated with high temperature grease according to the maintenance schedule (**Table 1**). Refer to Chapter Nine.

BRAKES

Inspect the brake function, brake fluid level and brake pad wear according to the maintenance schedule (**Table 1**). The disc brakes automatically compensate for wear and require no periodic free play adjustment.

Brake Function

Check for a solid feel at the lever and pedal. If the hydraulic brake feels spongy, perform

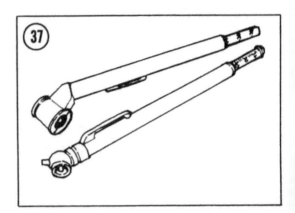

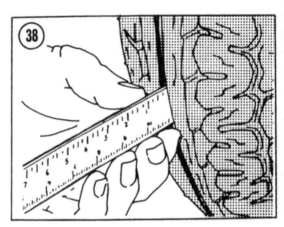

Brake System Bleeding as described in Chapter Nine.

Brake Pedal Height

CAUTION
The brake pedal free play must be properly adjusted or the rear brake pads may drag causing excessive friction and pad wear.

The master cylinder pushrod is adjusted until the brake pedal is the specified height below the top of the foot rest (**Figure 39**). Set the pedal height as follows:
   a. KZ750E and L models (U.S. and Canada)—8-12 mm (5/16-1/2 in.).
   b. KZ750E and L models (except U.S. and Canada)—13-17 mm (9/16-3/4 in.).
   c. KZ750H—4-8 mm (1/8-5/16 in.).
To adjust the brake pedal height, perform the following steps.

1. Remove the brake pedal, rear footpeg and the footpeg bracket mounting bolts (A, **Figure 40**).

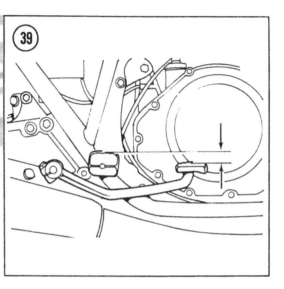

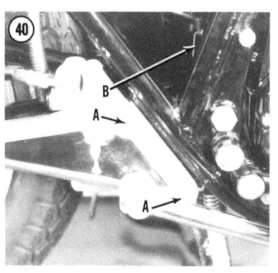

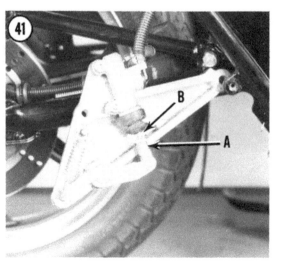

2. Disconnect the rear brake light switch spring at the switch (B, **Figure 40**), then carefully turn the footpeg bracket over, without twisting the brake hose excessively.

3. Loosen the brake pushrod locknut (A, **Figure 41**) and turn the adjuster (B) as required for fine adjustment of pedal height. Check the adjustment by temporarily installing the bracket and pedal.

4. Tighten the pushrod locknut and install the footpeg bracket, rear footpeg, brake light switch spring and the brake pedal. Tighten all bolts securely.

Brake Light Switch Adjustment

1. Turn the ignition switch ON.

2. Depress the brake pedal. The light should come on just as the brake begins to work.

3. To make the light come on earlier, hold the switch body and turn the adjusting nut (**Figure 42**) to move the switch body *up*. Move the switch body *down* to delay the light.

CAUTION
Do not turn the brake light switch body or you may twist the wires off.

NOTE
Some riders prefer having the light come on a little early. This way, they can tap the pedal without braking to warn drivers who follow too closely.

Brake Fluid Level Inspection

Check that the fluid level is between the upper and lower level lines (**Figure 43**). If you can not see the fluid level, clean the outside of the reservoir and remove the cap (**Figure 44**).

NOTE
To minimize fluid spillage, hold the handlebar as close to horizontal as possible when removing the front reservoir cap.

Adding Brake Fluid

1. Clean the outside of the reservoir cap thoroughly with a dry rag and remove the cap. Remove the diaphragm under the cap.

2. The fluid level in the reservoir should be up to the upper level line. Add fresh brake fluid as required.

WARNING
Kawasaki recommends DOT 3 brake fluid only. Lower grades may vaporize and cause brake failure. Never use old brake fluid or fluid from a container that has been left unsealed for a long time. Do not leave the reservoir cap off too long or the fluid will absorb moisture from the air and will vaporize more easily.

WARNING
Brake fluid is an irritant. Keep it away from your skin and eyes.

CAUTION
Be careful not to spill brake fluid on painted or plastic surfaces or it will destroy the finish. Wash spills immediately with soapy water and thoroughly rinse.

3. Reinstall the diaphragm (and washer on the rear master cylinder) and cap. Make sure that the cap is tight.

Brake Pad Wear Inspection

Inspect the disc brake pads for wear according to the maintenance schedule (**Table 1**).

1. Apply the brake and hold it tight.

2. Shine a light between the caliper and the disc to inspect the brake pads. The brake caliper is shown removed in **Figure 45** for clarity.
3. If either pad has worn thinner than 1/16 inch (1 mm) (**Figure 46**), replace both pads as a set; see *Brake Pad Replacement* in Chapter Nine.

Brake Seal Replacement

The rubber cup inside the master cylinder, the rubber piston seal inside the wheel caliper and their dust seals should be replaced every 2 years regardless of the mileage put on the bike. Replacement of the seals should be accompanied by inspection and, if necessary, rebuilding of the master cylinder and calipers. Because of

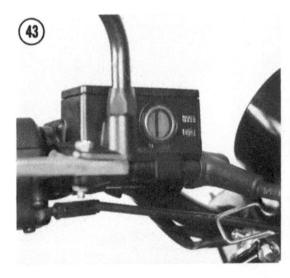

the special tools required for this kind of work, we recommend you have the job done by a Kawasaki dealer or qualified specialist. Brake system repair is critical work; see *Brakes* in Chapter Nine before attempting to rebuild the cylinder and calipers.

Brake Hose Replacement

The hydraulic brake hoses should be replaced every 4 years regardless of the mileage put on the bike; see *Brake Hose Replacement* in Chapter Nine.

NUTS, BOLTS AND FASTENERS

Check all exposed nuts, bolts, cotter pins, safety clips and circlips. Pay particular attention to:

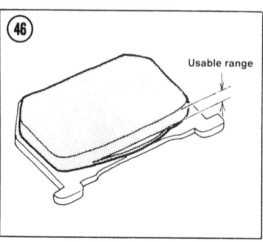

Usable range

a. Control lever, pedal and linkage pivots
b. Engine mounting bolts
c. Handlebar clamp bolts
d. Top triple clamp bolts
e. Bottom triple clamp bolts
f. Front axle clamp and axle nuts
g. Shock absorber mounts
h. Swing arm pivot
i. Rear brake torque link
j. Rear axle nut

This check is *especially* important on high mileage machines.

ENGINE TUNE-UP

The following list summarizes routine engine tune-up procedures. Detailed instructions follow the list. These tune-up procedures are arranged so that you start with the jobs that require a cold engine and finish with the jobs that call for a fully warmed-up engine. If you follow the sequence, you won't waste time waiting for your bike to cool down when required. If you aren't giving the bike a complete tune-up, backtrack through the procedures to make sure you've installed all parts. Consult Chapter Two for troubleshooting procedures when you suspect more serious trouble. Refer to **Table 2** at the end of the chapter for tune-up specifications.

1. Inspect the air filter and clean it or install a new one.
2. Clean the fuel system. Inspect the fuel lines for cracks or leakage.
3. Inspect the spark plugs; clean them and adjust the gap or replace them if necessary.
4. Inspect valve clearance and adjust if necessary.
5. Inspect the ignition timing and adjust if necessary.
6. Adjust the carburetors if required: throttle cable play, idle mixture (non-U.S. models), idle speed and synchronization.
7. Check and record cylinder compression.

AIR FILTER

A clogged air filter will cause a rich fuel/air mixture, resulting in power loss and poor gas mileage. Never run the bike without an air filter. Even minute particles of dust can cause severe internal engine wear and clogging of carburetor passages.

> *NOTE*
> *The air filter element is a dry paper type. Do not oil the filter or you will ruin it and cause the engine to run too rich.*

Twist the air cleaner cap (A, **Figure 47**) to the left and remove it, then remove the filter (B) from the housing.

Tap the filter against a solid surface to remove the heavy particles, then blow it clean from the inside out with compressed air, if available. Kawasaki recommends cleaning the paper-type filter in a non-oily solvent, then allowing it to dry.

Install a new air filter every 5 cleanings or every 6,000 miles (10,000 km) or any time the element or gaskets are damaged. Clean the filter more frequently in dusty areas and after riding in the rain.

FUEL SYSTEM

As water and dirt accumulate in the fuel tank or carburetor float bowls, engine performance will deteriorate. The fuel system should be cleaned when the engine is cold so gasoline doesn't spill on hot surfaces.

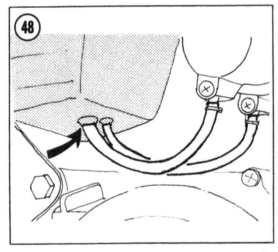

> *WARNING*
> *Some fuel may spill during these procedures. Work in a well-ventilated area at least 50 feet from any sparks or flames, including gas appliance pilot lights. Do not smoke in the area. Keep a BC rated fire extinguisher handy.*

Fuel System Cleaning

1. Check that the ignition switch is OFF.
2. Turn the fuel tap to PRI (prime).
3. Pull the carburetor overflow tubes out of the air cleaner housing (**Figure 48**) and put their ends into a container suitable for gasoline.
4. Loosen one float bowl drain screw one or two turns (**Figure 49**). Any accumulated water will flow out of the attached overflow hose. When clean gasoline comes out of the tube,

tighten the drain screw. Repeat for the other carburetors, then reconnect the overflow tubes.

5. If any dirt came out of the carburetors, inspect and clean the carburetors, fuel tap and fuel tank as described in Chapter Seven.

6. Make sure there are no leaks and that the fuel lines are not cracked or worn out.

SPARK PLUGS

Heat Range and Reach

The proper spark plug is very important for maximum performance and reliability. The proper heat range requires that a plug operate hot enough to burn off unwanted deposits, but not hot enough to burn up or cause preignition. A spark plug of the correct heat range will show a light tan color on the portion of the insulator within the cylinder after the plug has been in service.

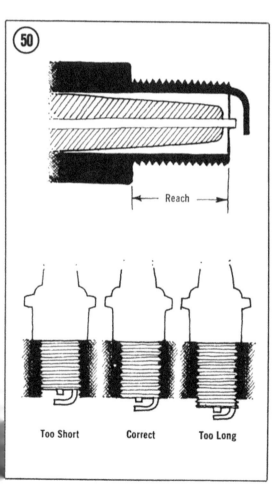

Too Short   Correct   Too Long

The spark plug recommended by the factory is usually the most suitable for your machine. For low speed riding or when riding in cold weather, a plug one step hotter may be preferable. Refer to **Table 2** at the end of the chapter for the recommended spark plug.

> *CAUTION*
> *Ensure the spark plug used has the correct thread reach (**Figure 50**). A thread reach too short will cause the exposed threads in the cylinder head to accumulate carbon, resulting in stripped cylinder head threads when the proper plug is installed. Too long a reach may cause plug/piston contact and serious damage.*

Spark Plug Inspection

1. Grasp the spark plug lead as near to the plug as possible and pull it off the plug. Clear away any dirt that has accumulated in the spark plug well.

> *CAUTION*
> *Dirt could fall into the cylinder when the plug is removed, causing serious engine damage.*

2. Remove the spark plug with a spark plug wrench.

3. Inspect the spark plug carefully. Look for broken center porcelain, excessively eroded electrodes and excessive carbon or oil fouling (See illustrations in Chapter Two). If deposits are light the plug may be cleaned with a wire brush or in a spark plug sandblast cleaner, but the price of a new plug is cheap insurance for high power and gas mileage. Check the spark plug gasket. If it's completely flattened, install a new one.

> *CAUTION*
> *Never sandblast an oily or wet plug. The grit will stick to the plug and later drop into the engine. After sandblasting a plug, clean it thoroughly.*

4. If the plug is reusable, file the center and side electrodes flat. Less voltage is required to jump the gap when the electrode corners are sharp.

5. Measure the gap with a round wire spark plug gauge (**Figure 51**). The gauge should just be able to pass through the gap. Adjust the gap by bending the side electrode only. The gap should be 0.028-0.032 in. (0.7-0.8 mm).

6. Apply a small amount of anti-seize compound to the plug threads. Don't use oil or grease—they'll turn to pure carbon and make the plug harder to get out the next time.

> *NOTE*
> *If you're going to adjust the valve clearance, leave the spark plugs out until you're finished. It will be easier to turn the engine.*

7. Clean the seating area on the cylinder head and thread the plug in by hand until it seats. Then tighten the plug 1/8 to 1/2 turn with a spark plug wrench. If you use a torque wrench, the proper torque is 20 ft.-lb. (2.8 mkg).

AIR SUCTION VALVES
(U.S. MODELS)

Suction Valve Operation

Models imported into the U.S. are equipped with a simple air suction system to minimize exhaust emissions. The air suction valves (reed valves) allow clean intake air to be sucked into the exhaust ports during exhaust vacuum pulses. This helps complete combustion of unburned hydrocarbons in the exhaust system.

The air suction valves also prevent exhaust gas from backing up and flowing into the air cleaner. If that happened, you would have an unplanned and unwanted EGR (exhaust gas recirculation) system. That would cut power a great deal and probably damage the engine. Periodic inspection of suction valve operation is very important.

Suction Valve Inspection

Check the suction valves with the engine OFF. You can check the air suction valve function by disconnecting the long suction hose at the air cleaner housing (**Figure 52**). You should be able to blow through this hose into the exhaust system and you should not be able to draw any air out of it because of the suction

valve reeds. If you can draw air out of the hose, one or more of the suction valves is faulty.

To inspect each valve individually, remove the suction valves.

Suction Valve
Removal/Installation

> *WARNING*
> *Some fuel may spill during these procedures. Work in a well-ventilated area at least 50 feet from any sparks or flames, including gas appliance pilot lights. Do not smoke in the area. Keep a BC rated fire extinguisher handy.*

1. Check that the ignition switch is OFF.

2. Remove the fuel tank; see *Fuel Tank Removal* in Chapter Seven.

3. Remove the 4 bolts and lift off the air suction valve cover (**Figure 53**).

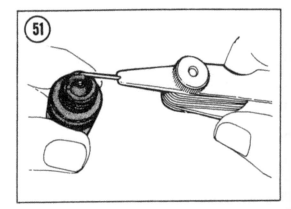

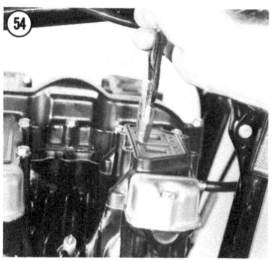

4. Remove the air suction valve assembly from the valve cover. There is a tab for pulling the valve up (**Figure 54**).

5. Check the reed valves for cracks, folds, warpage or any other damage (**Figure 55**).

6. Check the sealing lip coating around the perimeter of the valve. It must be free of grooves, scratches or signs of separation from the metal holder.

NOTE
The valve assembly cannot be repaired. It must be replaced if damaged.

7. Wash off any carbon deposits between the reed and the reed contact area with solvent.

CAUTION
Do not scrape deposits off or the assembly will be damaged.

8. Install by reversing the removal steps.

NOTE
If you're going to adjust valve clearance next, leave the air suction valve covers and fuel tank off.

9. Install the air suction valves; see *Suction Valve Installation* in Chapter Seven.

10. Install the fuel tank; see *Fuel Tank Installation* in Chapter Seven.

VALVE CLEARANCE

Normal wear of the valves and valve seats decreases valve clearance and alters valve timing slightly. Insufficient valve clearance can lead to burnt valves and seats and will eventually cause serious engine damage. Excessive clearance causes noisy operation and more rapid valve train wear.

Engines that use overhead camshafts with shims to adjust valve clearance should not require frequent valve clearance *adjustment*, but the clearance must be *inspected* according to the maintenance schedule (**Table 1**). Inspection is a simple, easy task compared to adjustment.

NOTE
Check and adjust valve clearance with the engine cool, at room temperature.

Valve Clearance Inspection

> *WARNING*
> *Some fuel may spill during these procedures. Work in a well-ventilated area at least 50 feet from any sparks or flames, including gas appliance pilot lights. Do not smoke in the area. Keep a BC rated fire extinguisher handy.*

1. Remove the fuel tank as described in *Fuel Tank Removal* in Chapter Seven.
2. Remove the spark plugs.
3. Remove the valve cover; see *Valve Cover Removal* in Chapter Four.
4. Check that all 16 camshaft cap bolts (**Figure 56**) are properly tightened to 8.5 ft.-lb. (1.2 mkg).
5. Remove the timing cover and gasket from the lower right side of the engine (**Figure 57**).
6. Turn the crankshaft with a 17 mm wrench on the outer advancer bolt until the "T" mark for cylinders No. 1 and 4 aligns with the index mark (**Figure 58**).

> *NOTE*
> *For clarity, the timing marks are shown with the pickup coil assembly removed. Observe the marks through the upper hole in the timing plate (Figure 59).*

7. Insert a feeler gauge between the cam lobe and the lifter for the pair of intake valves (rear cam) that has some clearance (**Figure 60**). This may be either cylinders No. 1 and 3 or cylinders No. 2 and 4, depending on whether the No. 1 piston is at TDC on its exhaust or compression stroke. The clearance is measured correctly when the feeler gauge drags slightly when it is inserted and withdrawn. Record the measurements and cylinder numbers.

> *NOTE*
> *The cylinders are numbered 1 through 4, starting at the left side of the engine.*

8. Turn the crankshaft to the right (clockwise) 1/2 turn until the "T" mark for cylinders No. 2 and 3 aligns with the index mark (**Figure 61**). Measure and record the clearance for the pair of exhaust valves (front cam) that has some clearance.
9. Turn the crankshaft to the right (clockwise) 1/2 turn until the "T" mark for cylinders No. 1

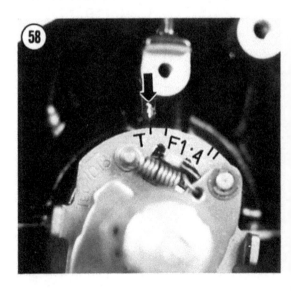

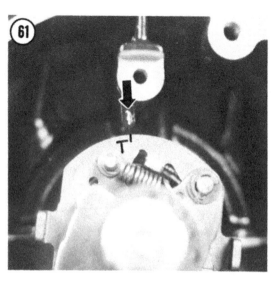

and 4 aligns and measure the clearance for the remaining pair of intake valves (rear cam).

10. Turn the crankshaft to the right (clockwise) 1/2 turn until the "T" mark for cylinders No. 2 and 3 aligns and measure the clearance for the remaining pair of exhaust valves (front cam).

11. If any valve's clearance is not within the range specified in **Table 2**, skip ahead to *Valve Clearance Adjustment*. If all the valve clearances are within specification, continue this procedure and install the removed parts.

12. Install the valve cover; see *Valve Cover Installation* in Chapter Four.

13. Install the fuel tank; see *Fuel Tank Installation* in Chapter Seven.

Valve Clearance Adjustment

To adjust the valve clearance, the camshaft must be removed and the shim under the valve lifter must be removed and replaced with one of a different thickness. The shims are available from Kawasaki dealers in increments of 0.05 mm and range in size from 2.0-3.2 mm. The dimension is marked on the bottom shim surface.

This procedure pertains only to valves that need adjustment. Do not remove any shims whose clearance falls within the specified range.

1. Remove the camshaft with valve clearances to be adjusted; see *Camshaft Removal* in Chapter Four.

2. Remove the lifter (**Figure 62**).

> *CAUTION*
> *Be very careful not to damage the outer surface of the lifter. Use a rubber suction cup, if available, to pull the lifter out.*

3. Remove the shim (**Figure 63**). Check the thickness marked on it or measure the shim thickness with a micrometer.

4. Calculate the correct shim thickness using this example. Refer to **Table 2** for the specified clearance.

> *NOTE*
> *The following numbers are for example only.*

Example:

Actual measured clearance	0.52 mm
Subtract specified clearance	-0.15 mm
Equals excess clearance	0.37 mm
Existing shim number	220
Add excess clearance	+37
Equals new shim number	257
(round up to the nearest	
shim number)	260

5. Insert the new shim under the valve lifter.

CAUTION
Never put shim stock under a shim. The shim could come loose at high rpm and cause extensive engine damage. Never grind the shim; this can remove the hardened outer shim surface and cause a fracture and extensive engine damage.

NOTE
*If the smallest available shim does not increase clearance to within acceptable limits, the valve seat is probably worn. If this is the case, repair the valve seat, grind the valve stem slightly and recheck the clearance. See **Valve/Seat Inspection** in Chapter Four.*

6. Install the camshaft; see *Camshaft Installation* in Chapter Four.
7. Install the valve cover; see *Valve Cover Installation* in Chapter Four.
8. Install the fuel tank; see *Fuel Tank Installation* in Chapter Seven.

IGNITION TIMING

All models have transistorized ignition. Ignition timing inspection is not *required* maintenance because transistorized ignition initial timing is very stable. Once it is set properly it should last the life of the motorcycle without adjustment. This optional procedure is provided in case of suspected trouble.

The ignition advance mechanism is a mechanical device that must be lubricated according to the maintenance schedule. If not maintained properly, the advance mechanism could stick and cause low power, overheating, spark knock or detonation. Transistorized ignition can not be checked statically. It must be inspected dynamically (engine running) with a strobe timing light.

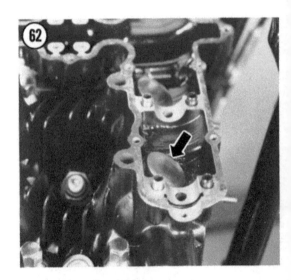

Check that the idle speed is within specification; see *Idle Speed* before inspecting dynamic timing. A too-high idle speed will begin the ignition advance process and give a faulty reading.

Dynamic Ignition Timing

1. Install the spark plugs; clean the seating area on the cylinder head and thread the plug in by hand until it seats. Then tighten the plug 1/8 to 1/2 turn with a spark plug wrench. If you use a torque wrench, the proper torque is 20 ft.-lb. (2.8 mkg).
2. Remove the timing cover and gasket from the lower right side of the engine.

3. Hook up a stroboscopic timing light to the No. 1 or No. 4 spark plug lead according to the manufacturer's instructions.
4. Start the engine and allow it to idle. Shine the light at the timing inspection marks. The "F" mark on the advancer should align with the index mark at idle (**Figure 64**). If the "F" mark does not align at idle, stop the engine and remove and inspect the ignition advance assembly; see *Ignition Advance* in Chapter Eight. Recheck the timing.

NOTE
*For clarity, the timing marks are shown with the pickup coil assembly removed. Observe the marks through the upper hole in the timing plate (**Figure 59**).*

5. Increase the engine speed to 3,800 rpm and check that the index mark falls between the double line advance mark (**Figure 65**). If the advancer does not work correctly, refer to *Ignition Advance* in Chapter Eight.
6. Stop the engine and install the timing cover and gasket.

CARBURETOR

The carburetor should be adjusted only when the engine is fully warmed up and all other tune-up operations are done.

Throttle Cable Play

Always check the throttle cables before you make any carburetor adjustments. Too much free play causes jerky throttle response; too little free play will cause unstable idling.

Check free play at the throttle grip flange (**Figure 66**). Kawasaki specifies about 1/8 in. (2-3 mm). If adjustment is required, loosen the throttle grip cable adjuster locknut (A, **Figure 67**) and turn the adjuster (B) as required to get your desired free play. Tighten the locknut.

If all the adjustment range is used up at the throttle grip, use the adjuster at the carburetor end of the throttle cable (**Figure 68**).

Idle Speed

Proper idle *speed* setting is necessary to prevent stalling and to provide adequate engine compression braking when you let off the

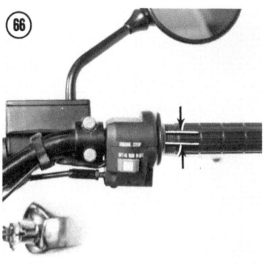

throttle, but you can't set it perfectly with the bike's tachometer—it's just not accurate at the low end. If you don't have a portable tachometer, you're about as well off setting idle by ear and feel. If it stalls, set idle faster; if you want more engine braking when decelerating, set idle slower.

1. Ride the bike to warm it up fully (about 10 minutes).

2. Stop the engine and attach a portable tachometer, following the instrument manufacturer's instructions.

3. Start the engine and turn the idle speed screw (**Figure 69**) to set idle at 1,000-1,100 rpm. If you have no tachometer, set the idle at the lowest speed at which the engine will idle smoothly.

4. Rev the engine a couple of times to see if it settles down to the set speed. Readjust, if necessary.

Idle Mixture
(Non-U.S. Models)

The idle fuel/air *mixture* affects low-speed emissions, as well as idling stability and response off idle. On motorcycles imported into the United States, the idle *mixture* screw is set and sealed at the factory and requires no adjustment. Adjust other models according to this procedure.

1. Adjust the idle speed to specification.

2. Stop the engine and remove the fuel tank; see *Fuel Tank Removal* in Chapter Seven. Hook up a temporary fuel supply.

> *WARNING*
> *When supplying fuel by temporary means, make sure the tank is secure and all fuel lines tight—no leaks.*

3. Turn each idle mixture screw (**Figure 70**) in until it seats lightly, then back it out 2 turns.

> *CAUTION*
> *Never turn the idle mixture screw in tight. You'll damage the screw or its soft seat in the carburetor.*

4. Start engine and turn each idle mixture screw in or out slightly to the setting that gives the highest stable idle speed.

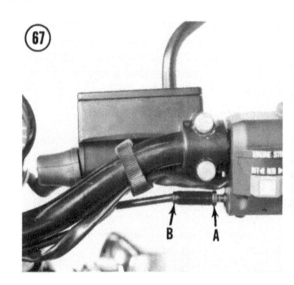

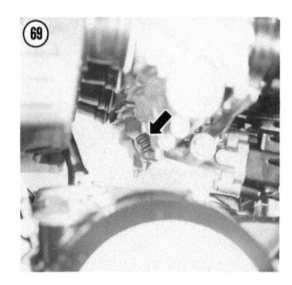

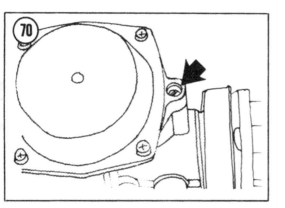

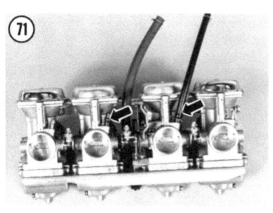

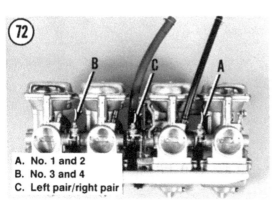

A. No. 1 and 2
B. No. 3 and 4
C. Left pair/right pair

5. Readjust idle speed to specification, if necessary.
6. Turn each mixture screw slightly again to see if the idle speed increases. Readjust idle speed, if necessary.

Carburetor Synchronization

Synchronizing the carburetors makes sure that one cylinder doesn't try to run faster than the others, cutting power and gas mileage. The only accurate way to synchronize the carbu-

retors is to use a set of vacuum gauges (a manometer) that measures the intake vacuum of all cylinders at the same time. A typical set of gauges is shown in Chapter One.

> *NOTE*
> *Before you try to synchronize the carburetors, make sure all of the following are checked or adjusted first; if not, you won't get a good synch: air filter, spark plugs, air suction valves (U.S. models), valve clearance, throttle cable play, carburetor holders and clamps air-tight.*

1. Ride the bike to warm it up fully, set the idle speed, then stop the engine.
2. Remove the fuel tank; see *Fuel Tank Removal* in Chapter Seven. Hook up a temporary fuel supply or use a long fuel line to connect the carburetors to the fuel tank. Plug the vacuum line that goes from the carburetor to the vacuum fuel tap.

> *WARNING*
> *When supplying fuel by temporary means, make sure the tank is secure and all fuel lines tight—no leaks.*

3. Remove the rubber caps or vacuum line from each carburetor's vacuum tap (**Figure 71**) and attach a set of vacuum gauges following the instrument manufacturer's instructions.

> *NOTE*
> *The carburetors are shown removed for clarity only. Do not remove the carburetors for this procedure.*

4. Turn the fuel tap to PRI (prime), start the engine and check that the vacuum difference between the cylinders is less than 0.80 in. Hg (20 mm Hg). Identical readings are desirable.
5. If the difference is greater than specified, loosen the locknut and turn the appropriate synchronizing screw located between the carburetors as required to equalize the vacuum in all cylinders (**Figure 72**). First match No. 1 and No. 2 carburetors, then match No. 3 and No. 4 and finally match the left pair of cylinders to the right pair.
6. Rev the engine once and check that all cylinders return to equal readings. Readjust if necessary and tighten the adjuster locknuts while holding the adjusters steady.

7. Reset the idle speed, stop the engine and install the vacuum lines and caps.

8. Install the fuel tank; see *Fuel Tank Installation* in Chapter Seven.

CYLINDER COMPRESSION

A cylinder cranking compression check is not *required maintenance,* but it is the quickest way to check the internal condition of the engine, i.e., rings, valves, head gasket, etc. It's a good idea to check compression at each tuneup, write it down and compare it with the reading you get at the next tune-up. This will help you spot any developing problems before they create a major breakdown.

1. Ride the bike to warm it up fully. Make sure the choke is OFF.

2. Remove the spark plugs.

3. Insert the tip of a compression gauge into the spark plug hole, making sure it seals fully (**Figure 73**).

4. Turn the kill switch off, hold the throttle wide open and crank the engine several revolutions until the gauge gives its highest reading. Record the number and repeat for the other cylinders.

When interpreting the results, the actual reading is not as important as the difference from the last check and the difference among cylinders. Individual gauge calibrations vary widely. A significant drop (more than 15 psi) since the last check (made with the same gauge) may indicate engine top end problems.

If the compression is 125 psi or more and there is less than a 15 psi difference among cylinders, compression is normal. If any cylinder reads less than about 110 psi, check your readings with a recently calibrated gauge. It may be time to rebuild the top end (rings and valves).

To tell the source of a problem, pour about a teaspoon of motor oil into the spark plug hole. Turn the engine over once to distribute the oil, then take another compression reading. If the compression increases significantly, the valves are good, but the rings are worn. If compression does not increase, the valves may be damaged.

STORAGE

Several months of inactivity can cause problems and a general deterioration of bike condition if proper care is neglected. This is especially true in areas of weather extremes.

During the winter months you should prepare your bike carefully for "hibernation."

Selecting a Storage Area

Most cyclists store their bikes in their home garages. If you do not have a garage, storage spaces are available for rent or lease in many areas. In selecting a building, consider the following points.

1. The storage area must be dry, free from excessive dampness. Heating is not necessary, but an insulated building is preferable.

2. Buildings with large window areas should be avoided or such windows should be masked if direct sunlight can fall on the bike (also a good security measure).

3. If you live near the ocean, make sure the area is sealed against salt spray and mist.

4. Select an area with minimum risk of fire or theft. Check your insurance to see if your bike is covered while in storage.

Preparing the Bike for Storage

Careful preparation will minimize deterioration and make it easier to restore the bike to service later. Use the following procedure.

1. Ride the bike until it is fully warmed up. Drain the oil, regardless of mileage since the

last oil change. Replace the oil filter and fill the engine with the normal quantity of fresh oil.

2. Wash the bike completely. Make certain to remove any road salt which may have accumulated during the first weeks of winter. Wax all painted and polished surfaces, including any chromed areas.

3. Remove the battery and coat the cable terminals with petroleum jelly. If there is evidence of acid spillage in the battery box, neutralize with baking soda, wash clean and repaint the damaged area. Store the battery in an area where it will not freeze and recharge it once a month.

4. Drain all gasoline from the fuel tank, interconnecting hoses and carburetors. As an alternative, a fuel preservative may be added

to the fuel and the tank should be filled to minimize water condensation. These preservatives are available from many motorcycle shops and marine equipment suppliers.

5. Remove spark plug and add a small quantity of oil to each cylinder. Crank the engine a few revolutions to distribute the oil and install the spark plug.

6. Check the tire pressures. Move the machine to the storage area and store it on the centerstand.

7. Cover the bike with material that will allow air circulation. Don't use plastic.

After Storage

A motorcycle that has been properly prepared and stored in a suitable area requires only light maintenance to restore it to service.

1. Thoroughly check all fasteners, suspension components and brake components.

2. Check the oil level and top up if necessary. If possible, change the engine oil and filter. The oil may have become contaminated with condensation.

3. Make sure the battery is fully charged and the electrolyte level is correct before installing the battery.

4. Fill the fuel tank with fresh gasoline. Check the fuel system for leaks. Drain and flush the fuel system if a preservative was used.

5. Before starting the engine, remove the spark plugs and spin the engine over a few times to blow out the excess storage oil. Place a rag over the cylinder head to keep the oil off the engine. Install new spark plugs and connect the spark plug leads.

6. Check safety items such as lights, horn, turn signals, etc., as oxidation of electrical contacts and/or sockets may make one or more of these electrical devices inoperative. Spray oxidized contacts with tuner cleaner available from most radio or electronic stores.

MODEL IDENTIFICATION

In the process of building motorcycles, the factory often introduces new models throughout the calendar year. New models, when introduced, are not necessarily identified by year but by model number suffix. See **Table 7** at the end of this chapter for model year and model suffix equivalents.

When you need to order parts for your motorcycle, make sure you get the right ones. Note the frame serial number on the steering head (**Figure 74**) and the engine serial number on the engine cases (**Figure 75**). Your dealer will often need these numbers to get the right parts for your bike.

GENERAL SPECIFICATIONS

See **Table 8** and **Table 9** for general specifications of all bikes covered by this manual.

Table 1   KZ750 MAINTENANCE SCHEDULE

Weekly/Gas Stop Maintenance	
Tire pressure	Check cold and adjust to suit load and speed
Brake function	Check for a solid feel
Throttle grip	Check for smooth opening and return
Clutch lever play	Check/adjust if necessary
Steering	Smooth but not loose
Drive chain	Lubricate every 200 miles (300 km) Check/adjust play if necessary
Nuts, bolts, fasteners	Check axles, suspension, controls and linkage/tighten if necessary
Engine oil	Check level/add oil if necessary
Lights and horn	Check operation, especially brake light
Engine noise and leaks	Check for any abnormality
Kill switch	Check operation
Monthly/3,000 Mile Maintenance (5,000 km)	
Battery electrolyte level	Check/add water if necessary Check more frequently in hot weather
Brake fluid	Check level/add if necessary
6 Month/3,000 Mile Maintenance (5,000 km)	
Air filter	Clean or replace
Air suction valves	Inspect
Fuel system	Drain float bowls
Spark plugs	Clean, set gap, replace if necessary
Valve clearance	Check/adjust if necessary
Carburetor	Check/adjust cable play, idle speed, mixture if necessary
Engine oil and filter	Change oil (and filter every other time)
General lubrication	Lube cables, levers, pedals, pivots, throttle grip
Clutch	Adjust clutch release
Tires	Check wear
Drive chain	Check wear
Brake pads	Check wear
Steering play	Check/adjust if necessary
Suspension	Check
Yearly/6,000 Mile Maintenance (10,000 km)	
Air filter	Replace
Brake fluid	Change
(continued)	

Table 1   KZ750 MAINTENANCE SCHEDULE (continued)

Yearly /6,000 Mile Maintenance (10,000 km)	
Fork oil	Change
Ignition advance	Lubricate
Nuts, bolts, fasteners	Check/tighten all
Swing arm	Grease pivot
2 year/12,000 Mile Maintenance (20,000 km)	
Speedometer gear housing	Grease
Wheel bearings	Grease
Steering bearings	Grease
2 Year Maintenance	
Brake master cylinder cup and dust seal	Replace
Brake caliper piston seal and dust seal	Replace
4 year Maintenance	
Brake hoses	Replace
Fuel hoses	Replace

Table 2   KZ750 TUNE-UP SPECIFICATIONS

Spark plug gap	0.028 – 0.032 in. (0.7-0.8 mm)
Spark plug type	
Normal conditions	
U.S.	NGK B8ES, ND WZ4ES-U
Others	NGK BR8ES, ND WZ4ESR-U
Cold weather (below 50° F/ 10° C) or low-speed riding	
U.S.	NGK B7ES, ND WZZES-U
Others	NGK BR7ES, ND WZZESR-U
Valve clearance (cold)	
Intake and exhaust	0.003 – 0.007 in. (0.08 – 0.18 mm)
Idle speed	1,000-1,100 rpm
Idle mixture (non-U.S. models)	2 turns out from seated, then adjust for highest idle speed

Table 3   KZ750 FORK AIR PRESSURE

Model	Standard	Range
KZ750-E, L	10 psi (70 kPa)	8.5-13 psi (60-90 kPa)
KZ750-H	8.5 psi (60 kPa)	7-14 psi (50-100 kPa)

Table 4   KZ750 STANDARD FORK OIL*

Model	Dry Capacity U.S. Fl. Oz. (cc)	Wet Capacity U.S. Fl. Oz. (cc)	Oil Level • Inches (mm)
KZ750-E,L			
U.S., Canada	8.4 (248)	7.8 (230)	12.0 (355)
Others	7.8 (232)	7.3 (215)	12.2 (362)
KZ750-H			
1980 (all)	9.5 (280)	8.8 (260)	14.7 (436)
1981 (all)	9.5 (280)	8.8 (260)	15.0 (445)

* Fork oil level is checked with forks fully extended and the fork spring removed. Use oil grade SAE 10W.

Table 5   KZ750 TIRES AND TIRE PRESSURE (U.S. AND CANADA)

Model/Tire Size	Pressure @ Load	
	0-215 lb. (0-97.5 kg)	215-364 lb. (97.5-165 kg)

KZ750-E (tubeless)
   Front 3.25H-19 4PR
   Rear 4.00H-18 4PR
KZ750-H (tubeless)
   Front 3.25H-19 4PR
   Rear 130/90-16 67H 4PR
	
	28 psi (200 kPa)
	32 psi (225 kPa)
	
	25 psi (175 kPa)
	22 psi (150 kPa)
		
		28 psi (200 kPa)
		36 psi (250 kPa)
		
		25 (175 kPa)
		25 (175 kPa)

Table 6   KZ750 TIRES AND TIRE PRESSURE (EUROPEAN)

Model/Tire Size	Pressure @ Load		
	0-210 lb. (0-95 kg)	210-300 lb. (95-136 kg)	300-397 lb. (136-180 kg)

Front
   KZ750-H,L 3.25H-19 4PR
      Up to 110 mph (180 kph)
      Over 110 mph (180 kph)
	
	25 psi (175 kPa)
	28 psi (200 kPa)
		
		25 psi (175 kPa)
		28 psi (200 kPa)
			
			25 psi (175 kPa)
			25 psi (175 kPa)

Rear
   KZ750-H 130/90-16 67H 4PR
   KZ750-L 4.00H-18 4PR
      Up to 110 mph (180 kph)
      Over 110 mph (180 kph)
	
	25 psi (175 kPa)
	28 psi (200 kPa)
		
		28 psi (200 kPa)
		32 psi (225 kPa)
			
			32 psi (225 kPa)
			36 psi (250 kPa)

Table 7   MODEL YEAR/SUFFIX DESIGNATION

1980	1981
KZ750-E1 KZ750-H1	KZ750-E2 KZ750-H2 KZ750-L1 (European)

Table 8   GENERAL SPECIFICATIONS
(1980 KZ750-E; 1980 KZ750-H, LTD MODEL)

General	
Engine type	4-stroke, 4-cylinder, DOHC, 2 valves/cyl
Lubrication system	Trochoid pump, wet sump
Clutch	Wet, 7 friction, 6 steel discs
Transmission	5-speed, constant mesh
Starting system	Electric starter
Ignition system	Transistorized
Charging system	Alternator, rectifier/regulator
Carburetion	4 Keihin CV34
Air filter	Dry paper
Dry weight	
E	463 lb. (210 kg)
H	465 lb. (211 kg)
Engine	
Displacement	45.0 cu. in. (738 cc)
Max. horsepower	74 @ 9,500 rpm
Max. torque	46 ft.-lb. (6.4 mkg) @ 7,500 rpm
Compression ratio	9.0:1
Valve timing	IN: Open 30° BTDC Close 60° ATDC EX: Open 60° BBDC Open 30° ATDC
Engine oil capacity	3.7 U.S. qt. (3.5 liters 3.1 Imp. qt.)

(continued)

Table 8   GENERAL SPECIFICATIONS1980 KZ750-E;
(1980 KZ750-H, LTD MODEL) (continued)

Transmission

Primary drive	Hy-Vo silent chain
Primary drive ratio	2.55 (27/23 x 63/29)
Transmission gears	Constant mesh, 5-speed
Gear ratios	
1st	2.33 (35/15)
2nd	1.63 (31/19)
3rd	1.27 (28/22)
4th	1.04 (26/25)
5th	0.88 (21/24)
Final drive ratio	
E	2.54 (33/13)
H	2.46 (32/13)

Drive chain	EK630S-T30, 84 links

Chassis

Type	Double cradle
Caster	
E	27°
H	30°
Trail	
E	4.2 in. (107 mm)
H	4.8 in. (121 mm)
Front suspension	Telescopic fork, oil damped, air/coil spring
Front wheel travel	
E	6.3 in. (160 mm)
H	7.1 in. (180 mm)
Rear suspension	Swing arm, dual shock absorber
Rear wheel travel	
E	3.7 in. (95 mm)
H	3.9 in. (100 mm)
Wheel type	Cast aluminum, tubeless
Wheel size	
E	Front MT1.85 x 19 Rear MT2.15 x 18
H	Front MT1.85 x 19 Rear MT3.00 x 16
Tire type	Tubeless
Front brake	Dual hydraulic disc, 8.9 in. (226 mm)
Rear brake	Hydraulic disc, 8.9 in. (226 mm)
Fuel tank capacity	
E	4.6 U.S. gal (17.3 liters, 3.8 Imp. gal.)
H	3.3 U.S. gal (12.4 liters, 2.7 Imp. gal.)

Electrical

Ignition system	Transistorized, mechanical advance
Initial timing	10° BTDC @ 1,050 rpm
Advance	40° BTDC @ 3,650 rpm
Charging system	Alternator, rectifier/regulator
Alternator output	238W @ 10,000 rpm
Regulator/rectifier	14.5 +/-0.5V
Battery	Furukawa FB12A-A
Capacity	12V, 12AH

(continued)

3

Table 8 GENERAL SPECIFICATIONS 1980 KZ750-E;
(1980 KZ750-H, LTD model) (continued)

Electrical (continued)	
Lighting	
Headlight	
E (U.S., Canada)	Sealed beam 60/50W
E (Europe)	Quartz halogen 60/50W
H (U.S., Canada)	Quartz halogen 60/55W
Tail/brakelights	
E (Europe, Australia)	8/21W
E (Others)	8/27W
H (Europe)	5/21W
H (Others)	8/27W
Turn signals	
Europe	21W
Others	23W
Turn signal/running light	23/8W
	3.4W
Meters & indicators	4W
City light (Europe)	

Table 9   GENERAL SPECIFICATIONS (1981 KZ750-E;
1981 KZ750-H, LTD MODEL; 1981 KZ750-L, EUROPEAN)

General	
Engine type	4-stroke, 4-cylinder, DOHC, 2 valves/cyl
Lubrication system	Trochoid pump, wet sump
Clutch	Wet, 7 friction, 6 steel discs
Transmission	5-speed, constant mesh
Starting system	Electric starter
Ignition system	Transistorized
Charging system	Alternator, rectifier/regulator
Carburetion	4 Keihin CV34
Air filter	Dry paper
Dry weight	
E	463 lb. (210 kg)
H,L	465 lb. (211 kg)
Engine	
Displacement	45.0 cu. in. (738 cc)
Max. horsepower	74 @ 9,500 rpm
Max. torque	46 ft.-lb. (6.4 mkg) @ 7,500 rpm
Compression ratio	9.0:1
Valve timing	IN: Open 30° BTDC Close 60° ATDC
	EX: Open 60° BBDC Open 30° ATDC
Engine oil capacity	3.7 U.S. qt. (3.5 liters, 3.1 Imp. qt.)
Transmission	
Primary drive	Hy-Vo silent chain
Primary drive ratio	2.55 (27/23 x 63/29)
Transmission gears	Constant mesh, 5-speed
Gear ratios	
1st	2.33 (35/15)
2nd	1.63 (31/19)
3rd	1.27 (28/22)
4th	1.04 (26/25)
5th	0.88 (21/24)

(continued)

Table 9   GENERAL SPECIFICATIONS (1981 KZ750-E;1981
KZ750-H, LTD MODEL; 1981 KZ750-L, EUROPEAN) (continued)

Transmission (continued)	
Final drive ratio	
E,L	2.54 (33/13)
H	2.46 (32/13)
Drive chain	EK630S-T30, 84 links
Chassis	
Type	Double cradle
Caster	
E,L	27°
H	30°
Trail	
E,L	4.2 in. (108 mm)
H	4.8 in. (121 mm)
Front suspension	Telescopic fork, oil damped, air/coil spring
Front wheel travel	
E,L	6.3 in. (160 mm)
H	7.1 in. (180 mm)
Rear suspension	Swing arm, dual shock absorber
Rear wheel travel	
E,L	3.7 in. (95 mm)
H	3.9 in. (100 mm)
Wheel type	Cast aluminum, tubeless
Wheel size	
E,L	Front MT1.85 x 19 Rear MT2.15 x 18
H	Front MT1.85 x 19 Rear MT3.00 x 16
Tire type	Tubeless
Front brake	Dual hydraulic disc, 8.9 in. (226 mm)
Rear brake	Hydraulic disc, 8.9 in. (226 mm)
Fuel tank capacity	
E	4.6 U.S. gal (17.3 liters, 3.8 Imp. gal.)
H	3.3 U.S. gal (12.4 liters, 2.7 Imp. gal.)
L	5.7 U.S. gal (21.7 liters, 4.8 Imp. gal.)
Electrical	
Ignition system	Transistorized, mechanical advance
Initial timing	10° BTDC @ 1,050 rpm
Advance	40° BTDC @ 3,650 rpm
Charging system	Alternator, rectifier/regulator
Alternator output	238W @ 10,000 rpm
Regulator/rectifier	14.5 +/-0.5V
Battery	Furukawa FB12A-A
Capacity	12V, 12AH
Lighting	
Headlight	Quartz halogen 60/55W
Tail/brakelights	
Europe, Australia	5/21W
Others	8/27W
Turn signals	
Europe	21W
Others	23W
Turn signal/running light	23/8W
Meters & indicators	3.4W
City light (Europe)	4W

3

CHAPTER FOUR

ENGINE

This chapter provides complete service and overhaul procedures for the Kawasaki 750 cc 4-cylinder engine. **Table 1** provides detailed specifications for the engine. **Table 2** provides tightening torques. All tables are at the end of the chapter. Routine inspections and adjustments, including a cranking compression test, are given in Chapter Three.

This chapter is written in a general teardown sequence. If you only need to remove one particular part, follow the *Disassembly* procedures until you have the part you want. Then refer to the *Inspection* procedure for that part and finally work through the *Assembly* procedures from installation of that part.

Although the clutch and transmission are located within the engine, they are covered in separate chapters to simplify the material.

Service procedures for all models are virtually the same. Where differences occur, they are identified. Right now, before you start any work, go back and read the *Service Hints* in Chapter One. You will save yourself a lot of mistakes with those hints fresh in your mind.

TOOLS

Several specialized tools will be helpful in the disassembly and inspection procedures in this chapter.

Although you may be able to make do without circlip or snap ring pliers (**Figure 1**), we highly recommend you have pliers with both straight and angled tips to prevent damage or loss of circlips.

Inspection measurements require a precision inside and outside micrometer, dial gauge or the equivalent (**Figure 2** and **Figure 3**). If you don't have the right tools, remove the parts and have your dealer or machine shop take the required measurements.

ENGINE DESIGN

The unit construction combines the power plant, clutch and transmission into one set of engine cases.

The crankcase is the front portion of the aluminum alloy engine cases, which are split horizontally. The forged one-piece crankshaft is mounted in 5 split insert main bearings. Power is delivered to a secondary shaft by a primary drive chain inside the crankcase. The secondary shaft drives the clutch ring gear and contains the starter clutch mechanism.

The 4 pistons operate inside an alloy cylinder block with pressed-in iron cylinder sleeves. The alloy cylinder head houses the 2 overhead camshafts. Both camshafts are driven by a single chain from a sprocket on the

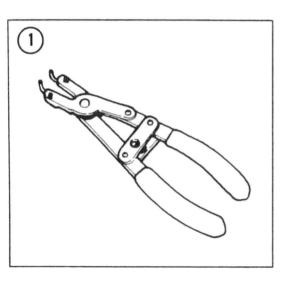

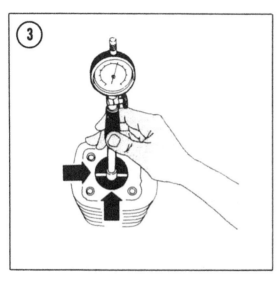

crankshaft (between cylinders 2 and 3). The cam lobes depress lifter cups fitted to the tops of the valve stems, opening the valves.

The alternator is mounted on the left end of the crankshaft. The ignition timing system (electronic pickups) and advance mechanism are on the right end of the crankshaft.

Engine lubrication is by wet sump, with the oil supply stored in the bottom of the crankcase. An oil pump, reached from the bottom of the engine, feeds the main and big-end bearings, primary chain, camshafts, valves and some of the bearings on the transmission shafts.

BREAK-IN

Following cylinder repair (boring, honing, new rings, etc.) and major lower end work, the engine should be broken in just as though it were new. The performance and service life of the engine depend greatly on a careful and sensible break-in. For the first 500 miles, no more than 1/3 throttle should be used and speed should be varied as much as possible within the 1/3 throttle limit. Avoid prolonged, steady running at one speed, no matter how moderate, as well as hard acceleration. Following the 500 mile service, increasingly more throttle can be used, but full throttle should not be used until the motorcycle has covered at least 1,000 miles and then it should be limited to short bursts until 1,500 miles have been logged.

SERVICING ENGINE IN FRAME

The engine has been laid out so that most "top end" repairs (camshaft, cylinder head, cylinder block and piston) can be done with the engine still in the frame. However, for repairs to the "bottom end" (crankshaft, cam chain, connecting rods and bearings), transmission and shift drum/forks, the engine must be removed from the frame for separation of the crankcases. Although the engine "top end" can be left attached for engine removal, we recommend that you remove it first. It makes the engine much lighter to handle and, while the engine is in the frame, you can use the rear brake to lock the drive train instead of resorting to makeshift or expensive tools.

Once the engine is removed from the frame, some parts (like the alternator rotor, engine sprocket, and clutch) can not be loosened without special tools or locking techniques.

CAM CHAIN AND TENSIONER

See **Figure 4**. Proper cam chain tension is essential for safe operation, quiet running and maximum power. The cam chain has no master link and wraps around the crankshaft,

so it can't be removed without splitting the crankcases; see *Crankshaft Removal* in this chapter. The chain guides are removed during "top end" disassembly.

The tensioner pushrod is held against the rear chain guide by the cam chain tensioner assembly. Two kinds of tensioners have been used on the KZ750 motorcycles: a "ball lock" tensioner on 1980 and early 1981 models and a "cross wedge" on late 1981 models.

The early ball lock tensioner sometimes was the source of a rattle or knock at low engine speeds, so Kawasaki cured the problem with the cross wedge tensioner. If your bike has the

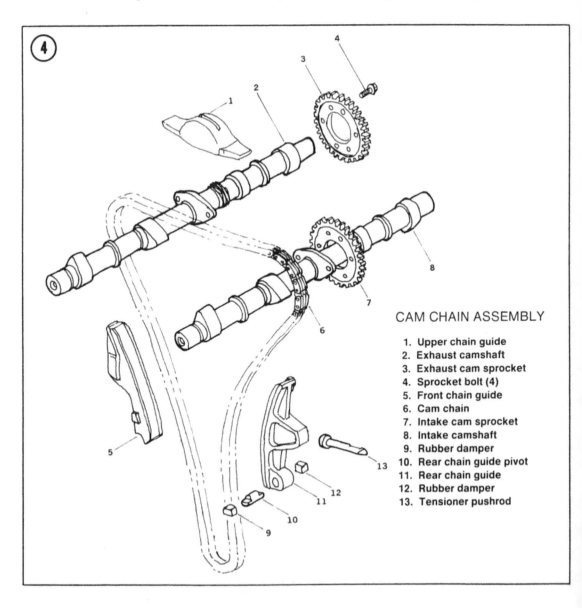

CAM CHAIN ASSEMBLY

1. Upper chain guide
2. Exhaust camshaft
3. Exhaust cam sprocket
4. Sprocket bolt (4)
5. Front chain guide
6. Cam chain
7. Intake cam sprocket
8. Intake camshaft
9. Rubber damper
10. Rear chain guide pivot
11. Rear chain guide
12. Rubber damper
13. Tensioner pushrod

ball lock tensioner and it is giving you no trouble, there is no need to change to the cross wedge type, but replacement parts are not available for the ball lock tensioner. If your ball lock tensioner is causing a knock or rattle, take your bike to a Kawasaki dealer for free installation of the improved cross wedge type.

BALL LOCK
CHAIN TENSIONER

See **Figure 5**. This automatic tensioner is continually self-adjusting. The bolt on the side is used only to lock the tensioner during engine disassembly/assembly. During normal operation, this short bolt doesn't touch the tensioner pushrod. The pushrod is free to move inward, but can't move out because of a one-way ball and retainer assembly.

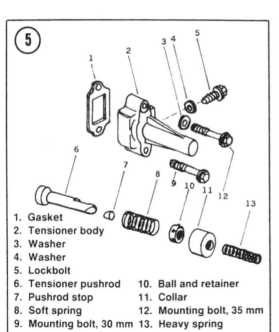

1. Gasket
2. Tensioner body
3. Washer
4. Washer
5. Lockbolt
6. Tensioner pushrod
7. Pushrod stop
8. Soft spring
9. Mounting bolt, 30 mm
10. Ball and retainer
11. Collar
12. Mounting bolt, 35 mm
13. Heavy spring

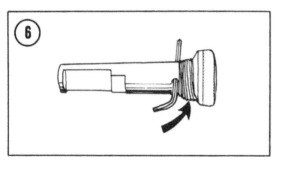

The tensioner should be locked before performing any work that slackens the cam chain. Special attention to tensioner installation is required during the *Valve Cover, Camshaft and Cylinder Head Removal/Installation* procedure in this chapter.

Locking the Tensioner

Remove the standard lock bolt and install a longer bolt to lock the tensioner. Any 6 mm diameter bolt about 16 mm or longer will do. After your top end assembly work is complete and the tensioner is in place, unlock the tensioner by removing the longer bolt and installing the original bolt and washer.

Removal

1. Remove the carburetor assembly (Chapter Seven). As an alternative, loosen the clamps at the front of the carburetors and slide the carburetor assembly back out of the front rubber holders. Push the front of the carburetors up for working room.
2. Lock the tensioner as described in this chapter.
3. Remove the 2 tensioner mounting bolts and the tensioner assembly.

> *CAUTION*
> *If the tensioner is not locked during removal, you must reset and lock the tensioner before installation. If this is not done, the pushrod will overextend and lock, damaging the cam chain when the mounting bolts are tightened. See **Resetting and Installation** in this chapter.*

Resetting and Installation

If the tensioner was properly locked during removal, start this procedure at Step 6. Steps 1-5 describe how to lock a tensioner that is not installed on the engine.
1. See **Figure 5**. Loosen the lock bolt and remove the pushrod, pushrod stop and fine spring.
2. Compress the spring against the pushrod head and hold it in place temporarily with a wire or awl (**Figure 6**).
3. Hold the tensioner rod up and drop the retainer and ball assembly onto the pushrod.

4. Insert the heavy spring and pushrod stop in the tensioner, then insert the pushrod and retainer into the tensioner body with the pushrod flat facing the lock bolt.

5. Push the pushrod in as far as it will go and hold it while you tighten the lock bolt. Remove the tool that kept the fine spring compressed. The pushrod shouldn't move. If it does, repeat Steps 1-5.

6. Install the locked tensioner assembly and gasket on the cylinder block. The upper bolt is longer than the lower and has an aluminum washer.

7. After engine top end assembly is complete, position the crankshaft with either No. 1 and 4 or No. 2 and 3 pistons at TDC; turn the crankshaft with the 17 mm bolt on the right end of the crankshaft, until either "T" mark aligns with the index mark (**Figure 7**).

CAUTION
Do not use the small inner bolt to turn the engine or you will damage the ignition advance mechanism.

8. Listen carefully as you loosen the lock bolt. You should hear the tensioner rod spring out against the cam chain and take up the chain slack. Do not start the engine until the tensioner is working properly.

CAUTION
Do not loosen the lock bolt if the valve cover is not installed. The pushrod will overextend and lock, damaging the cam chain when the other parts are tightened.

9. Tighten the lock bolt; it should turn freely all the way until the bolt head seats against the tensioner body. If the lock bolt doesn't screw in all the way, do not start the engine. Repeat this procedure starting with Step 1.

10. Install the carburetor assembly; see *Carburetor Installation* in Chapter Seven.

CAUTION
Make sure there are no air leaks where the carburetors join with the front rubber holders. You should be able to feel the carburetors bottom out in the holders on both sides. Any leakage here will cause a lean fuel mixture and engine damage.

CROSS WEDGE CHAIN TENSIONER

See **Figure 8**. This automatic tensioner is continually self-adjusting. The small bolt on the side (A, **Figure 9**) is used only to keep the tensioner pushrod from falling out during tensioner installation. During normal operation, this short bolt doesn't touch the tensioner pushrod. The pushrod is free to move inward, but can't move out because of the one-way cross wedge assembly.

The cross wedge cap (B, **Figure 9**) should be removed before performing any work that slackens the cam chain. Special attention to

1. Pushrod and spring	4. Cross wedge and spring
2. Tensioner body	5. Cross wedge cap
3. Bolt	

tensioner installation is required during the *Valve Cover, Camshaft and Cylinder Head Removal/Installation* procedure in this chapter.

Removal

1. Remove the carburetor assembly (Chapter Seven). As an alternative, loosen the clamps at the front of the carburetors and slide the carburetor assembly back out of the front rubber holders. Push the front of the carburetors up for working room.

2. Remove the tensioner cross wedge cap and spring (**Figure 10**).

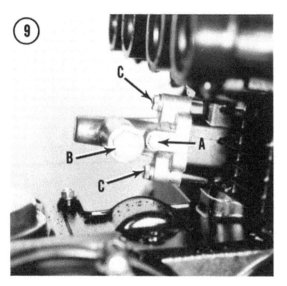

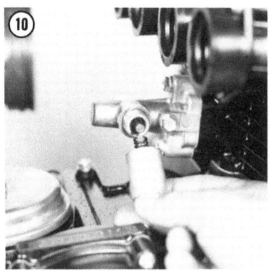

3. Remove the 2 tensioner mounting bolts (C, **Figure 9**) and the tensioner assembly.

> *CAUTION*
> *Do not loosen the tensioner mounting bolts without removing the cross wedge cap. The pushrod will overextend and lock, damaging the cam chain when the mounting bolts are tightened.*

Installation

1. See **Figure 8**. Remove the cross wedge cap and spring.
2. Insert the large spring and pushrod into the tensioner body with the pushrod flat facing the assembly bolt on the side of the tensioner.
3. Install and tighten the assembly bolt.
4. Install the tensioner body and gasket on the cylinder block. The upper bolt is longer than the lower and has an aluminum washer. There is no washer on the bottom bolt.
5. After engine top end assembly is complete, position the crankshaft with either No. 1 and 4 or No. 2 and 3 pistons at TDC; turn the crankshaft with the 17 mm bolt on the right end of the crankshaft, until either "T" mark aligns with the index mark (**Figure 7**).

> *CAUTION*
> *Do not use the small inner bolt to turn the engine or you will damage the ignition advance mechanism.*

6. Grease the tensioner cross wedge and push it into the body lightly by hand so the flat on the end of the cross wedge faces the flat on the end of the pushrod. The end of the cross wedge should stick out about 3/8 in. (10 mm) from the tensioner body (**Figure 11**). If the cross wedge sticks out much more or less then specified, recheck for proper engine and tensioner assembly. Also, the cam chain, guides or sprockets may be excessively worn.
7. Install the cross wedge spring, check that the aluminum washer is in place and install the cap.
8. Install the carburetor assembly; see *Carburetor Installation* in Chapter Seven.

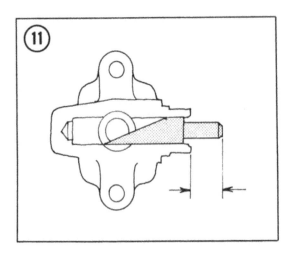

CAUTION
Make sure there are no air leaks where the carburetors join with the front rubber holders. You should be able to feel the carburetors bottom out in the holders on both sides. Any leakage here will cause a lean fuel mixture and engine damage.

VALVE COVER

Removal

The valve cover must be removed to inspect and adjust valve clearance.

1. Check that the ignition switch is OFF.
2. Remove the fuel tank; see *Fuel Tank Removal* in Chapter Seven.
3. Remove the spark plugs; grasp the spark plug leads as near to the plug as possible and pull them off the plugs. Clear away any dirt that has accumulated in the spark plug wells. Remove the spark plugs with a spark plug wrench.
4. Disconnect the leads to the ignition coils, then remove the coil bracket mounting bolts and the coils (**Figure 12**).
5. *U.S. models*: Slide up the lower hose clamps and pull the hoses off the air suction valve covers (**Figure 13**). Swing the vacuum switch and air hoses up out of the way. Remove the 8 bolts securing the air suction valve covers (**Figure 13**). Remove the covers and pull the suction valves up out of the valve cover (**Figure 14**).
6. Lock the cam chain tensioner (ball lock type) or remove the tensioner cross wedge cap (cross wedge type); see *Cam Chain Tensioner* in this chapter.

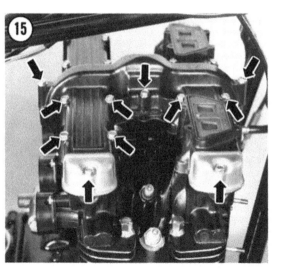

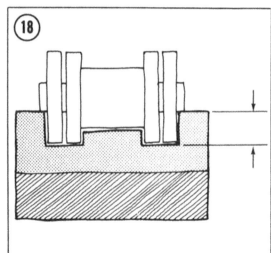

7. Remove the 20 bolts (U.S. model) or 24 bolts (others) securing the valve cover (**Figure 15**). Note which bolts have spark plug cable clamps attached. Tap around the cover's edge with a plastic or rubber mallet to loosen it and remove the cover.

Installation

1. Check that the cam chain tensioner is locked (ball lock type) or remove the tensioner cross wedge cap (cross wedge type); see *Cam Chain Tensioner* in this chapter.
2. Check that the 4 rubber plugs at the ends of the camshafts are in place and in good condition (**Figure 16**). When installing new plugs, coat the curved sides of the plugs with gasket sealer.

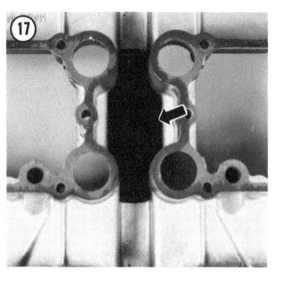

NOTE
If the rubber plugs leak oil, you may be able to stop leakage by removing and cleaning the plugs, then wrapping them around their circumference with 5 or 6 layers of Teflon tape, such as that used for plumbing pipe thread sealing.

3. Inspect the top cam chain guide in the valve cover (**Figure 17**) and replace it if worn deeper than the service limit in **Table 1** (**Figure 18**).
4. Place the valve cover gasket and the valve cover on the cylinder head. The arrow on the cover points to the front (**Figure 19**).

5. Install the valve cover bolts (**Figure 15**). Torque the valve cover bolts as specified in **Table 2**.

6. *U.S. models*: Install the 2 air suction valves in the valve cover (**Figure 14**). Install the suction valve covers (**Figure 13**). Install the hoses on the suction covers and slide the hose clamps into place (**Figure 13**).

7. Unlock the cam chain tensioner if locked (ball lock type) or install the tensioner cross wedge cap (cross wedge type); see *Cam Chain Tensioner* in this chapter.

8. Install the spark plugs.

9. Install the ignition coils and connect the primary coil wires (**Figure 20**). The *red* leads go to the positive (+) coil terminal. The *green* wire goes to the left coil and the *black* wire goes to the right coil.

10. Install the fuel tank; see *Fuel Tank Installation* in Chapter Seven.

11. After the engine has run and cooled off, retighten the valve cover bolts.

CAMSHAFTS

The exhaust camshaft is at the front of the engine and the intake camshaft is at the rear. The camshaft journals turn in bearing surfaces machined directly into the cylinder head. There are no separate bearings or bushings. If the bearing surfaces become damaged, the cylinder head must be replaced.

The cam lobes push down on valve lifter cups which move the valves off their seats. There is a shim under each cup that controls working valve clearance. The shims are available from Kawasaki in a wide range of thicknesses to adjust valve clearance. As the valve and valve seat wear, the working valve clearance decreases and the proper clearance must be restored by removing the original shim and installing a thinner one. See *Valve Clearance Inspection* in Chapter Three.

Removal

The camshafts can be removed after removing the valve cover as described in this chapter.

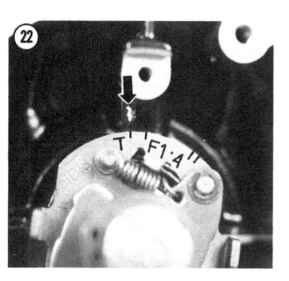

1. Remove the ignition timing cover and gasket from the lower right side of the engine (**Figure 21**).

2. Turn the crankshaft with the 17 mm bolt on the right end of the crankshaft, until the "T" mark next to the No. 1 and No. 4 "F" mark aligns with the fixed pointer (**Figure 22**). No. 1 and 4 cylinders are now at top dead center (TDC).

> *CAUTION*
> *Do not use the small inner bolt to turn the engine or you will damage the ignition advance mechanism.*

3. Remove the 16 camshaft cap bolts securing the camshaft caps (**Figure 23**). Gently tap the caps with a soft mallet to loosen them and lift them off. There is a hollow dowel pin at each cap.

4. Remove the camshafts. Tie the cam chain up to the frame with wire or place a tool through the chain loop to keep it from falling into the crankcase.

> *CAUTION*
> *If the crankshaft must be rotated when the cam chain is off the sprockets, pull up on the cam chain and keep it taut while turning the crankshaft. If the chain is slack, it may jam up at the crankshaft and damage the chain and the sprocket.*

> *CAUTION*
> *Do not install the camshafts on a cylinder head that has been removed from the engine. You may bend the valves; and the cams will have to be removed again before installing the cylinder head.*

Inspection

> *NOTE*
> *Camshaft journal/bearing surface clearance is measured with Plastigage while installing the camshafts. See* **Camshaft, Installation**.

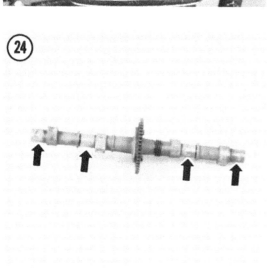

1. Check the camshaft journal outside diameters (OD) for wear and scoring (**Figure 24**). If any journal outside diameter is smaller than the wear limit in **Table 1**, install a new camshaft.

2. Measure the cam lobe height (**Figure 25**). Replace the camshaft if any lobe height is less than the wear limit in **Table 1**. The lobes should not be scored and the edges should be square. Slight damage may be removed with a silicon carbide oilstone. Use No. 100-120 grit initially, then polish with a No. 280-320 grit.

3. Check the camshaft sprockets for damaged teeth or other obvious faults (cracks, etc.). If damage can be seen, remove the sprocket from the camshaft. Sprocket installation is described under *Camshaft, Installation* in this chapter.

Camshaft Journal/Bearing Clearance Inspection

To check camshaft journal/bearing surface clearance with Plastigage, follow the *Camshaft Installation* procedure, but leave the cam and bearing surfaces *dry*. It is very important not to turn the cams while assembled dry. After checking clearance, remove and lubricate the cams.

Installation

Whenever a camshaft has been removed, the camshaft rotation must be timed in relation to crankshaft rotation or the engine will not develop normal power. If the timing is too far off, the valves could try to open when the piston is at TDC (top dead center). Bent or damaged valves and pistons could result.

1. Check that the cam chain tensioner was locked (ball lock type) or remove the tensioner cross wedge cap (cross wedge type); see *Cam Chain Tensioner* in this chapter.

2. The intake and exhaust sprockets are identical. If removed, install the sprocket on the camshaft, using the proper holes for each cam (**Figure 26**). The exhaust camshaft has a tachometer drive gear machined into it (**Figure 27**). The marked side of the sprocket faces the end of the cam with a notch in it. Use a locking agent such as Loctite Lock N' Seal on the sprocket bolts and torque them as specified in **Table 2**.

3. Remove the tachometer bolt and pinion holder stop(s) at the cylinder head (**Figure 28**).

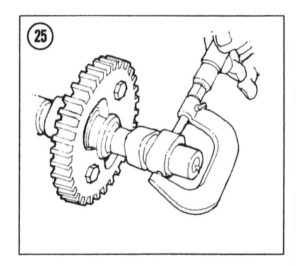

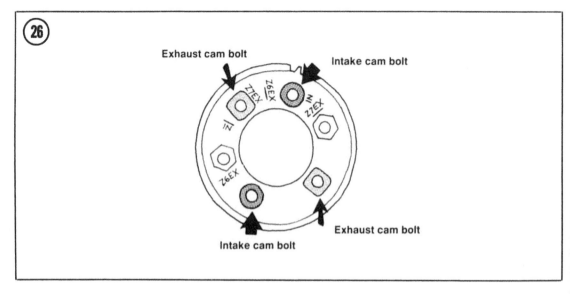

Exhaust cam bolt

Intake cam bolt

Exhaust cam bolt

Intake cam bolt

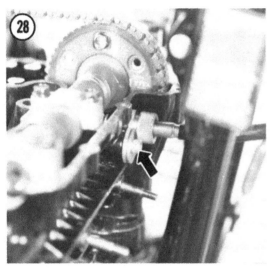

Remove the cable guide and tachometer drive gear from the cylinder head.

4. Make sure the No. 1 and 4 pistons are still at TDC; the "T" mark next to the No. 1 and 4 "F" mark should align with the index mark (**Figure 22**).

> *CAUTION*
> *Pull up on the cam chain and keep it taut while turning the crankshaft. If the chain is slack, it may jam up at the crankshaft and damage the chain and the sprocket.*

5. *If you are checking cam/bearing clearance with Plastigage*, cut strips of Plastigage and lay them lengthwise across each camshaft cap. Do not lubricate the journals or bearing caps until after you remove the Plastigage.

6. *If you are not checking cam/bearing clearance*, coat the camshaft journals and lobes with clean engine oil. Use molybdenum disulfide grease if new parts are being installed.

7. Insert the exhaust camshaft through the cam chain into the cylinder head. The exhaust camshaft has a tachometer worm gear machined into it (**Figure 27**). The notched end of the camshaft faces the right side of the engine.

8. Without turning the crankshaft, align the "Z7EX" line on the exhaust cam sprocket with the front cylinder head surface (**Figure 29**). Pull the cam chain tight at the front and fit it onto the sprocket.

> *NOTE*
> *Do not rotate either camshaft if Plastigage is in place. Align the marks by lifting and turning the camshaft.*

9. Without turning the crankshaft, insert the intake camshaft through the cam chain into the cylinder head. Align the "IN" line a little above the rear cylinder head surface (**Figure 30**) and fit the chain to the sprocket. The notched side of the camshaft faces the right side of the engine.

> *NOTE*
> *When the camshafts are correctly installed with the No. 1 piston at TDC, the notches on the right ends of the camshafts will point toward each other (**Figure 31**).*

10. Locate the cam chain pin on the exhaust sprocket in line with the "Z7EX" line (**Figure 32**). Beginning with this pin as zero, count off 45 pins toward the intake cam. The "IN" line on the intake sprocket must lie between the 45th and 46th pins. If it does not align, recheck your pin count and reposition intake camshaft if required.

11. Check that the cam chain is properly seated in the front and rear cam chain guides.

12. Check that the camshaft cap dowel pins are in place and loosely install the camshaft caps in their original location.

> *NOTE*
> *Each of the caps is numbered to match its location on the cylinder head and marked with an arrow that must point to the front of the engine.*

13. Tighten the bolts of the left inside cap on both camshafts just enough to seat the camshafts, then tighten the remaining cap bolts gradually. Torque the cap bolts as specified in **Table 2**.

14. If you are checking cam/bearing clearance with Plastigage:

    a. Remove the camshaft cap bolts. Remove the caps and measure the width of the Plastigage with the Plastigage wrapper (**Figure 33**). The material may stick to the camshaft journal or the cap.

    b. If any bearing clearance is larger than the limit in **Table 1**, install the camshaft caps on the cylinder head without the

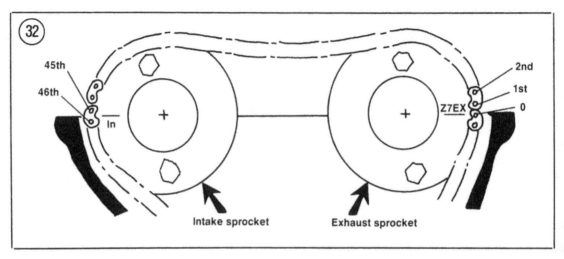

Intake sprocket   Exhaust sprocket

camshafts and measure the vertical bearing surface inside diameters (ID). If any bearing surface ID is larger than the wear limit in **Table 1**, install a new cylinder head. The bearing caps and cylinder head are machined as one piece and can not be replaced separately. If the bearing surface ID does not exceed the limit in **Table 1**, check clearance with a *new* camshaft. If a new camshaft will not bring the clearance within specification, the cylinder head must be replaced.

c. Remove the camshafts, clean all Plastigage from the camshafts and caps and reinstall the camshafts beginning at Step 6.

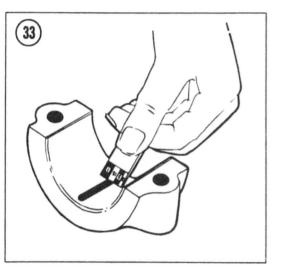

15. Make sure the No. 1 and 4 pistons are still at TDC; the "T" mark next to the No. 1 and 4 "F" mark should align with the index mark (**Figure 22**).

16. Slowly turn the crankshaft to the right (clockwise) 2 full turns, using the 17 mm bolt on the right end of the crankshaft. Check again that all timing marks align as shown in **Figure 32**. When you tighten the camshaft caps, the "IN" mark should align with the rear cylinder head surface. If so, the cam timing is correct.

> *CAUTION*
> *If there is any binding while turning the crankshaft, **stop**. Recheck the camshaft timing marks. Improper cam timing can cause valve and piston damage.*

17. *If the valve seats were ground or if a new valve, cylinder head, valve lifter or camshaft was installed*, check and adjust the valve clearance; see *Valve Clearance* in Chapter Three.

18. Apply molybdenum disulfide grease to the tachometer drive gear shaft and install the gear, holder, stops and Allen screw in the cylinder head (**Figure 28**).

19. Install the valve cover and gasket; see *Valve Cover Installation* in this chapter.

CYLINDER HEAD

The alloy cylinder head has cast-in valve seats and pressed-in valve guides. Each valve operates against 2 coil springs, one inside the other.

As the valves and valve seats wear, the valves move closer to the camshaft, decreasing the clearance. When the clearance can no longer be restored by fitting thinner valve shims, the end of the valve stem can be ground down slightly.

Removal

The camshafts must be removed as described in this chapter before removing the cylinder head.

1. Remove the exhaust system; see *Exhaust Removal* in Chapter Seven.

2. Remove the carburetor assembly; see *Carburetor Removal* in Chapter Seven.

3. Remove the 2 cylinder head bolts (A, **Figure 34**) and the 12 nuts and washers.

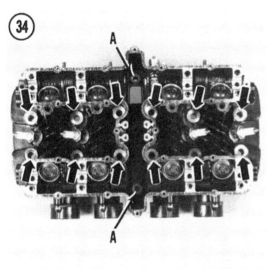

4. Remove the valve lifters and shims (**Figure 35**) now to avoid accidental mixup if they should fall out while removing the head. Remove lifters and shims one cylinder at a time and place them into a container (like an egg carton—see **Figure 36**) marked with the specific cylinder and "intake" or "exhaust." The No. 1 cylinder is on the left side of the bike.

> *CAUTION*
> *For minimum wear, the lifters must be reinstalled in their original location during assembly.*

5. Loosen the head by tapping around the edge with a rubber or plastic mallet. If necessary, *gently* pry the head loose with a broad-tipped screwdriver only in the ribbed areas of the fins.

> *CAUTION*
> *Remember, the cooling fins are fragile and may be damaged if tapped or pried too hard. Never use a metal hammer to loosen the cylinder head.*

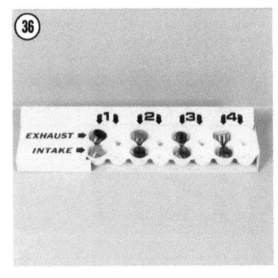

6. Lift the cylinder head straight up and off the studs and remove it.

7. Keep track of the 2 hollow dowel pins at the outer front cylinder block studs.

8. Keep the cam chain tied up (**Figure 37**) and place a clean shop rag into the cam chain opening in the cylinder block to keep out dirt.

> *NOTE*
> *If you are going to remove the valves yourself, go on to **Valve Removal**. Otherwise, take the cylinder head to a Kawasaki dealer or qualified specialist for valve and valve seat work.*

Inspection

1. Remove all traces of gasket or sealant from the head and cylinder block mating surfaces.

2. Remove all deposits from the combustion chambers and intake and exhaust ports with a wire brush or *soft* metal scraper. Be careful not to gouge the soft aluminum surfaces. Burrs will create hot spots which can cause preignition and heat erosion. Clean the spark plug hole threads with a fine wire brush, then clean the head thoroughly with solvent.

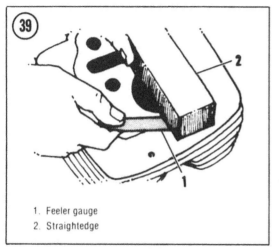

1. Feeler gauge
2. Straightedge

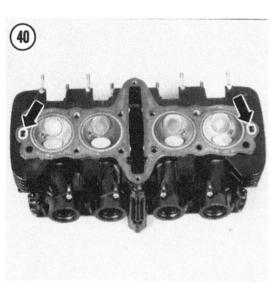

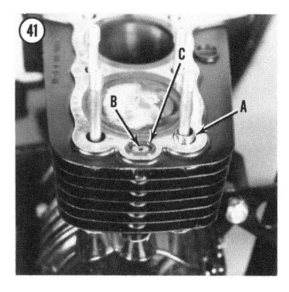

NOTE
If one or more of the combustion chambers contains unusually large carbon deposits, check the valve guides and oil seals for those combustion chambers very carefully.

3. *U.S. models*: Clean the air suction passages in the cylinder head exhaust ports (**Figure 38**).
4. Inspect the combustion chambers for cracks, especially between the valve seats. It may be possible to salvage a cracked head with Heliarc welding; check with a qualified specialist.
5. Place a straightedge across the gasket surface at several points. Measure warp by inserting a feeler gauge between the straightedge and cylinder head at each location (**Figure 39**). There should be less warp than the limit in **Table 1**. If a small amount is present, the head can be resurfaced. Have the job done by a machine shop.

Installation

1. Check that the top surface of the cylinder block and the bottom surface of the cylinder head are clean before installing a new gasket. Make sure the oil passages are clear (**Figure 40**).
2. Check that the 2 dowel pins (A, **Figure 41**) are installed at the outer front cylinder block studs and that the oil control orifices (B) are in place with their small hole pointing up.
3. Install 2 new O-rings (C, **Figure 41**).

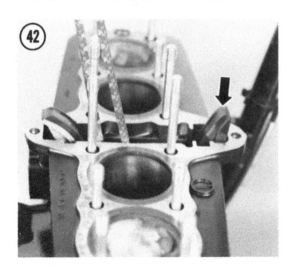

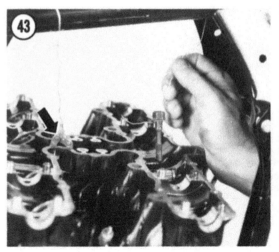

4. Apply a small amount of liquid gasket sealer to the top and bottom of a new cylinder head gasket *only* around the 4 outer corner cylinder block stud holes. Install the head gasket. If one side of the gasket is marked "TOP," that side faces up.

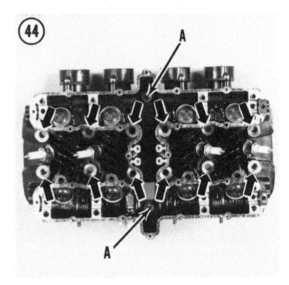

> NOTE
> *The cylinder head gasket is not syme-trical. Make sure the gasket matches all holes in the head and the center cam chain tunnel cutaway.*

5. Check that the front cam chain guide is correctly installed (**Figure 42**).

6. Lower the cylinder head onto the cylinder block studs, threading the cam chain up through the head. Tie the chain up (**Figure 43**) or stick a screwdriver through it to keep it from falling down into the crankcase.

7. Check that the cylinder head is fully seated against the cylinder block all around, then loosely install the 12 washers and nuts and the 2 bolts (A, **Figure 44**).

> NOTE
> *To keep from dropping the cylinder head bolts down into the crankcase, loop a piece of soft wire around the bolt heads as you lower them into place (**Figure 45**).*

8. Tighten the cylinder head nuts evenly in 2 stages to the torque specified in **Table 2**. Follow the sequence shown in **Figure 46**.

9. Torque the 2 cylinder head bolts as specified in **Table 2**.

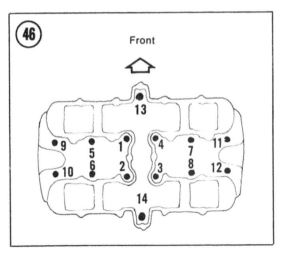

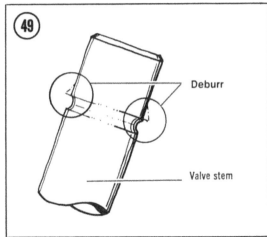

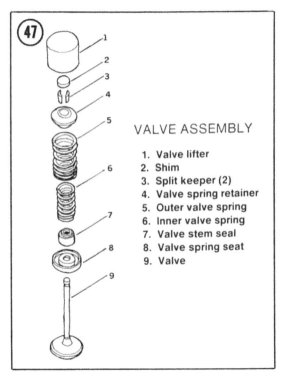

VALVE ASSEMBLY

1. Valve lifter
2. Shim
3. Split keeper (2)
4. Valve spring retainer
5. Outer valve spring
6. Inner valve spring
7. Valve stem seal
8. Valve spring seat
9. Valve

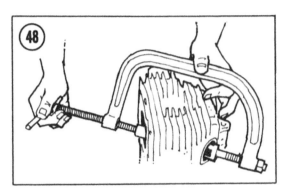

10. Oil and install the valve shims and lifters in the same locations from which they were removed. Apply molybdenum disulfide grease if new parts are being used.

VALVES AND GUIDES

Valve Removal

See **Figure 47**. The valves can be removed after removing the cylinder head as described in this chapter. Use this procedure if you are going to disassemble the valves yourself. Otherwise, take the cylinder head to a Kawasaki dealer or qualified specialist for valve and valve seat work.

1. Fit a valve spring compressor to the valve spring retainer and the bottom of the valve head (**Figure 48**). Use the tool to press down the valve spring retainer and expose the split keepers on the valve stem. Remove the keepers.

2. Retract the compressor tool and remove the valve.

> *CAUTION*
> *Remove any burrs from the valve stem grooves before removing the valve (**Figure 49**). Otherwise the valve guides will be damaged when the valves come out.*

3. Remove the valve seal and valve spring seat (**Figure 50**) with needlenose pliers. Discard the old valve seal and install a new one; rubber seals harden and crack with age and should be replaced whenever the valves are removed.

Valve/Seat Inspection

1. Clean valves with a wire brush and solvent. Inspect the contact face of each valve for burning (**Figure 51**). Minor roughness and pitting can be removed by lapping the valve as described in this chapter. Excessive unevenness of the valve face shows that the valve is not serviceable. The valve face may be ground lightly on a valve grinding machine, but it is best to replace a burned or damaged valve with a new one.

2. Measure the vertical runout of the valve stem with a V-block and dial indicator (**Figure 52**) or by rolling the stem on a piece of plate glass and measuring any gap with a feeler gauge. The runout should not exceed the service limit in **Table 1**.

3. Measure valve stems for wear (**Figure 53**). Replace the valve if the stem OD (outside diameter) is less than the wear limit in **Table 1**.

4. Measure the valve head thickness (**Figure 54**). If the head is thinner than the limit in **Table 1**, install a new valve.

5. Remove all carbon and varnish from the valve guides in the cylinder head with a stiff spiral wire brush.

6. Inspect the valve/guide clearance; insert each valve in its guide. Hold the valve just slightly off its seat and rock it sideways, then at a right angle to the first check (**Figure 55**). If it rocks more than the limit in **Table 1**, the guide is probably worn and should be replaced.

7. Take the cylinder head to a dealer and have the valve guides measured and replaced if necessary. Installation of new guides requires special installation and reaming tools.

8. Inspect the valve seats in the cylinder head. If worn or burned, they must be reconditioned with special cutting or grinding tools. This work should be done by your dealer or local machine shop. If you are performing the work yourself, see **Table 1** at the end of this chapter for valve seat specifications (**Figure 56**).

Valve Lapping

Valve lapping is a simple operation which can restore the valve seal without machining if the amount of wear or distortion is not too great.

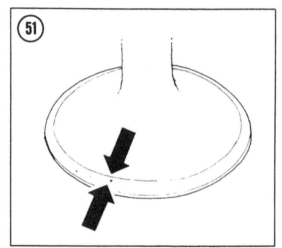

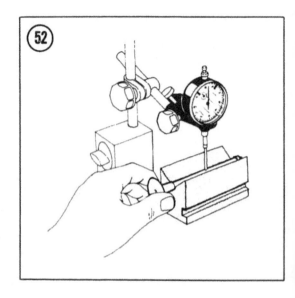

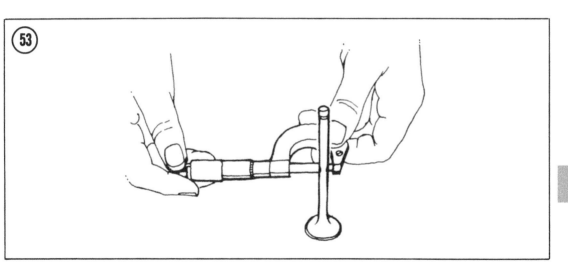

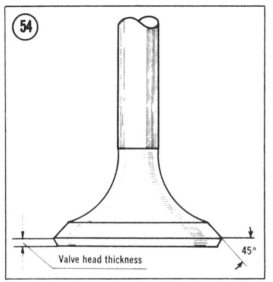

Valve head thickness
45°

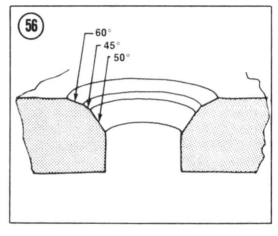

60°
45°
50°

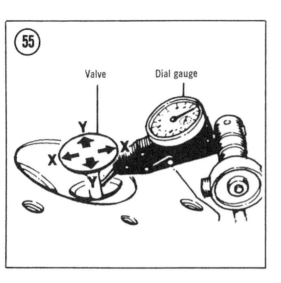

Valve   Dial gauge

NOTE
Valve lapping is not a substitute for precision grinding or cutting of valves and their seats. Get a professional opinion on whether lapping will do the job before you settle for it.

1. Coat the valve seating area in the head with a lapping compound such as Carborundum or Clover Brand.

2. Insert the valve into the combustion chamber.

3. Wet the suction cup of a lapping stick (**Figure 57**) and stick it onto the head of the valve. Lap the valve to the seat by rotating the lapping stick back and forth by hand. Every 5 to 10 seconds, turn the valve 120° in the seat; continue lapping until the contact surfaces of the valve and the seat are a uniform grey. Do not remove too much material.

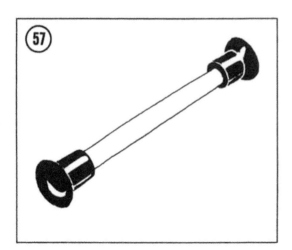

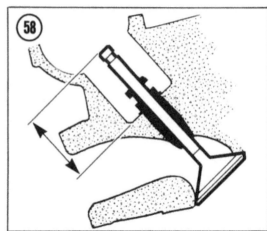

4. Thoroughly clean the valves and cylinder head in solvent to remove all grinding compound. Any compound left on the valves or the cylinder head will end up in the engine and cause serious damage.

5. After the lapping is finished and the valves have been reinstalled in the head, the valve seal should be tested. Check the seal of each valve by pouring solvent into the intake and exhaust ports. There should be no leakage past the valve seat. If fluid leaks past any of the seats, disassemble that valve assembly and repeat the lapping procedure until there is no leakage.

Valve Stem Height

If the valve faces or seats were reground or recut, the valves will drop deeper into the cylinder head than before. Check valve stem installed height before you assemble the valves in the cylinder head; otherwise, you may not be able to get proper valve clearance with the available shims.

1. Insert the valve into the cylinder head and measure valve stem height (**Figure 58**).

2. If the stem height is more than specified in **Table 1**, the valve is too long for that cylinder. Swap valves between cylinders to see if another valve will bring the stem height within tolerance; if not, install a new cylinder head.

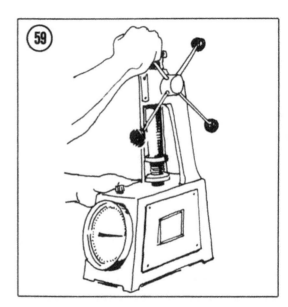

CAUTION
*If the valve stub (**Figure 58**) is ground to less than 0.165 in. (4.2 mm) thickness, the valve lifter may hit the spring retainer and drop the valve into the engine while running.*

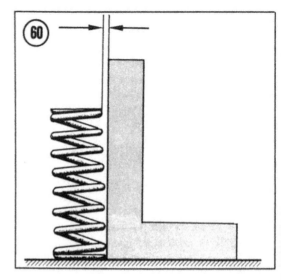

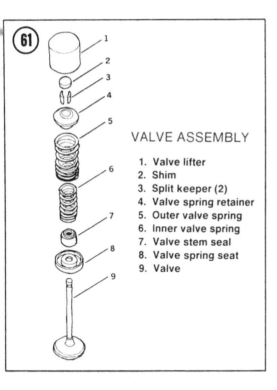

VALVE ASSEMBLY

1. Valve lifter
2. Shim
3. Split keeper (2)
4. Valve spring retainer
5. Outer valve spring
6. Inner valve spring
7. Valve stem seal
8. Valve spring seat
9. Valve

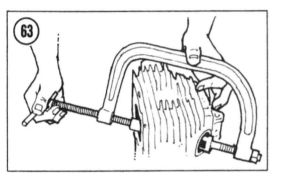

Valve Spring Inspection

As the valve springs wear, they become weaker. The valve springs must be checked for tension *while compressed* (**Figure 59**). See **Table 1** at the end of this chapter for compression specifications.

Measure each valve spring (inner and outer) for squareness by standing it on a flat surface and butting it against the vertical edge of a square (**Figure 60**). Install a new spring if the gap at the top is more than the limit in **Table 1**.

Check the valve spring retainer and valve keepers. If they are in good condition, they can be reused.

Valve Installation

1. See **Figure 61**. Install the spring seats and new oil seals on the valve guides.
2. Coat the valve stems with molybdenum disulfide grease and insert the valves into the cylinder head.
3. Install the 2 valve springs and the valve spring retainer.

NOTE
*Install variable pitch springs with the closely-wound coils toward the cylinder head (**Figure 62**).*

4. Fit a valve spring compressor to the valve spring retainer and the bottom of the valve head (**Figure 63**). Use the tool to press down the retainer and expose the keeper groove on the valve stem. Install the keepers and remove the spring compressor. Make sure the keepers are securely seated. Tap the stem end lightly with a hammer if necessary to jar the keepers into place.

CYLINDER BLOCK

The alloy cylinder block has pressed-in cylinder sleeves, which can be bored to 0.020 in. (0.5 mm) or 0.040 in. (1 mm) oversize.

The cylinder block can be removed after removing the cylinder head as described in this chapter.

Removal

1. Pull the front cam chain guide up out of the block (**Figure 64**).

2. Loosen the cylinder block from the crankcase; there is a cast-in pry point at the cylinder block base (**Figure 65**). Use the widest tool that will fit the slot. Do not hammer into the opening.

3. Pull the cylinder block straight up and off the pistons and studs.

4. Stuff a clean rag into the crankcase under each piston to keep dirt or small parts from entering.

5. Remove the 2 oil control orifices and O-rings from both ends of the crankcase, under the cylinder block.

6. Remove the O-rings from the studs to the right of the cam chain tunnel, if present.

7. Lift the rear cam chain guide up out of the crankcase.

Inspection

1. Do not remove the carbon ridge at the top of the cylinder bore unless you are going to install new piston rings or bore or hone the cylinder. The ridge helps the top ring's compression seal.

2. Check the cylinder walls for scratches; if evident, the cylinders should be rebored.

3. Measure the cylinder bores, with a cylinder gauge or inside micrometer, at the points shown in **Figure 66**. Measure 2 ways—in line with the piston pin and at a right angle to the pin. If any measurement exceeds the wear limit in **Table 1** or if the taper or out-of-round is greater than 0.002 in. (0.05 mm), the cylinders must be rebored to the next oversize and new pistons and rings installed. The liner can be bored oversize to 0.020 in. (0.5 mm) or 0.040 in. (1.0 mm). Rebore all cylinders even though only one may be faulty.

NOTE
Get the new pistons before you have the cylinders bored. You'll need them to achieve proper piston/cylinder clearance.

4. Inspect the cam chain guides (**Figure 67**). Replace them if visibly damaged or if worn deeper than the service limit in **Table 1** (**Figure 68**).

5. Make sure the 2 rubber dampers on the rear chain guide pivot shaft are bonded in place.

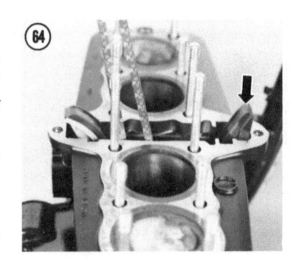

The "UP" marks face up, away from the shafts (**Figure 69**).

NOTE
The best adhesive to bond these dampers in place is one of the cyanoacrylate "super glues" like Loctite Super Bonder. Follow the manufacturer's instructions; this adhesive will stick your fingers together if you are not careful.

Installation

1. Check that the top surface of the crankcase and the bottom surface of the cylinder block are clean before installing a new gasket. Make sure the oil passages at either end of the cylinder block are clear (**Figure 70**).

2. Install the rear cam chain guide assembly (A, **Figure 71**).

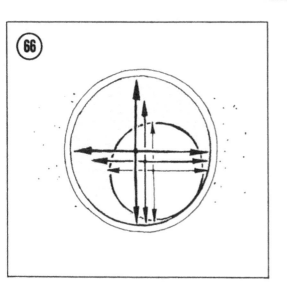

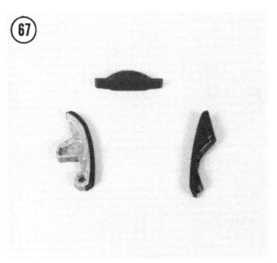

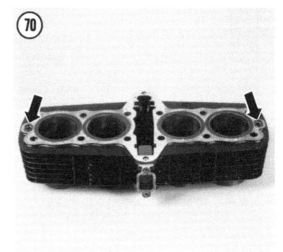

4

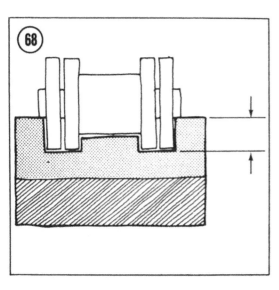

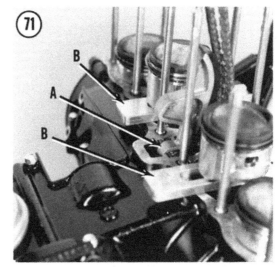

3. Install the 2 oil control orifices and new O-rings at either end of the crankcase (**Figure 72**). The end of the orifice with the small hole faces up.

4. Install new O-rings on the studs to the right of the cam chain tunnel, if present when disassembled. Not all models have these O-rings.

5. Install a new cylinder block base gasket on the crankcase. Apply a small amount of liquid gasket sealant to the top and bottom of the cylinder block base gasket *only* at the rear of the cam chain tunnel.

6. Install a piston holding fixture under the 2 inner pistons (B, **Figure 71**).

> *NOTE*
> *You can easily make a simple piston holding fixture out of wood. See **Figure 73**.*

7. Lightly oil the piston rings and cylinder walls.

8. Carefully slide the cylinder block down over the cylinder studs, threading the cam chain up through the block (**Figure 74**). Tie the chain up or stick a screwdriver through it to keep it from falling down into the crankcase.

9. Rock the cylinder block and slide it down over the inner 2 pistons and rings. Compress each piston ring with your fingers as the cylinder starts to slide over it, then slide the block down over the outer pistons.

> *NOTE*
> *If the rings are hard to compress, you can use a large hose clamp as a cheap, but effective piston ring compressor (**Figure 75**).*

10. Remove the piston holding fixture and push the cylinder block down until it seats on the crankcase. Check that the cylinder block is fully seated against the crankcases all around.

PISTONS AND RINGS

The pistons can be removed after removing the cylinder block as described in this chapter. With the pistons off, the connecting rods can be examined without separating the crankcase halves.

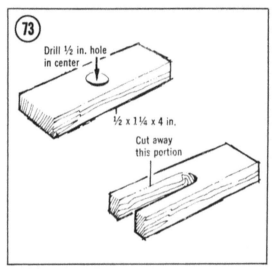

Drill ½ in. hole in center

½ x 1¼ x 4 in.

Cut away this portion

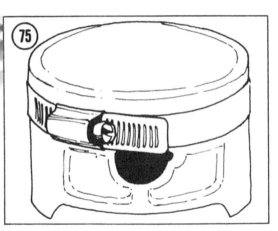

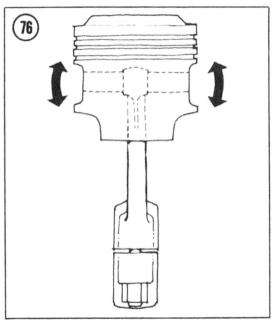

Each piston has 3 rings. The top 2 rings are compression rings, to prevent compression blowby into the crankcase. The bottom ring is an oil control ring, to keep excess oil out of the combustion chamber. Some models have a 1-piece oil control ring; others have a 3-piece oil control ring consisting of 2 flat rails with an expander in between.

Piston Removal

1. Lightly mark the top of the piston with a "1," "2," "3" or "4" so that each will be reinstalled into the correct cylinder. The No. 1 cylinder is on the left.

2. Before removing the piston, hold the rod tightly and rock the piston as shown in **Figure 76**. Any rocking motion (do not confuse with the normal side-to-side sliding motion) indicates wear on the piston, piston pin, connecting rod small bore or more likely a combination of all three. If there is detectable rocking, install new pistons and pins in sets.

3. Pry out one or both piston pin circlips with an awl (**Figure 77**). The circlip can spring out forcefully, so protect your face.

4. Push the piston pin out the side of the piston from which you removed the circlip. The pin will probably slide right out. Remove the piston and keep each piston pin inside its piston, so they can be reassembled in the original sets.

CAUTION
Do not try to hammer the pin out. You could bend the connecting rod or damage the rod bearings.

NOTE
It may be necessary to heat the piston slightly with a rag soaked in hot water or use a homemade tool to push the pin out, as shown in Figure 78.

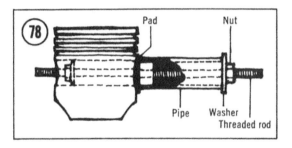

5. After you remove the piston, remove the top ring with a piston ring expander (**Figure 79**) or spread the ends with your thumbs just enough to slide it up over the piston (**Figure 80**). Repeat for the remaining rings.

> *WARNING*
> *The rail portions of a 3-piece oil scraper can be very sharp. Be careful when handling them to avoid cut fingers.*

Piston Inspection

1. Carefully clean the carbon from the piston crown with a chemical remover or with a *soft* scraper.
2. Clean the carbon and gum from the ring grooves with a broken ring or a groove cleaner (**Figure 81**). Any deposits left in the grooves will prevent the rings from seating correctly and may result in piston damage. Inspect the grooves carefully for burrs, nicks or broken or cracked lands. Recondition or replace the piston if necessary. Examine each ring groove for burrs, dented edges and wear. Pay particular attention to the top compression ring groove, as it usually wears more than the others.
3. Check piston wear; measure the OD (outside diameter) of the piston with a micrometer. Take the measurement 3/16 in. (5 mm) above the bottom of the piston skirt, at a right angle to the pin bore (**Figure 82**). If the diameter of the piston measures less than the wear limit in **Table 1**, install a new piston.
4. Measure piston-to-cylinder clearance as described in this chapter.

Piston/Cylinder Clearance

The most accurate way to check piston/cylinder clearance is to measure the inside diameter of the cylinder just above its bottom edge (where it will have undergone the least amount of wear), then subtract the piston diameter as measured in *Piston Inspection*. The clearance should be within the range specified in **Table 1**.

You can also measure installed piston/cylinder clearance with a feeler gauge near the bottom of the cylinder (**Figure 83**). The piston

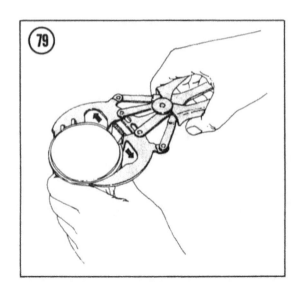

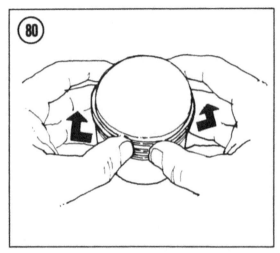

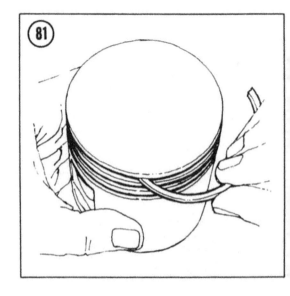

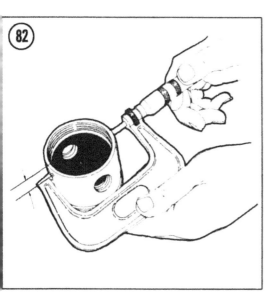

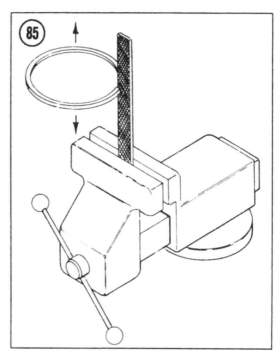

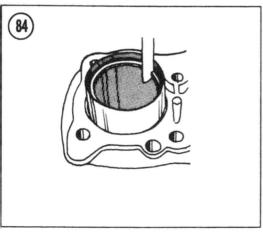

(with no rings) should be just free enough to slide with a light push. This method is *not* as accurate as micrometer measurement calculation.

If a cylinder has not worn past the acceptable inside diameter limit and installing a new piston will bring the clearance within tolerance, the cylinder block need not be bored. However, in no case should the piston/cylinder clearance be less than the minimum.

Piston Ring Inspection

Measure the top 2 rings for wear by inserting each into the *bottom* of the cylinder where the cylinder is least worn. Seat the ring squarely in the cylinder by pushing it in slightly with a piston. Measure the installed end gap with a feeler gauge (**Figure 84**). A new ring's gap should be no smaller than the limit in **Table 1**.

If the gap is smaller than specified, hold a small file in a vise, grip the ends of the ring with your fingers and enlarge the gap to the required minimum (**Figure 85**). As old rings wear, the gap will increase. Discard any rings whose installed gap exceeds the limit in **Table 1**. Always install new rings when installing new pistons or when you have any doubt about the condition of the rings.

Roll each ring around in its piston groove to make sure there is no binding (**Figure 86**). Check the side clearance of each ring with a feeler gauge (**Figure 87**). Refer to **Table 1**. If the clearance is incorrect, replace the pistons, rings or both.

Connecting Rod Inspection

After removing the pistons, the connecting rods can be inspected without removing them from the engine. If any rod must be replaced, see *Crankshaft Disassembly* in this chapter.

1. Measure the ID (inside diameter) of the small end of the connecting rod with a snap gauge and micrometer (**Figure 88**). If the ID is larger than the limit given in **Table 1**, install a new connecting rod.

2. Check the rod for obvious damage such as cracks and burns.

3. Check connecting rod big-end side clearance with feeler gauges (**Figure 89**). If the clearance exceeds the limit in **Table 1**, replace the connecting rod. If the clearance still exceeds the wear limit with a new connecting rod, the crankshaft should be replaced.

4. Check the connecting rod big end radial (up-and-down) clearance. You can make a preliminary inspection of the connecting rod big end radial clearance by turning the crankshaft until the crankpin is at TDC (top dead center). Grasp the connecting rod firmly and pull up on it. Tap sharply on the top of the rod with your free hand. If the bearing and crankpin are in good condition, there should be no movement felt in the rod. If movement is felt or if there is a sharp metallic click, measure the radial clearance with Plastigage as described in this chapter in *Crankshaft Inspection*.

Piston Installation

1. Spread the ends of the rings with your thumbs, or use a ring expander tool, and install them in the proper piston groove. Note the following.

   a. The side of the ring with a letter or number mark always faces *up* (**Figure 90**). If there is no mark, either side can face up.

   b. The top ring has chamfered outer edges and the second ring has a notched upper inner edge (**Figure 91**).

2. Coat the connecting rod holes and piston pin holes with clean engine oil.

> *NOTE*
> *When assembling used parts, be sure to install each piston on the same rod from which it was removed.*

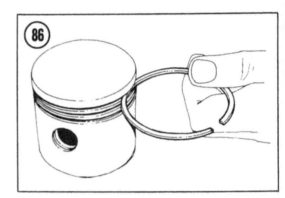

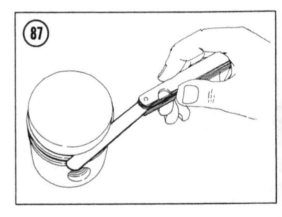

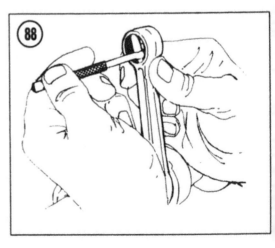

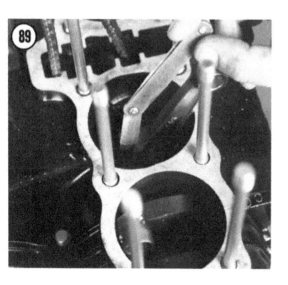

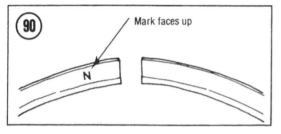

Mark faces up

N

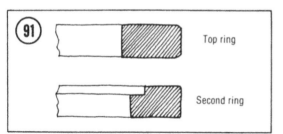

Top ring

Second ring

3. Oil the piston pin and push it into the piston until the end extends slightly beyond the inside of the boss.

4. Place the piston over the connecting rod with the arrow on the top of the piston pointing forward (**Figure 92**). Line up the pin with the rod small end and push the pin into the piston until it is even with the circlip grooves.

CAUTION
Do not try to hammer the pin in. You could bend the connecting rod or damage the rod bearings.

NOTE
*It may be necessary to heat the piston slightly with a rag soaked in hot water or use a homemade tool to push the pin in (**Figure 78**).*

5. Install new circlips where removed. After installing each circlip, rotate it so that the gap lies at the bottom or top (**Figure 93**). Make sure that the clips are fully seated in the grooves.

CAUTION
Never reuse an old circlip. It becomes weak in the process of removal and could work loose and cause serious engine damage.

6. To minimize blowby, rotate the top and bottom ring on each piston so that the ring openings face forward. Rotate the second ring

so that its opening faces the rear of the engine. With 3-piece oil rings, the 2 rails should each be offset about 30° to either side of the top ring (**Figure 94**).

ENGINE REMOVAL/INSTALLATION

Engine removal and crankcase separation is required for repair of the "bottom end" (crankshaft, connecting rods and bearings), transmission and shift drum/forks. Although the following parts can be left attached for engine removal, we recommend that you remove them first; it makes the engine much lighter to handle and , while the engine is in the frame, you can use the rear brake to lock the drive train when required instead of resorting to makeshift or expensive tools:

   a. Top end (camshafts, cylinder head, cylinder, pistons).
   b. Alternator rotor and electric starter (refer to Chapter Eight).
   c. Secondary shaft nut (see *Secondary Shaft Removal* in this chapter).
   d. Clutch hub and plates (refer to Chapter Five).
   e. Shift linkage (refer to Chapter Six).
   f. Engine sprocket (refer to Chapter Six).

Once the engine is removed from the frame, some parts (like the alternator rotor, secondary shaft nut, clutch hub nut and engine sprocket) can not be removed from the engine without special tools or locking techniques.

If you only need to repair the transmission, removal of the engine "top end" is not necessary.

Engine Removal

1. Drain the engine oil and discard it. Don't re-use old oil.
2. Disconnect the breather hose that runs from the air cleaner housing to the breather cap.
3. Remove the engine sprocket cover dowel pins for clearance when removing the engine (A, **Figure 95**).

4. To keep from bending the shift linkage, remove the shift linkage cover and linkage (B, **Figure 95**); see *Shift Linkage Removal* in Chapter Six.
5. Remove the clutch pushrod (C, **Figure 95**)
6. Disconnect the alternator leads and the neutral switch lead (**Figure 96**)
7. Remove the ignition timing cover at the right end of the crankshaft, then remove the 3 timing plate screws (**Figure 97**) and the pickup coil assembly.
8. Disconnect the oil pressure switch lead behind the ignition timing mechanism (**Figure 98**).
9. Remove the rear brake light switch (**Figure 99**).
10. Disconnect the batter ground lead at the right rear side of the engine (**Figure 100**).
11. Remove the right footpeg.
12. Remove the brake pedal.
13. Take a final look all over the engine to make sure everything has been disconnected.
14. Loosen, then remove, all engine mounting nuts and bolts on both sides of the engine (**Figure 101**).

> *WARNING*
> *Keep your hands clear as you remove the bolts. The engine can easily smash your fingers.*

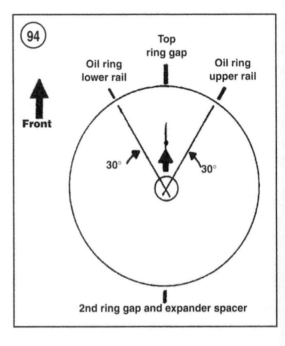

(94) Front

Top ring gap
Oil ring lower rail
Oil ring upper rail
30°   30°
2nd ring gap and expander spacer

NOTE
Bolt removal will be easier if you jack or lever the engine up slightly to take the weight off the bolts.

15. Pull the engine up slightly and to the right side so that it clears the lower front and rear right mounting support brackets. Take the engine to a workbench for further disassembly.

WARNING
If the recommended parts have not been removed, 2 people are required to safely remove the engine from the frame.

Engine Installation

If you have removed the recommended parts listed at the beginning of *Engine Removal*, leave them off until you have installed the bare engine in the frame. It will be easier to handle the engine and to tighten the rotor bolt, clutch hub nut, secondary shaft nut and engine sprocket nut.

1. Install the engine through the right side of the frame.
2. Install the two right and one left engine mounting brackets loosely with lockwashers.
3. Lift the engine and install the 6 long bolts with locknuts. Use a locking agent such as Loctite Lock N' Seal on all engine mounting bolts and nuts. The upper rear bolt has a spacer.
4. Torque the mounting bracket nuts, then the engine mounting nuts as specifed in **Table 2**.

NOTE
Some engines have shims at the engine mounting bolts to take up play resulting from manufacturing variations. Check all mating surfaces after you torque the engine mounting bolts. If there is a gap between the engine and frame, add shims to take up any slack and provide a rigid engine/frame assembly.

5. Install the brake pedal.
6. Install the output shaft O-ring (**Figure 102**).
7. Install the sprocket cover dowel pins (A, **Figure 95**) and the clutch pushrod (C).
8. Install the shift linkage (B, **Figure 95**) and the linkage cover; see *Shift Linkage Installation* in Chapter Six.

9. Push the output shaft sleeve through the linkage cover onto the output shaft (**Figure 103**).
10. Install the engine sprocket and sprocket cover; see *Engine Sprocket Installation* in Chapter Six. You can now use the rear brake to lock the engine while tightening the sprocket nut, alternator rotor bolt, clutch hub nut and secondary shaft nut.
11. Install the clutch; see *Clutch Installation* in Chapter Five.
12. Install the alternator and electric starter (refer to Chapter Eight).
13. Install the "top end" (pistons, cylinder, cylinder head, camshafts and valve cover) as

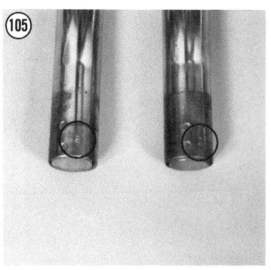

described in this chapter. Follow the *Installation* procedures when given seperately or reverse the *Removal* procedures when separate installation procedures are not given. Do not omit any steps. Note the following:

a. Secure the ignition timing leads in the clamps attached to the front and rear bottom clutch cover screws.

b. Route the alternator and starter leads behind the shift mechanism cover tabs (**Figure 104**).

c. After connecting multiple-pin plastic connectors, make sure none of the male pins have popped out of place. Tug on the wires to find a loose pin and push loose pins back into place until you feel the locking tang seat fully.

d. The oil pressure switch lead is *blue/red*. Install the oil pressure switch lead with the connector pointing to the rear.

e. The neutral switch lead is *green*.

f. The rear brake light switch leads are *blue* and *brown*.

g. When installing the footpeg assemblies, lubricate the damper rubbers with soapy water to ease installation.

h. *Ignition coils*: The No. 1 and 4 coil goes on the left. Connect the *red* leads to the positive (+) coil terminals. The *black* lead goes to the left coil and the *green* lead goes to the right coil.

i. Don't tighten the air cleaner housing bolts until after the carburetors are installed.

j. The No. 2 and the No. 3 exhaust pipes have identifying numbers (**Figure 105**).

k. Check that the engine oil drain plug is torqued as specified in **Table 2** and add engine oil to the crankcase; see *Engine Oil* in Chapter Three.

l. Adjust the throttle cables, clutch, drive chain, rear brake and rear brake light switch as described in Chapter Three.

m. Adjust the throttle cables, clutch, drive chain, rear brake and rear brake light switch as described in Chapter Three.

WARNING
Make sure none of the control cables or wires are stretched or pinched when the handlebars are turned from lock to lock.

CAUTION
Do not hurry to start or ride the motor-cycle yet. You have invested a lot of time, energy and money at this point. Do not take a chance on serious injury or mechanical damage. Thoroughly check and recheck all parts and controls on the motorcycle. Make sure all cables are correctly routed, adjusted and secured and all bolts and nuts are properly tightened. Position all cables away from the exhaust system.

n. Start the engine and check for leaks.

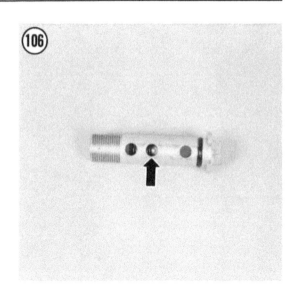

OIL FILTER
BYPASS VALVE

The oil filter bypass valve is inside the oil filter mounting bolt (**Figure 106**). If the oil filter becomes so dirty that it blocks oil flow to the engine, the bypass valve routes unfiltered oil directly to the engine.

To remove the bypass valve, remove the oil filter. The bypass valve can not be disassembled without damaging it, but you can push on the end of the bypass ball inside the valve to make sure it moves freely inside the valve body. If it does not move freely, replace it.

OIL PUMP

Operation

The trochoid oil pump is geared to the secondary shaft. The double rotor pump pulls oil through a coarse mesh screen and pushes it through the oil filter to trap fine particles. If the filter becomes clogged, a bypass valve routes the oil—still dirty—around the filter.

From the filter, one oil line goes to the oil pressure relief valve in the oil pan and then to the oil pressure switch behind the ignition timing plate (**Figure 107**). The pressure switch turns on the low oil pressure warning light when the oil pressure drops below a safe minimum level. The relief valve limits engine oil pressure by dumping oil back into the crankcase when the pump pressure exceeds the limit in **Table 1**.

The filtered oil follows several distribution paths:

a. A route goes to the crankshaft main bearings and big-end bearings, the spray from which lubricates the cylinder walls and the piston pins.

b. A route goes through an orifice and leads to the camshaft journals; after the oil leaves the camshaft journals it lubricates the cam lobes and valve lifters.

c. A route goes through a spray nozzle to the primary drive chain.

d. A route goes through an orifice to the secondary shaft and starter clutch.

e. A route goes through an orifice to the transmission main shaft and output shaft

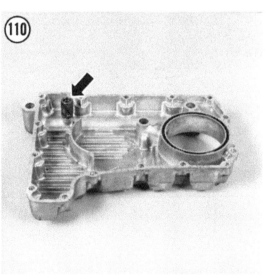

bearings, then through the shafts to the gear bushings.

Oil Pressure Check

To check the operation of the oil pump and relief valve, start with a cold engine. Remove the plug from the oil pressure check point on the right side of the crankcase (**Figure 108**) and attach an oil pressure gauge (Kawasaki has a special gauge available).

Check the relief valve by starting the engine and observing the cold oil pressure as you increase engine speed. If the cold oil pressure exceeds the upper limit in **Table 1**, the relief valve may be stuck closed. If the cold oil pressure is less than the lower limit when the engine speed is above about 5,000 rpm, the relief valve may be stuck open. If you suspect the relief valve is faulty, remove it as described in *Oil Pump Removal*. Check that the internal ball slides smoothly when pushed away from its seat and that the ball seats fully when released.

Check the oil pump operation by running the engine long enough to warm the oil to normal operating temperature. The oil pressure should be as specified in **Table 1**. If the oil pressure is very low, inspect the oil pump, then the oil distribution routes.

Removal

The oil pump can be removed without removing the engine from the frame. For clarity, the engine is shown removed.

1. Drain the engine oil and remove the oil filter.

2. Remove the exhaust system; see *Exhaust Removal* in Chapter Seven.

3. Remove the 15 bolts and washers that mount the oil pan to the crankcase (**Figure 109**). Remove the oil pan, gasket and 3 O-rings.

4. Remove the oil pressure relief valve from the oil pan (**Figure 110**).

5. Remove the clutch assembly; see *Clutch Removal* in Chapter Five.

6. Remove the one bolt and two screws that mount the oil pump to the right side of the crankcase (**Figure 111**). Pull out the oil pump assembly.

Disassembly/Inspection/Assembly

1. See **Figure 112**. Inspect the pump body for cracks.
2. Remove the circlip and the washer from the pump shaft (A, **Figure 113**).
3. Remove the 3 pump cover screws (B, **Figure 113**) and the cover and gasket.
4. With feeler gauges, measure the minimum clearance between the inner rotor and the outer rotor (**Figure 114**).
5. Measure the clearance between the outer rotor and the pump body (**Figure 115**).
6. Measure the rotor side clearance with a straightedge and feeler gauge (**Figure 116**).
7. Replace any parts whose clearance has worn larger than the limits specified in **Table 1**.
8. To assemble the oil pump, reverse the disassembly procedure. Note the following:
   a. Check that the inner dowel pin is in place (**Figure 117**).
   b. Use a new gasket.
   c. After assembly, rotate the shaft to make sure it turns freely.

Installation

1. Check that the 2 dowel pins are in place in the right side of the crankcase (**Figure 118**).
2. Fill the pump with engine oil to prime it.
3. Install the pump in the engine. If it will not seat easily, rotate the drive gear enough to

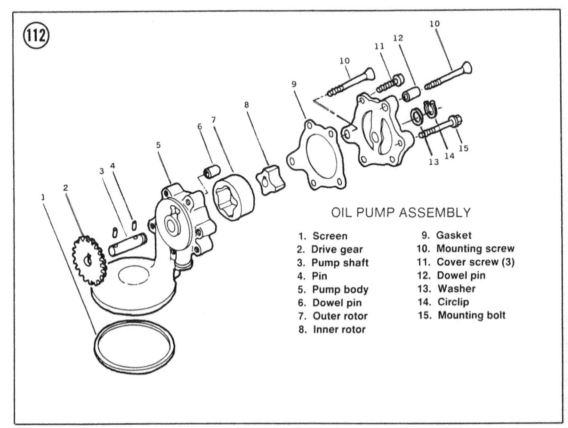

OIL PUMP ASSEMBLY

1. Screen
2. Drive gear
3. Pump shaft
4. Pin
5. Pump body
6. Dowel pin
7. Outer rotor
8. Inner rotor
9. Gasket
10. Mounting screw
11. Cover screw (3)
12. Dowel pin
13. Washer
14. Circlip
15. Mounting bolt

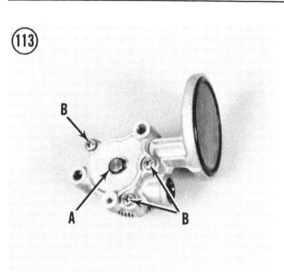

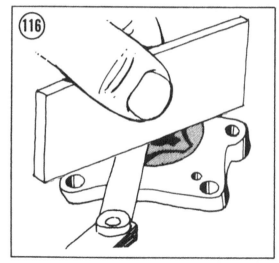

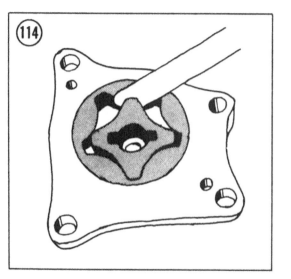

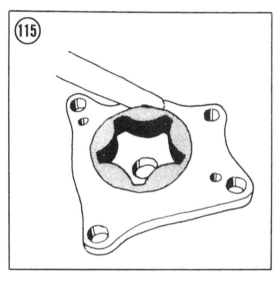

4

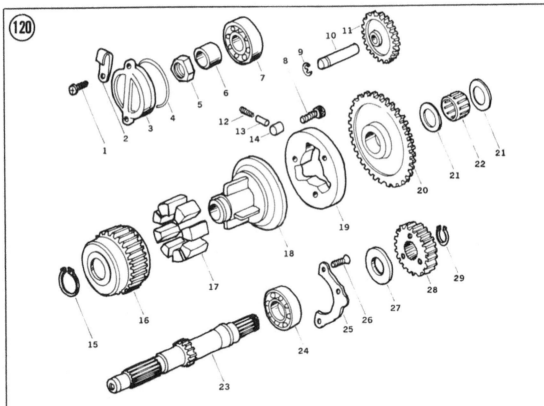

SECONDARY SHAFT/STARTER CLUTCH

1. Cover screw
2. Wiring clamp
3. Bearing cap
4. O-ring
5. Nut
6. Sleeve
7. Ball bearing
8. Starter clutch screw (3)

9. Circlip
10. Idler shaft
11. Starter idler gear
12. Spring (3)
13. Spring cap (3)
14. Roller (3)
15. Circlip

16. Driven sprocket
17. Rubber damper
18. Inner coupling
19. Starter clutch
20. Starter clutch gear
21. Washer
22. Needle bearing

23. Secondary shaft
24. Ball bearing
25. Bearing retainer
26. Screw (2)
27. Thrust washer
28. Secondary drive gear
29. Circlip

mesh it with the secondary shaft gear. Stake the oil pump mounting screws with a punch to lock them in place.

4. Install the oil pressure relief valve in the oil pan. Use a locking agent such as Loctite Lock N' Seal on the relief valve threads.

5. Check the oil passage O-rings (**Figure 119**). Install new ones if damaged. The flat side of the O-ring faces the engine.

6. Install the oil pan with a new gasket. Tighten the mounting bolts in a crisscross pattern to the torque specified in **Table 2**.

7. Install the clutch assembly; see *Clutch Installation* in Chapter Five.

8. Install the exhaust system; see *Exhaust Installation* in Chapter Seven.

9. Install the oil filter and add engine oil; see *Engine Oil and Filter* in Chapter Three.

SECONDARY SHAFT AND STARTER CLUTCH

Refer to **Figure 120**. The secondary shaft is an intermediate shaft driven by a chain from the crankshaft. The secondary shaft transmits power to the clutch ring gear; it also turns the crankshaft when driven by the electric starter.

The driven sprocket on the secondary shaft has an internal rubber damper to smooth engine torque peaks. The starter drive clutch is mounted to the sprocket damper.

Secondary shaft removal is required before the crankcase halves can be separated. The secondary shaft can be removed while the engine is in the frame in order to repair the starter drive clutch.

Primary Chain Inspection

The primary chain can be inspected after removing the oil pan. See *Oil Pan Removal* in this chapter. The primary chain should be replaced when it has more than 1.06 in. (27 mm) of up-and-down slack (**Figure 121**). Inspect the secondary shaft sprocket carefully when you replace the chain. It is likely that both will need to be replaced at the same time.

Secondary Shaft/Clutch Removal

The secondary shaft can be removed without removing the engine from the frame, after removing the oil pump. For clarity, the engine is shown removed.

1. Remove the remaining screw in the clutch-side bearing retainer (**Figure 122**).

2. Remove the engine sprocket cover; see *Sprocket Cover Removal* in Chapter Six.

3. Remove the bearing cap from the alternator side of the engine (**Figure 123**).

4. Remove the secondary shaft nut (**Figure 124**).

NOTE
While the engine is in the frame, you can shift the engine into gear and lock the drive train with the brake pedal. There will be considerable play in the rubber damper, but the nut will loosen.

5. Tap on the left end of the secondary shaft with a soft driver until the right end pushes the right bearing out **(Figure 125)**.

6. Pull the secondary shaft out of the clutch side of the engine.

7. Remove the starter clutch/secondary sprocket from the crankcase.

8. Remove the circlip and remove the starter idler gear and shaft from the crankcase.

Starter Clutch Disassembly/Assembly

1. Pull the starter clutch gear out of the starter clutch.

2. Remove the 3 rollers, spring caps and springs from the clutch **(Figure 126)**.

3. Remove the 3 Allen screws that mount the clutch body to the rotor.

4. The secondary shaft internal rubber dampers can be replaced after removing the circlip on the end of the driven sprocket.

5. To assemble, reverse this procedure. Use a locking agent such as Loctite Lock N' Seal on the 3 starter clutch Allen bolts.

Secondary Shaft/Clutch Installation

1. Install the starter idler gear with the gear hub boss pointing to the clutch side of the engine **(Figure 127)**. Install the circlip on the idler gear shaft.

2. Install the starter clutch/secondary sprocket through the primary chain **(Figure 128)**.

3. Turn the starter clutch gear while pushing it into the clutch and roller assembly **(Figure 129)**.

4. Put the thrust washer, needle bearing and second thrust washer on the secondary shaft **(Figure 130)**, then insert the secondary shaft through the clutch side of the engine, aligning the shaft splines with the secondary sprocket splines. You will probably have to tap on the end of the shaft to seat the bearing in the crankcase. Before the shaft is all the way in, insert the bearing retainer **(Figure 131)**.

5. Tap the secondary shaft into place, until the clutch-side bearing bottoms in the crankcase. Install the bearing retainer screw and stake it in place **(Figure 122)**.

6. Check that the alternator-side bearing sleeve is in place **(Figure 132)**.

NOTE
The alternator-side bearing should be 0.42-0.44 in. (10.7-11.3 mm) inside the outer surface of the crankcase boss.

7. Install the secondary shaft nut and torque the nut as specified in **Table 2**.

NOTE
While the engine is in the frame, you can shift the engine into gear and lock the drive train with the brake pedal. There will be considerable play in the rubber damper, but the nut will tighten.

8. Install the alternator-side bearing cap with wiring clamp (**Figure 123**).

9. Install the engine sprocket cover; see *Sprocket Cover Installation* in Chapter Six.

CRANKCASE

Separation

If you only need to repair the transmission, removal of the engine "top end" is not necessary before separating the case halves.
1. Remove the engine from the motorcycle; see *Engine Removal*.
2. Remove the ignition pickup coil assembly and the ignition advance unit as described in Chapter Eight.
3. Check that all parts outside of the crankcases have been removed from the engine (**Figure 133** and **Figure 134**).

NOTE
Secondary shaft removal is part of this crankcase separation procedure.

4. Remove the 13 upper crankcase bolts (**Figure 135**).
5. Turn the engine over and remove the oil pump as described in this chapter.

NOTE
*Before you proceed, inspect the primary chain play. The primary chain should be replaced when it has more than 1.06 in. (27 mm) of slack (**Figure 121**). Inspect the secondary shaft sprocket carefully when you replace the chain. It is likely that both will need to be replaced at the same time.*

6. Remove the secondary shaft, starter clutch and idle gear; see *Secondary Shaft/Clutch Removal* in this chapter.

7. Remove the 17 lower crankcase bolts (**Figure 136**). Don't miss the bolt at the rear of the engine.

> *CAUTION*
> *Make sure that you have removed all the fasteners. If the cases are hard to separate later, check for any fasteners you may have missed.*

8. Pry the crankcase halves apart, using the largest tool that will fit in the 4 pry points (**Figure 137**). There are pry points on both sides of the engine. If you encounter resistance, check for bolts you may have missed. Don't overlook the upper bolt under the starter motor.

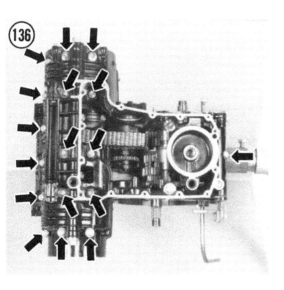

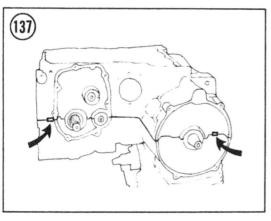

9. Lift the bottom crankcase half off. See Chapter Six for transmission removal and inspection.

> *NOTE*
> *Keep each crankshaft bearing insert in its original place in the crankcases. If you are going to assemble the engine with the original inserts, they must be installed exactly as removed in order to prevent rapid wear.*

Inspection

Check the crankcase halves for cracks or fractures in the stiffening webs, around the bearing bosses and at threaded holes. While such damage is rare, it should be checked for, particularly following a major failure (such as piston breakage, bearing failure or gear breakage) or after a collision or hard spill in which the engine suffers external damage.

If cracks or fractures are found, they should be repaired immediately by a reputable shop experienced in and equipped to perform repairs on precision aluminum castings.

The upper and lower crankcase halves are machined as a pair and if one half is not usable, both must be replaced.

Bearing and Seal Replacement

The crankshaft oil seals should be replaced whenever the engine is disassembled. Seals should be installed with their marked side facing out of the engine. The secondary shaft and transmission bearings should be replaced if there is any doubt about their condition.

Most bearings and seals are installed with an interference fit. If possible, heat the part in an oven to about 212° F (100° C) to aid seal or bearing removal and installation.

> *CAUTION*
> *Heating should be done in an oven and not with a torch; it's hard to get uniform heating with a torch and the likelihood of warping the case is great.*

If the bearings are hard to remove, they may need to be tapped out with a socket or piece of pipe the same size as the bearing outer race. Install the new bearing while the parts are still

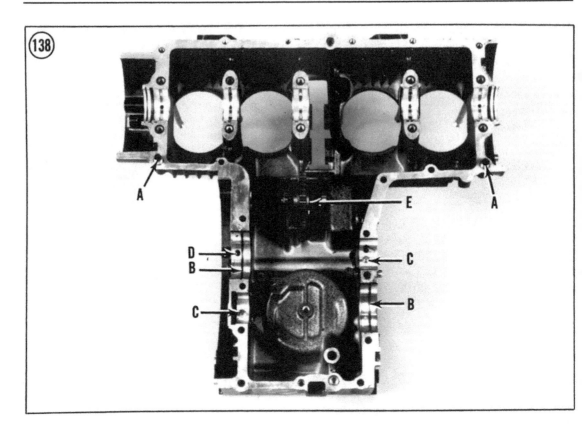

hot. The marked side of a bearing faces out. If necessary, freeze the bearing or seal before installing, to provide extra clearance.

If bearings are hard to remove or install, don't take a chance on expensive case damage. Have the work done by a Kawasaki dealer or competent machine shop.

Assembly

The crankcases should be assembled with the engine upside down.

> *CAUTION*
> *Set the upper crankcase half on wood blocks to protect the cylinder studs.*

1. In the upper crankcase half, check that the 2 crankcase dowel pins (A, **Figure 138**), 2 transmission bearing alignment 1/2 rings (B), 2 transmission dowel pins (C) and the oil passage plug (D) are in place.
2. Check that the starter idler gear is installed in the upper case half with the gear hub boss pointing to the clutch side of the engine (E, **Figure 138**). Install the circlip on the idler gear shaft.

3. In the lower crankcase half, check that all oil passages are clear and that the 2 oil control nozzles are in place (**Figure 139**).
4. Install the main bearing inserts in the upper and lower crankcase halves. Make sure they are locked in place (**Figure 140**).

> *NOTE*
> *If the old inserts are reused, be sure they are installed in their original positions to prevent rapid wear.*

5. Install new left and right oil seals on the ends of the crankshaft (**Figure 141**); apply high-temperature grease to the seal lips. The arrow on the seal must face out and must point in the direction of crankshaft rotation (clockwise when viewed from the right side of the engine).

6. Oil the main bearing inserts and crankshaft main journals.

7. Position the cam chain and primary chain on the crankshaft.

8. Install the crankshaft assembly with primary chain and cam chain in the upper crankcase half (**Figure 142**). The oil seal ribs must fit into the grooves in the crankcase.

9. Install the transmission shafts in the upper crankcase half as described in *Gear/Shaft Installation* in Chapter Six. Align the transmission gears in the NEUTRAL position (**Figure 143**).

10. Rotate the shift drum to the NEUTRAL position (**Figure 144**).

11. *Transmission test*: This is an optional procedure to make sure the transmission has been assembled properly.

   a. Carefully place the lower case half in position, fitting the shift forks in their proper gear grooves (**Figure 145**).

   b. Seat the upper case half onto the lower and tap lightly with a plastic or rubber mallet—do not use a metal hammer or it will damage the cases.

c. Carefully rotate the shift drum while turning the output shaft. Check that all gears engage smoothly and that each gear position can be identified. You will have to spin the output shaft quickly to keep the "neutral finder" from locking you out of 2nd and higher gears. Refer to Chapter Six if the transmission does not work correctly.

d. Remove the lower case half.

12. Make sure the crankcase mating surfaces are completely clean and dry.

13. Apply a light coat of gasket sealer to the sealing surfaces of the bottom case half. Cover only flat surfaces, not curved bearing surfaces. Make the coating as thin as possible.

> *CAUTION*
> *Do not block any oil passage with sealant.*

14. Carefully place the lower case half in position, fitting the shift forks in their proper gear grooves (**Figure 145**).

15. Seat the upper case half onto the lower and tap lightly with a plastic or rubber mallet—do not use a metal hammer or it will damage the cases.

16. Apply oil and loosely install the 17 lower crankcase bolts (**Figure 146**). Don't miss the bolt at the rear of the engine.

> *NOTE*
> *When installing crankcase bolts, check that each one sticks up the same amount before you screw them all in. If not, you've got a short bolt in a long hole and vice versa.*

17. Torque the 10 large bolts in 2 stages, as specified in **Table 2**, following the sequence numbers on the bottom of the engine (**Figure 147**).

18. Torque the 7 smaller bolts as specified in **Table 2**.

19. Rotate the crankshaft to make sure the bearings are not too tight.

20. Turn the transmission input and output shafts to see that they are free and, while spinning the output shaft, shift the transmission (turn the shift drum) through all gears to make sure the transmission is working right.

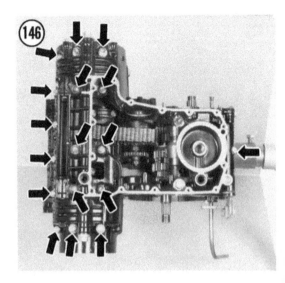

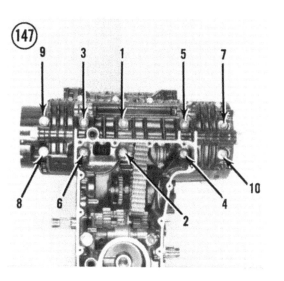

NOTE
These models have neutral locating balls inside output 4th gear. When operating correctly, the neutral locator will not allow the transmission to shift up from NEUTRAL as long as the output shaft is not turning. You must spin the output shaft to shift up from NEUTRAL.

21. Install the secondary shaft and starter clutch assembly; see *Secondary Shaft/Clutch Installation* in this chapter.

NOTE
*Before proceeding, inspect the primary chain play as described in **Crankcase Separation**.*

22. Install the oil pump and oil pan; see *Oil Pump Installation* in this chapter.
23. Turn the engine over and install the 13 upper crankcase bolts (**Figure 148**). Torque them as specified in **Table 2**.
24. Install the ignition advance assembly and pickup coil assembly as described in Chapter Eight.

CRANKSHAFT AND CONNECTING RODS

Once the crankshaft is removed, the cam chain and primary chain can be removed. The crankshaft is a 1-piece design that uses plain bearing inserts at the crankshaft main bearings and at the connecting rod big-end bearings.

Crankshaft Disassembly

1. Split the engine cases; see *Crankcase Separation* in this chapter.

NOTE
Keep each bearing insert in its original place in the crankcases. If you are going to assemble the engine with the original inserts, they must be installed exactly as removed in order to prevent rapid wear.

2. Before removing the crankshaft assembly from the upper crankcase, measure the crankshaft side clearance with feeler gauges, as shown in **Figure 149**. Pry the crankshaft toward one end, then toward the other, and measure the gap between the main bearing

boss thrust face and the crankshaft flyweight thrust face. If play exceeds the limit in **Table 1**, the crankcase halves should be replaced as a set.

3. Lift the crankshaft assembly out of the engine case. Take off the cam chain and primary chain.

4. Remove the oil seals from the ends of the crankshaft.

5. Check connecting rod big-end side clearance with feeler gauges (**Figure 150**). If clearance exceeds the limit in **Table 1**, the crankshaft and connecting rods should be replaced.

6. Before removing the rods, mark the rods and caps. Number them 1, 2, 3 and 4 starting at the alternator end of the crankshaft.

7. Remove each connecting rod's cap nuts (A, **Figure 151**), the cap and the rod. Keep each cap with its original rod, with the weight mark on the end of the cap matching the mark on the rod (B, **Figure 151**).

CAUTION
After removing the rod caps, remove the rods carefully to keep the rod studs from scratching the crank journals.

NOTE
Keep each bearing insert in its original place in the crankcase, rod or rod cap. If you are going to assemble the engine with the original inserts, they must be installed exactly as removed in order to prevent rapid wear.

Crankshaft Inspection

1. Clean the oil holes with pipe cleaners (**Figure 152**); flush them thoroughly and dry. Lightly oil all bearing journals immediately to prevent rust.

2. Carefully inspect all the main bearing and connecting rod bearing journals for scratches, ridges, scoring, nicks, etc. Very small nicks and scratches may be removed with fine emery cloth. More serious damage must be removed by grinding—a job for a machine shop.

3. If these checks are satisfactory, take the crankshaft and connecting rod assembly to

your dealer or local machine shop. They can check connecting rod bend and twist and inspect the parts for cracks. Check against the measurements given in **Table 1** at the end of this chapter.

NOTE
When obtaining new connecting rods, make sure each pair (No. 1 and 2 or No. 3 and 4) has the same weight mark stamped on the side of the big end (B, Figure 151).

4. Check all the main bearing and connecting rod big-end bearing clearances as follows.

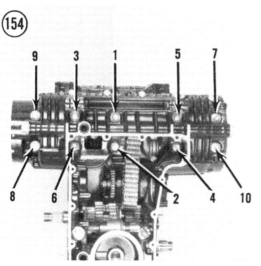

Main Bearing Clearance

To check crankshaft journal/bearing clearance with Plastigage, the crankshaft is installed with the journals and inserts *dry*. It is very important not to turn the crankshaft while assembled dry. After checking clearance, remove the crankshaft and clean and lubricate the journals.

1. Make a note of the bearing size (if any) stamped on the back of the insert. If the inserts are color-coded with paint on the edge, note the color.
2. Check the inside and outside surfaces of the bearing inserts for wear, a bluish tint (burned), flaking, abrasion or scoring. If the inserts are good, they may be reused if clearance is satisfactory. If any insert is questionable, replace the entire set.
3. Wipe any oil from the crankshaft main journals and the inserts.
4. Set the upper crankcase upside down on the workbench, on wood blocks, to prevent damage to the cylinder studs.
5. Install the existing inserts in the upper crankcase.
6. Install the crankshaft in the upper crankcase.
7. Place a strip of Plastigage over each main bearing journal parallel to the crankshaft (**Figure 153**).

NOTE
Do not rotate the crankshaft while the Plastigage strips are in place.

8. Install the main bearing inserts in the lower crankcase half. Make sure they are locked in place correctly (**Figure 140**).
9. Carefully turn the lower crankcase over and install it on the upper crankcase.
10. Apply oil to the crankcase bolt threads and install the 10 large lower crankcase bolts (**Figure 154**). Tighten the bolts in 2 stages, to the toque specified in **Table 2**, following the sequence numbers on the bottom of the crankcase next to the bolts.
11. Remove bolts in the same sequence you tightened them. Remove the lower case half.
12. Measure the width of the flattened Plastigage according to manufacturer's in-

structions (**Figure 155**). The standard crank-shaft main bearing clearance is 0.001 in. (0.025 mm). The maximum (service limit) is 0.004 in. (0.11 mm).

13. Compare both ends of the Plastigage strip. A difference of 0.001 in. (0.025 mm) or more indicates a tapered journal which should be reground. Confirm with micrometer mea-surement of the journal OD.

14. If the clearance is larger than the service limit, measure the OD (outside diameter) of the crankshaft main journal with a microme-ter. The minimum main journal OD (service limit) is given in **Table 1**.

    a. If the OD is smaller than the service limit, the crankshaft should be replaced.

    b. If the OD is larger than the service limit, install new bearing inserts. Always re-place all 10 inserts at the same time. Several different insert thicknesses are available and they are color-coded for size. Refer to **Table 1** for the insert thickness.

    c. Recheck the clearance with the new in-serts. The clearance should be less than the service limit and as close to standard as possible.

NOTE
You can install the thinnest new inserts, recheck clearance and install thicker inserts if the new ones aren't thick enough or you can calculate the required insert thickness: assemble the crankcase halves without the inserts or crankshaft and measure the crankcase ID; subtract the main journal OD from the crankcase ID, then subtract the standard clearance—0.001 in. (0.025 mm). The number you get will be twice the thickness of the required inserts; divide by 2 for the appropriate insert thickness.

15. Clean and oil the main bearing journals and insert faces.

16. Install the bearing inserts in the crankcase halves. Make sure they are locked in place correctly.

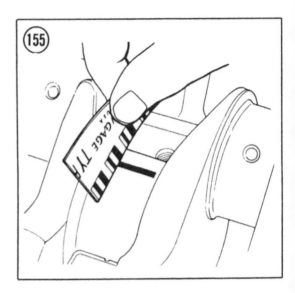

NOTE
If the old inserts are reused, be sure they are installed in their original positions for minimum wear.

Connecting Rod
Big-end Bearing Clearance

1. Make a note of the bearing size (if any) stamped on the back of the insert. If the insert is color-coded with paint on the edge, note the color.

2. Check the inside and outside surfaces of the bearing inserts for wear, a bluish tint (burned), flaking, abrasion and scoring. If the inserts are good, they may be reused if clearance is satis-factory. If any insert is questionable, replace the entire set.

3. Wipe oil from the crankpin journals and the inserts.

4. Place a piece of Plastigage on one crankpin parallel to the crankshaft.

5. Install the rod cap and torque the nuts as specified in **Table 2**.

NOTE
Do not rotate connecting rod while Plastigage is in place.

6. Remove the rod cap.

7. Measure the width of the flattened Plastigage according to manufacturer's instructions. The standard connecting rod big-end bearing clearance is 0.002 in. (0.051 mm). The maximum (service limit) is 0.004 in. (0.1 mm).

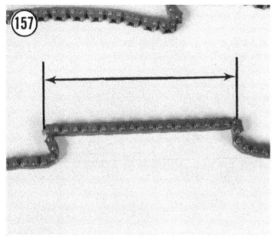

c. Recheck the clearance with the new inserts. The clearance should be less than the service limit and as close to standard as possible.

NOTE
*You can install the thinnest new inserts, recheck clearance and install thicker inserts if the new ones aren't thick enough or you can calculate the required insert thickness: assemble the connecting rod and cap without the inserts and measure the rod big-end ID; the cap must be with its original rod, with the weight mark on the end of the cap matching the mark on the rod (**Figure 151**). Install the cap nuts finger-tight, then torque them as specified in **Table 2**; subtract the crankpin journal OD from the rod big-end ID, then subtract the standard clearance—0.002 in. (0.051 mm). The number you get will be twice the thickness of the required inserts; divide by 2 for the appropriate insert thickness.*

10. Repeat for each connecting rod.
11. Clean and oil the crankpin journals and insert faces.
12. Install the bearing inserts in each connecting rod and cap. Make sure they are locked in place correctly (**Figure 156**).

NOTE
If the old inserts are reused, be sure they are installed in their original positions to prevent rapid wear.

Cam Chain
Inspection

Pull the cam chain taut with about a 10 lb. (5 kg) tension and measure the length from the 1st pin to the 21st pin (**Figure 157**). If the chain is longer than the service limit given in **Table 1**, install a new chain.

Primary Chain
Inspection

The primary chain wear is inspected by measuring chain play with the crankshaft and secondary shaft installed in the engine; see *Crankcase Separation* in this chapter.

8. Compare both ends of the Plastigage strip. A difference of 0.001 in. (0.025 mm) or more indicates a tapered journal which should be reground or a bent or twisted rod. Confirm with micrometer measurement of the journal OD.
9. If the clearance is larger than the service limit, measure the OD of the crankpin journal with a micrometer. The minimum crankpin journal OD (service limit) is given in **Table 1**.
  a. If the OD is smaller than the service limit, the crankshaft should be replaced.
  b. If the OD is larger than the service limit, install new bearing inserts. Always replace all 8 inserts at the same time. Several different insert thicknesses are available and they are color-coded for size. Refer to **Table 1** for the insert thickness.

Crankshaft Assembly

1. Install the rod bearing inserts in each connecting rod and cap. Make sure they are locked in place (**Figure 156**).

> *NOTE*
> *If the old inserts are reused, be sure they are installed in their original positions to prevent rapid wear.*

2. Lubricate the rod bearing inserts and crankpins with engine oil.

3. Install each rod, one at a time, on the proper crankpin, with No. 1 starting at the alternator end of the crank. The cap must be with its original rod, with the weight mark on the end of the cap matching the mark on the rod (B, **Figure 151**).

4. Install the cap nuts finger-tight, then torque them as specified in **Table 2**.

5. Install the remaining rods in the same manner (**Figure 158**).

Table 1   KZ750 ENGINE WEAR LIMITS

Cam chain (20 link length)	5.07 in.	(128.9 mm)
Cam chain guide rubber wear depth		
Upper	0.14 in.	(3.5 mm)
Front	0.09 in.	(2.2 mm)
Rear	0.14 in.	(3.5 mm)
Cam chain tensioner spring		
free length		
Cross wedge pushrod spring	1.42 in.	(36 mm)
Cross wedge lock rod spring	1.73 in.	(44 mm)
Cam lobe height	1.404 in.	(35.65 mm)
Camshaft journal/		
bearing cap clearance	0.007 in.	(0.19 mm)
Camshaft bearing ID	0.871 in.	(22.12 mm)
Camshaft journal OD	0.864 in.	(21.93 mm)
Camshaft runout	0.004 in.	(0.1 mm)
Compression	110-170 psi	$(7.7\text{-}12.0 \text{ kg/cm}^2)$
Combustion chamber volume (from		
bottom of cylinder head)	1.51 cu. in.	(24.8 cc)
Connecting rod bend and twist per		
4 in. (100 mm)	0.008 in.	(0.20 mm)
Connecting rod big end clearance		
Radial clearance: Limit	0.004 in.	(0.1 mm)
Standard	0.002 in.	(0.05 mm)
Side clearance	0.018 in.	(0.45 mm)
Connecting rod small end ID	0.593 in.	(15.05 mm)
Connecting rod journal OD	1.377 in.	(34.97 mm)
Connecting rod bearing		
insert thickness (standard)		
Green	0.0586 in.	(1.488 mm)
Black	0.0584 in.	(1.483 mm)
Brown	0.0582 in.	(1.478 mm)
Crankshaft main journal/		
bearing clearance		
Limit	0.004 in.	(0.11 mm)
Standard	0.001 in.	(0.025 mm)
Crankshaft main journal O.D.	1.416 in.	(35.96 mm)
Crank bearing insert thickness		
Brown	0.0587 in.	(1.492 mm)
Black	0.0589 in.	(1.496 mm)
Blue	0.0591 in.	(1.500 mm)
Crankshaft side clearance	0.016 in.	(0.40 mm)
Crankshaft runout	0.002 in.	(0.05 mm)
Cylinder head warp	0.002 in.	(0.05 mm)
Cylinder bore ID	2.602 in.	(66.10 mm)
Oil pressure (4,000 rpm @		
195° F/90° C)		
Standard	28-36 psi	$(2.0\text{-}2.5 \text{ kg/cm}^2)$
Relief valve (cold oil)	63-85 psi	$(4.4\text{-}6.0 \text{ kg/cm}^2)$
Oil pump		
Inner/outer rotor clearance	0.012 in.	(0.30 mm)
Outer rotor/body clearance	0.012 in.	(0.30 mm)
Rotor end play	0.005 in.	(0.12 mm)
Piston/cylinder clearance (standard)	0.0016 – 0.0026 in.	(0.040-0.067 mm)
Piston OD	2.591 in.	(65.80 mm)
Piston pin hole ID	0.5933 in.	(15.07 mm)
Piston pin OD	0.5925 in.	(15.05 mm)
Piston ring/groove clearance	0.006 in.	(0.15 mm)
Piston ring thickness	0.055 in.	(1.40 mm)
Piston ring groove width		
Top, second	0.063 in.	(1.60 mm)
Oil	0.102 in.	(2.60 mm)
Piston ring installed gap	0.008 – 0.027 in.	(0.2 – 0.7 mm)

(continued)

Table 1   KZ750 ENGINE WEAR LIMITS (continued)

Primary chain play	1.06 in.	(27 mm)
Valve stem runout	0.002 in.	(0.05 mm)
Valve stem OD	0.271 in.	(6.89 mm)
Valve guide ID	0.279 in.	(7.08 mm)
Valve/guide clearance (rocking method)		
Intake	0.009 in.	(0.24 mm)
Exhaust	0.007 in.	(0.19 mm)
Valve head thickness	0.020 in.	(0.5 mm)
Valve seat width (standard)	0.030 in.	(0.75 mm)
Valve seat angles:	30, 45, 60°	
Valve spring tension		
Inner	34 lb. @ 0.93 in.	(15.6 kg @ 23.6 mm)
Outer	63 lb. @ 1.01 in.	(28.5 kg @ 25.6 mm)
Valve spring squareness	0.06 in.	(1.5 mm)
from vertical @ top of spring		
Valve stem installed height (max.)	1.478 in.	(37.54 mm)

Table 2   KZ750 ENGINE TORQUES

Item	Ft.-lb.	mkg
Alternator rotor bolt	50	7.0
Cam chain tensioner cap		
(cross wedge type)	18	2.5
Camshaft cap bolts	8.5	1.2
Camshaft sprocket bolts	11	1.5
Clutch hub nut	100	13.5
Clutch spring bolts	90 in.-lb.	1.0
Connecting rod cap nuts	27	3.7
Crankcase bolts		
Small	90 in.-lb.	1.0
Large	18	2.5
Cylinder head		
Bolts	22	3.0
Nuts	30	4.0
Engine mounting nuts	30	4.0
Engine mounting bracket bolts	17.5	2.4
Engine sprocket nut	60	8.0
Ignition advance bolt	18	2.5
Neutral indicator switch	11	1.5
Oil drain plug	27	3.8
Oil filter mounting bolt	14.5	2.0
Oil pan bolts	90 in.-lb.	1.0
Oil pressure switch	11	1.5
Oil pressure relief valve	11	1.5
Secondary shaft nut	45	6.0
Shift return spring pin	18	2.5
Spark plugs	20	2.8
Starter clutch Allen bolts	25	3.5
Valve cover bolts	70 in.-lb.	0.8

NOTE: If you own a 1982 or later model, first check the Supplement at the back of the book for any new service information.

CHAPTER FIVE

5

CLUTCH

The clutch can be serviced with the engine in the motorcycle or on a workbench. Generally, most service operations are easier with the engine in the motorcycle because the engine is held firmly.

Clutch wear limit specifications are given in **Table 1** at the end of the chapter.

OPERATION

The clutch is a wet multi-plate type which operates immersed in the engine oil. It is mounted on the right end of the transmission input shaft. The inner clutch hub is splined to the input shaft; the outer housing can rotate freely on the input shaft when the clutch release is actuated. The outer housing is driven by a gear on the end of the secondary shaft. The clutch release is mounted in the sprocket cover on the left side of the engine.

> *CAUTION*
> *The clutch friction plates are bathed in the same oil you put in the engine. Do not use oil additives or you may cause clutch slippage.*

Between the clutch inner hub and outer housing is a sandwich of clutch plates. Every other plate (including the bottom and top friction plates) is locked to the outer housing and must turn whenever it turns. The remaining metal plates are locked to the inner hub; when they turn, it turns. Outside this sandwich of clutch plates is the pressure plate. There are coil springs that push the pressure plate in against the rest of the plates. This pressure jams the plates together and friction locks the clutch hubs together; then the crankshaft will turn the transmission input shaft.

To disengage the clutch, a clutch release mechanism lifts the pressure plate outward from the clutch. With the pressure gone, the outer housing and the friction plates locked to it continue to turn, but the metal plates and inner hub stop turning. All of the clutch parts can be removed while the engine is mounted in the frame.

CLUTCH RELEASE

Routine cable free play and release adjustment are described in Chapter Three.

The clutch release mechanism is mounted inside the engine sprocket cover. The release consists of a release arm that rides on ball bearings inside the release housing. The clutch cable is attached to the arm; when pulled, it rotates the arm which moves toward the

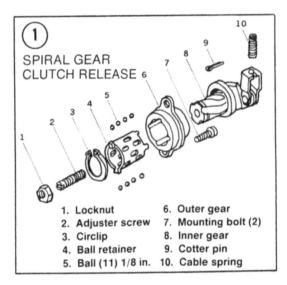

① SPIRAL GEAR CLUTCH RELEASE

1. Locknut
2. Adjuster screw
3. Circlip
4. Ball retainer
5. Ball (11) 1/8 in.
6. Outer gear
7. Mounting bolt (2)
8. Inner gear
9. Cotter pin
10. Cable spring

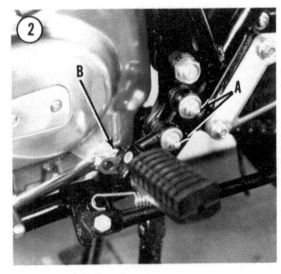

clutch. The adjusting screw in the arm then pushes against the clutch pushrod, which extends through the hollow transmission input shaft and releases the clutch pressure plate from the pack of friction and steel plates.

Disassembly/Assembly

Early models use a clutch release with 2 spiral gears riding on 11 ball bearings (**Figure 1**). Late models use a clutch release with 3 ball bearings riding up ramps in the release arm.

1. Place the bike on its centerstand.

2. Remove the 2 cap nuts that mount the left front footpeg assembly to the frame (A, **Figure 2**). Remove the footpeg assembly from its mounting studs.

3. Remove the shift pedal (B, **Figure 2**); remove the bolt and spread the slot open with a screwdriver if necessary. If your bike has linkage between the shift pedal and shift shaft, remove the circlip at the pedal and remove all the linkage.

4. Remove the 2 bolts securing the starter motor cover and remove the cover and gasket (**Figure 3**).

5. Remove the 4 bolts securing the engine sprocket cover and remove the cover (**Figure 4**). Be careful not to damage the shift shaft oil seal.

6. Remove the clutch release lever cotter pin (**Figure 5**). Remove the cable tip from the lever.

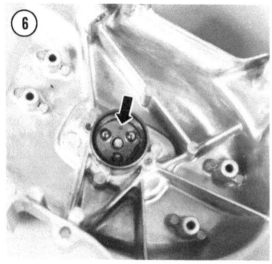

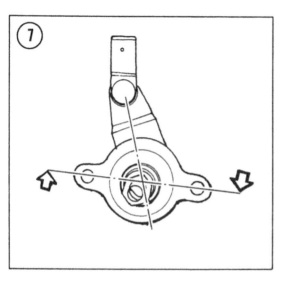

7A. On late models perform the following:
   a. Pull the release arm out of the release housing.
   b. Remove the 3 ball bearings and their cage (**Figure 6**).

7B. On early models, refer to **Figure 1** and perform the following:
   a. Remove the 2 screws that mount the release housing to the sprocket cover.
   b. Remove the circlip and separate the outer and inner spiral release gears. Be careful not to lose the 11 release balls; in case you do, they are 1/8 in. diameter.

8. To assemble, reverse the disassembly steps. Note the following:
   a. Apply grease to all parts before assembly.
   b. On early models, the inner spiral gear must be installed so that when the 2 gears are fully meshed, the clutch release lever will be positioned as shown in **Figure 7**, with the machined side of the housing facing up.
   c. Use a new cotter pin to secure the cable in the release lever. Spread its ends.
   d. Make sure the 2 sprocket cover dowel pins are in place.
   e. Align the shift pedal with the top of the footpeg.
   f. Adjust the clutch as described under *Clutch Adjustment* in Chapter Three.

Inspection

Clean all parts in solvent and dry them. Check the balls for wear or pitting; replace if damaged. On models with spiral release gears, inspect the grooves in the inner gear and outer housing. If they show signs of wear, replace the entire assembly. Upon reassembly, push the inner gear back and forth in the direction of the shaft without turning it. If there is excessive play, replace the entire assembly.

CLUTCH

All of the clutch parts can be removed while the engine is mounted in the frame.

Disassembly

1. See **Figure 8**. Put the bike up on its center-stand and drain the engine oil, then place the oil drain pan under the clutch cover.

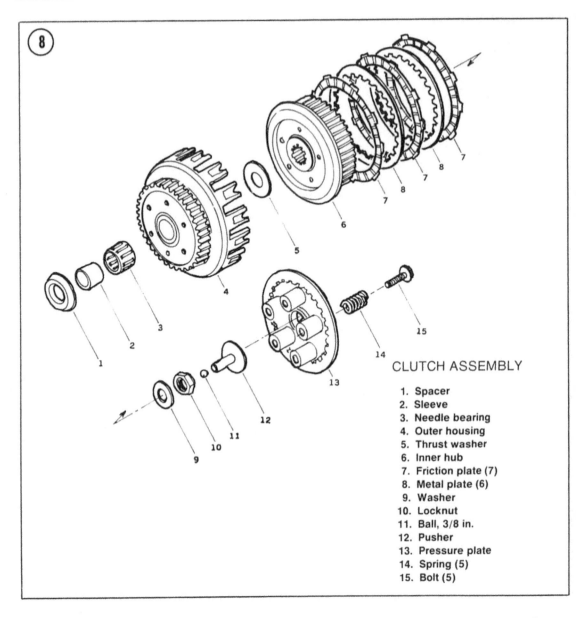

CLUTCH ASSEMBLY

1. Spacer
2. Sleeve
3. Needle bearing
4. Outer housing
5. Thrust washer
6. Inner hub
7. Friction plate (7)
8. Metal plate (6)
9. Washer
10. Locknut
11. Ball, 3/8 in.
12. Pusher
13. Pressure plate
14. Spring (5)
15. Bolt (5)

2. Remove the 2 cap nuts that mount the right front footpeg assembly to the frame. Remove the footpeg assembly from its mounting studs.

3. Remove the 10 clutch cover screws (**Figure 9**).

4. Free the cover by tapping it gently with a soft mallet. Remove the cover. Be careful not to damage the gasket.

CAUTION
Make sure that you have removed all the fasteners. If the cover is hard to remove, check for any fasteners you may have missed. Do not try to pry the cover off the

engine case or you will damage the sealing surfaces.

5. Loosen the 5 clutch pressure plate bolts gradually in a crisscross pattern.

6. Remove the bolts, springs and the pressure plate.

7. Remove the pressure plate pusher and the steel ball behind it.

8. Pull the clutch release pushrod out of the center of the transmission input shaft.

9. Remove the clutch plates from the hub.

10. With a socket on a breaker bar, remove the clutch hub nut from the input shaft (**Figure 10**).

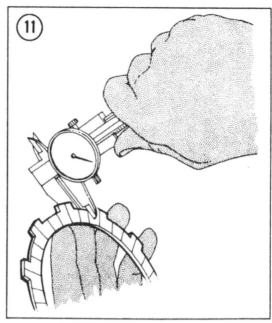

5

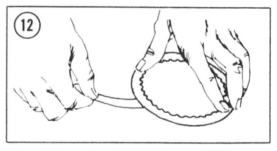

NOTE
To remove the clutch hub nut, you must hold the clutch hub (which is mounted on the transmission input shaft) steady. You can lock the input shaft by stuffing a clean rag or a copper penny between the secondary shaft gear and clutch ring gear teeth. A special tool, the "Grabbit," is available from Precision Mfg. and Sales Co., Box 149, Clearwater FL 33517.

11. Remove the spring washer, inner clutch hub and thrust washer.
12. Remove the input shaft sleeve, bearing and spacer.

Inspection

1. Inspect the friction plates for signs of overheating or a burnt smell. Replace them if damaged.

2. Measure the thickness of each friction plate at several places around the plate as shown in **Figure 11**. Replace any plate that is worn below the wear limit in **Table 1**.

3. Lay each plate on a flat surface. If there is a gap between any part of the clutch plate and the flat surface, measure the warpage with feeler gauges (**Figure 12**). Replace any plate with a warp greater than the service limit in **Table 1**.

4. Insert a friction plate into the outer housing and rotate the plate until one side of each tab on it is butted up against the housing. With feeler gauges, measure the clearance between the other side of each tab and the housing (**Figure 13**). Replace any plate with clearance greater than 0.040 in. (1.0 mm). Ideal clearance is 0.014-0.026 in. (0.35-0.65 mm). Too much clearance will cause a clutch rattle.

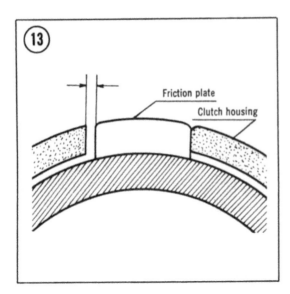

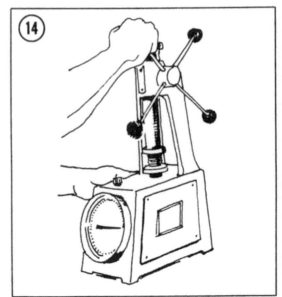

5. Visually inspect the outer splines that mount the clutch plates to the inner hub and outer housing. If the splines are chewed up or badly worn, install new parts.

6. Inspect the clutch springs. The spring tension must be checked *while compressed* in a special spring tester (**Figure 14**). See **Table 1** at the end of this chapter for compression specifications. Replace all springs if one has sagged below the limit.

7. Roll the clutch pushrod on a flat surface to check for bends or damage. Examine the rounded ends of the pushrod for damage. Replace it if bent or damaged.

8. Examine the clutch housing and ring gear and check for excessive wear or loose rivets. If the condition is marginal, replace the housing. Clutch failure at high rpm can cause expensive engine damage.

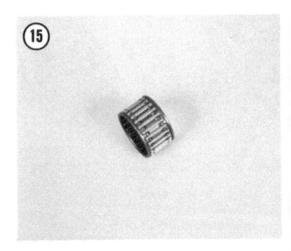

9. Check the clutch shaft bearing (**Figure 15**), spacers and thrust washers. Replace any parts that are cracked or excessively worn, loose or galled.

10. Measure the clutch housing inside diameter (ID) with an inside micrometer or vernier caliper. If the bushing ID is larger than the limit in **Table 1**, install a new clutch housing and bearing.

11. Measure the clutch sleeve outside diameter (OD) with an outside micrometer or vernier caliper. If the sleeve OD is smaller than the limit in **Table 1**, install a new clutch housing and bearing.

Assembly

1. See **Figure 8**. Install the input shaft spacer with the flat side facing out and install the bearing sleeve (**Figure 16**).
2. Install the clutch bearing (**Figure 17**).
3. Install the outer housing and thrust washer (**Figure 18**).
4. Install the inner clutch hub and install the washer (**Figure 19**).

> *NOTE*
> *If one side of the washer is marked "OUTSIDE," be sure that side faces out.*

5. Use a new clutch hub locknut when possible (**Figure 10**) and torque it to 100 ft.-lb. (13.5 mkg).

> *NOTE*
> *To install the clutch hub nut, you must hold the clutch hub (which is mounted on the transmission input shaft) steady. You can lock the input shaft by stuffing a clean rag or a copper penny between the secondary shaft gear and clutch ring gear teeth. A special tool, the "Grabbit," is available from Precision Mfg. and Sales Co., Box 149, Clearwater FL 33517.*

6. Install the clutch release pushrod (**Figure 20**).
7. Apply molybdenum disulfide grease to the steel ball and the end of the pusher. Install them in the end of the input shaft (**Figure 21**).

8. Install the clutch plates (**Figure 22**). When installing the clutch plates, the sequence is friction plate, metal plate, friction plate, etc., starting and ending with a friction plate. Take care to align the plate tabs carefully with the housing teeth.

NOTE
If you are installing new dry plates, first wet them with oil to prevent clutch seizure.

9. Install the pressure plate, aligning its splines with the clutch hub splines. Install the 5 springs and bolts (**Figure 23**). Tighten the bolts gradually in a crisscross pattern.
10. Check that the 2 clutch cover dowel pins are in place and install the clutch cover and gasket.
11. Install the cover screws and tighten them by hand. Note the ignition timing lead clamps at the front and rear bottom cover screws.

NOTE
When installing cover screws, check that each one sticks up the same amount before you screw them all in. If not,

you've got a short screw in a long hole and vice versa. Do not use an impact driver to install cover screws. They'll be too tight.

12. Install the right footpeg assembly.
13. Add engine oil and adjust the clutch release and cable play as described in Chapter Three.

Table 1   KZ750 CLUTCH WEAR LIMITS

Clutch housing ID	1.458 in.	(37.03 mm)
Clutch sleeve OD	1.258 in.	(31.96 mm)
Disc tab/housing clearance		
Wear limit	0.040 in.	(1.0 mm)
Standard:	0.014 – 0.026 in.	(0.35 – 0.65 mm)
Friction disc thickness	0.14 in.	(3.5 mm)
Disc/plate warp	0.016 in.	(0.4 mm)
Housing/secondary gear backlash		
Wear limit	0.005 in.	(0.12 mm)
Standard	0.0012 – 0.004 in.	(0.03 – 0.10 mm)
Spring tension	52 lb. @ 0.95 in.	(23.5 kg @ 24.1 mm)

5

NOTE: If you own a 1982 or later model, first check the Supplement at the back of the book for any new service information.

CHAPTER SIX

TRANSMISSION

This chapter covers all the parts that transmit power from the clutch to the drive chain: the engine sprocket, the transmission gears, the shift drum and forks that slide the gears and the shift linkage that turns the shift drum.

The shift linkage can be repaired while the engine is mounted in the frame, but repair of the transmission gears, shift drum and shift forks requires engine removal and crankcase separation.

Table 1 at the end of the chapter lists transmission wear limit specifications. Many inspection measurements require a precision inside and outside micrometer, dial gauge or the equivalent. If you don't have the right tools, have your dealer or machine shop take the required measurements.

SPROCKET COVER

The clutch release mechanism is mounted on the inside of the sprocket cover. The sprocket cover and shift linkage cover underneath it must be removed for access to the shift linkage.

Removal/Installation

1. Place the bike on its centerstand.
2. Remove the 2 cap nuts that mount the left front footpeg assembly to the frame (A, **Figure**

1). Remove the footpeg assembly from its mounting studs.
3. Remove the shift pedal (B, **Figure 1**); remove the bolt and spread the slot open with a screwdriver if necessary. If your bike has linkage between the shift pedal and shift shaft, remove the circlip at the pedal and remove all the linkage (**Figure 2**).
4. Remove the 2 bolts securing the starter motor cover and remove the cover and gasket (**Figure 3**).

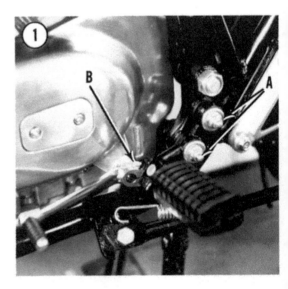

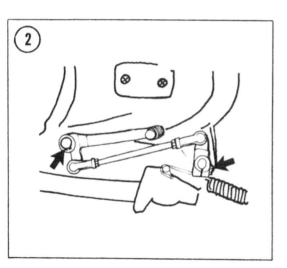

5. Remove the 4 bolts securing the engine sprocket cover and remove the cover (**Figure 4**). Be careful not to damage the shift shaft oil seal.

6. To detach the sprocket cover, remove the clutch release lever cotter pin (**Figure 5**). Remove the cable tip from the lever.

7. To install the sprocket cover, reverse the removal steps. Note the following:

   a. Use a new cotter pin to secure the clutch cable in the release lever. Spread its ends.

   b. Make sure the 2 sprocket cover dowel pins are in place.

   c. Align the shift pedal with the top of the footpeg.

ENGINE SPROCKET

The engine sprocket is on the left end of the transmission output shaft, behind the sprocket cover. The drive chain is endless—it has no master link. To remove the drive chain, remove the swing arm; see *Swing Arm Removal* in Chapter Ten.

Removal

1. Remove the engine sprocket cover; see *Sprocket Cover, Removal/Installation.*

2. Flatten the lockplate behind the sprocket nut, then lock the rear wheel with the brake and remove the engine sprocket nut and lockplate.

3. Slide the sprocket off the output shaft.

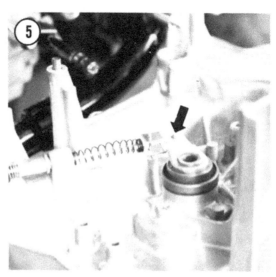

NOTE
*You may have to loosen the drive chain
to allow sprocket removal. See* **Drive
Chain Play** *in Chapter Three.*

Installation

1. Inspect the engine sprocket for wear. If the teeth are undercut as shown in **Figure 6**, install a new sprocket. A worn sprocket will quickly wear out a new drive chain.

2. Position the drive chain on the sprocket, then slide the sprocket onto the output shaft (**Figure 7**).

3. Install the splined lockplate (A, **Figure 8**) and sprocket nut.

4. Lock the rear wheel with the brake and torque the engine sprocket nut to 60 ft.-lb. (8.0 mkg). Fold up a side of the lockplate against the nut (**Figure 9**).

CAUTION
Do not fold a lockplate more than once in the same spot. The metal is likely to break off, with possible loosening of the part and engine damage. Install a new lockplate when possible.

5. Install the sprocket cover; see *Sprocket Cover Removal/Installation.*

6. Adjust the drive chain if it was loosened. See *Drive Chain Play* in Chapter Three.

NEUTRAL SWITCH

The neutral indicator light is activated by a switch mounted in the shift linkage cover, under the sprocket cover (B, **Figure 8**). When the shift drum end plate is at its NEUTRAL position (**Figure 10**), the insulated switch pin that rides against the end plate is grounded against the metal portion of the plate, completing the indicator light circuit.

SHIFT LINKAGE

Refer to **Figure 11**. The shift linkage can be repaired without separating the crankcases, but repair of the shift drum and forks requires engine removal and crankcase separation.

Inside the transmission, gears are moved by shift forks, which are moved from side to side by the camming slots in the cylindrical shift drum. The linkage (outside the engine cases)

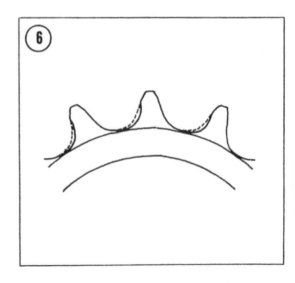

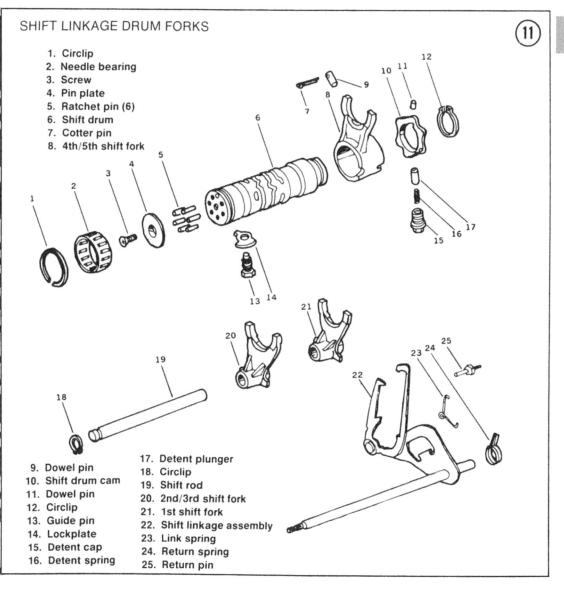

SHIFT LINKAGE DRUM FORKS

11

6

1. Circlip
2. Needle bearing
3. Screw
4. Pin plate
5. Ratchet pin (6)
6. Shift drum
7. Cotter pin
8. 4th/5th shift fork

9. Dowel pin
10. Shift drum cam
11. Dowel pin
12. Circlip
13. Guide pin
14. Lockplate
15. Detent cap
16. Detent spring

17. Detent plunger
18. Circlip
19. Shift rod
20. 2nd/3rd shift fork
21. 1st shift fork
22. Shift linkage assembly
23. Link spring
24. Return spring
25. Return pin

that converts up-and-down motions of the gearshift pedal into rotation of the shift drum is the gear shift mechanism.

The shift pedal is mounted on one end of the shift shaft. At the other end of the shaft are a shift arm and an overshift limiter under the arm. The shift arm pawls (fingers) rest against pins in the end of the shift drum. When the shift shaft is rotated, the pawls grasp the pins and rotate the shift drum. The overshift limiter hooks keep the shift drum from moving more than one gear at a time.

The 2 legs of the strong hairpin return spring on the shift shaft rest against a stationary centering pin. When the shift pedal is released, the return spring brings the shift shaft back to its center position.

Removal

The shift linkage can be repaired without separating the crankcases.

1. Remove the engine sprocket cover; see *Sprocket Cover Removal/Installation.*

2. Remove the engine sprocket; see *Engine Sprocket Removal.*

3. Place a drain pan under the shift linkage cover and remove the neutral indicator lead (B, **Figure 8**).

4. Remove the 5 screws and 2 bolts (**Figure 12**), then remove the shift linkage cover and gasket. Tap the cover loose with a soft mallet, if necessary. Use care; the cover is positioned with dowel pins.

5. Remove the output shaft collar.

6. Note carefully how the shift link pawls engage the shift drum (A, **Figure 13**). Move the shift linkage arms out of engagement with the shift drum and pull the shift linkage out of the crankcase.

> *CAUTION*
> *Do not pull the shift fork rod (B, **Figure 13**) out. If it is pulled out 1 1/2 in. (40 mm), the shift forks within the crankcase will fall off the rod. This would require removal and disassembly of the engine to reposition the forks.*

7. To expose the pins on the shift drum, remove the screw from the cover plate and remove the plate.

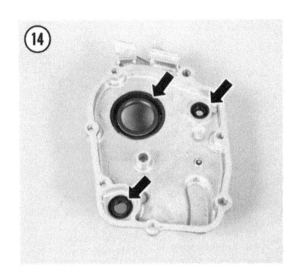

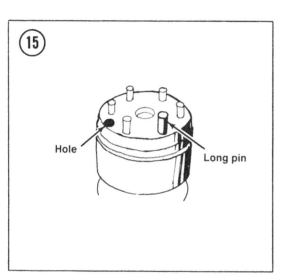

Hole

Long pin

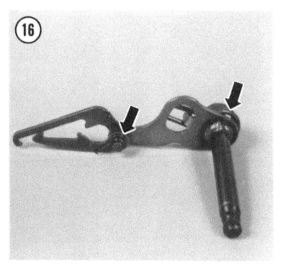

Inspection

1. Inspect the seals in the shift linkage cover (**Figure 14**). Replace any damaged seals; heat the cover in an oven to about 212° F and tap the old seals out. Install new seals flush with the surface of the case, with their numbered side out.

2. If the transmission fails to shift gears, check for a weak pawl spring; bent, worn or binding pawls; worn shift drum pins; a broken return spring; or a broken return spring pin.

3. If the transmission undershifts or over-shifts, check for a binding, bent or worn detent; a weak detent spring; bent or worn pawls; worn shift drum pins; a loose return spring pin; or a bent or weak return spring.

4. If the transmission jumps out of gear, check for a binding, bent or worn detent or a weak detent spring.

5. Replace any other broken, bent, or worn parts, including shift drum pins.

Installation

CAUTION
Use a locking agent such as Loctite Lock N' Seal on all shift linkage screws. Loose linkage will cause serious engine damage.

1. Check that the shift drum ratchet pins are in good condition. If the pins are removed, install the long pin (which times the NEUTRAL light) as shown in relation to the hole on the end of the drum (**Figure 15**). Secure the pin plate screw with a locking agent such as Loctite Lock N' Seal.

2. Check that the shift shaft return spring pin is tight (C, **Figure 13**). If loose, remove it, apply a locking agent such as Loctite Lock N' Seal and tighten it securely.

3. Be sure the return spring and link springs are in place on the shift linkage (**Figure 16**), then spread the shift linkage pawls as you install the linkage in the crankcase, mating the pawls to the shift drum.

4. Make sure the return spring is positioned correctly on the pin (C, **Figure 13**).

5. Make sure the output shaft O-ring is in place (A, **Figure 17**).

6. Check that the linkage cover dowel pins are in place (B, **Figure 17**) and install the cover and gasket. Install the 2 bolts and 5 screws (**Figure 12**).

7. Install the output shaft sleeve (**Figure 18**) after installing the shift linkage cover.

8. Install the neutral indicator lead (B, **Figure 8**).

9. Install the engine sprocket; see *Engine Sprocket Installation*.

10. Install the engine sprocket cover; see *Sprocket Cover Removal/Installation*.

SHIFT DETENT

The shift drum has a cam on the end opposite the pins and linkage (**Figure 19**). A spring-loaded detent is mounted on the bottom of the crankcases, riding on the face of the shift drum cam. The detent locks the shift drum in position after a shift has been made, to help keep the transmission from jumping out of gear.

Remove and inspect the detent assembly whenever the transmission will not stay in gear or if it is very hard to shift; remove the shift drum detent bolt, washer, spring and pin (**Figure 20**).

Check that the plunger slides freely inside the bolt. Measure the free length of the detent spring (**Figure 21**). If the spring is shorter than the limit in **Table 1**, install a new spring. Install the shift drum detent pin, spring, washer and bolt.

TRANSMISSION

Repair of the transmission gears, shift drum and shift forks requires engine removal and case separation. If the transmission fails to shift properly or jumps out of gear, check the condition of the shift linkage before splitting the engine cases. See *Shift Linkage Inspection* and *Shift Detent* in this chapter.

Transmission Operation

The basic transmission has 5 pairs of constantly meshed gears on the input and output shafts. Each pair of meshed gears gives one gear ratio. In each pair, one of the gears is locked to its shaft and always turns with it. The other gear is not locked to its shaft and can spin freely on it. Next to each free spinning gear is a third gear which is splined to the same shaft, always turning with it. This third gear can slide from side to side along the shaft splines. The side of the sliding gear and the free spinning gear have mating "dogs" and "slots." When the sliding gear moves up against the free spinning gear, the 2 gears are locked together, locking the free spinning gear to its shaft. Since both meshed input and output gears are now locked to their shafts, power is transmitted at that gear ratio.

Neutral "Finder"

In 4th gear on the output shaft are 3 steel balls, spaced 120° apart, which help keep the transmission from overshooting to 2nd gear when the rider wants to shift from 1st to NEUTRAL. As long as the bike is moving and the output shaft is turning, the balls are thrown away from the shaft and will allow upshifting to 2nd. When the bike stops, the ball on top falls into a groove in the shaft and keeps the gear from sliding into position for higher gears.

Shift Drum And Fork Operation

Each sliding gear has a deep groove machined around its outside. The curved shift fork arm rides in this groove, controlling the side-to-side sliding of the gear and therefore the selection of different gear ratios.

Each shift fork slides back and forth on a guide shaft or on the shift drum and has a pin that rides in a groove on the face of the shift drum. When the shift linkage rotates the shift drum, the zigzag grooves move the shift forks and sliding gears back and forth.

A spring-loaded plunger rides in a cam on the end of the shift drum. This detent helps keep the drum in the selected gear or in NEUTRAL.

SHIFT DRUM AND FORKS

Removal

Refer to **Figure 22**.
1. Remove the engine from the motorcycle and separate the case halves as described in Chapter Four, *Engine Removal* and *Crankcase Separation.*
2. In the lower engine case, pull out the shift fork rod and remove the 2 smaller shift forks.
3. Remove the large shift fork cotter pin and guide pin.
4. Turn the engine case over. Flatten the lock plate and unscrew the shift drum guide bolt (A, **Figure 23**).
5. Remove the shift drum detent bolt, spring and plunger (B, **Figure 23**).
6. On the clutch end of the shift drum, remove the detent cam circlip and the cam.
7. Pull the shift drum out of the engine case, taking off the third shift fork as you go.

Inspection

1. Inspect each shift fork for wear on the fork arms (**Figure 24**) and for signs of burning or cracking. See **Table 1** for the minimum shift fork thickness specification.
2. Make sure the forks slide smoothly on their shaft and check that the shaft is not bent.
3. Check the grooves in the shift drum for wear, chipping or roughness (**Figure 25**). The fork pins should fit in the drum grooves without excessive play. The maximum shift drum groove width and minimum shift fork guide pin diameter are specified in **Table 1**.

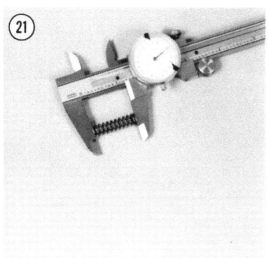

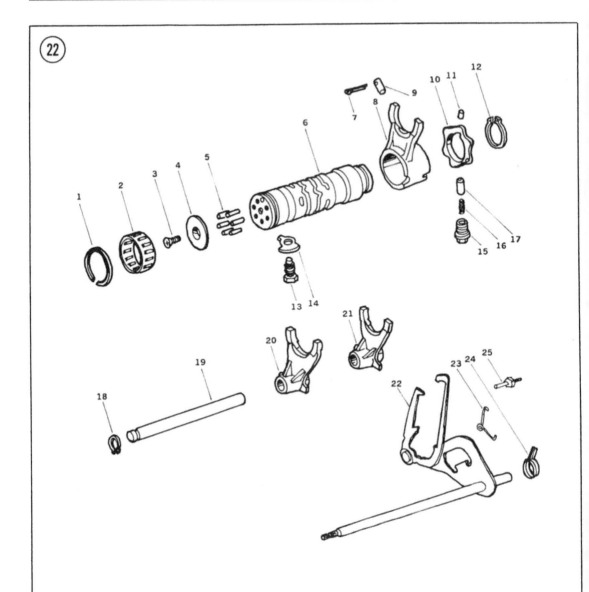

SHIFT LINKAGE, DRUM
AND FORKS

1. Circlip
2. Needle bearing
3. Screw
4. Pin plate
5. Ratchet pin (6)
6. Shift drum
7. Cotter pin
8. 4th/5th shift fork
9. Dowel pin
10. Shift drum cam
11. Dowel pin
12. Circlip
13. Guide pin

14. Lockplate
15. Detent cap
16. Detent spring
17. Detent plunger
18. Circlip
19. Shift rod
20. 2nd/3rd shift fork
21. 1st shift fork
22. Shift linkage assembly
23. Link spring
24. Return spring
25. Return pin

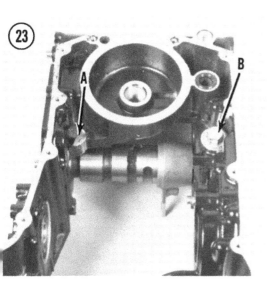

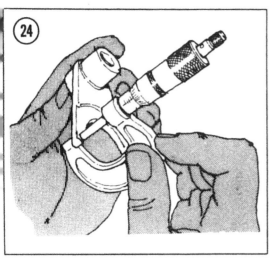

4. Measure the free length of the detent spring (**Figure 21**). Install a new spring if it is shorter than the limit in **Table 1**.

5. Inspect the detent cam for wear. Replace it if visibly worn.

6. Check that the shift drum ratchet pins are in good condition. If the pins are removed, install the long pin as shown in relation to the hole on the end of the drum (**Figure 15**). Secure the pin plate screw with a locking agent such as Loctite Lock N' Seal.

Installation

Refer to **Figure 22**.

1. Apply molybdenum disulfide grease to any new parts. Oil all parts before assembly.

2. Push the shift drum into the engine case. Position the large shift fork so that the drum will enter the short end of the fork boss first (**Figure 26**).

3. Check that the detent cam dowel pin is in place (**Figure 27**). Install the detent cam and circlip (**Figure 28**). Use a new circlip when possible. An old clip may have lost its tension during removal.

4. Install a new lockplate and the shift drum guide bolt. The lockplate tab must seat in the crankcase hole (**Figure 29**). Fold up a side of the lockplate.

5. Rotate the shift drum to the NEUTRAL position and install the neutral plunger, spring, washer and bolt (**Figure 30**).

6. Install the large shift fork guide pin with a new cotter pin as shown (**Figure 31**). The guide pin fits in the middle shift drum groove.

7. Install the 2 smaller shift forks and the shift rod (**Figure 32**), fitting the fork pins into the shift drum grooves.

NOTE
The 2 smaller shift forks are identical.

TRANSMISSION GEARS

Removal/Disassembly

Refer to **Figure 33**.

1. Remove the engine from the motorcycle and separate the case halves as described in Chapter Four, *Engine Removal* and *Crankcase Separation*.

2. Carefully lift out the input and output shaft gear clusters. Carefully note the location of thrust washers and bearings on the ends of the shafts.

3. To disassemble the transmission shafts, remove the circlips with circlip pliers. Carefully lay out all the clips, washers and gears in the order you remove them.

NOTE
To remove the output shaft 4th gear (on the end away from the engine sprocket splines), hold the shaft vertically with 4th gear up, spin the shaft and lift off 4th gear. Do not lose the 3 "neutral finder" balls. You may have to spin the shaft several times before the balls centrifuge out and unlock 4th gear.

Inspection

1. Inspect the transmission shaft bearings. Check for roughness, noise or excessive play. To replace a bearing, use a gear puller or heat the shaft in an oven to about 212° F and tap the old bearing off.

2. Check each gear for missing teeth, chips and excessive wear of the shift fork grooves. See **Table 1** for the maximum gear groove width specification.

NOTE
When one gear is replaced, inspect the mating gear on the opposite shaft very closely. Any damage is likely to affect both gears.

3. Check that the mating gear dogs and slots (**Figure 34**) are in good condition. Worn dogs and slots can cause the transmission to jump out of gear.

4. Check that the gears slide smoothly on the shaft splines.

Assembly/Installation

Refer to **Figure 33**.

1. Follow these general steps during assembly:
   a. Apply molybdenum disulfide grease to any new parts. Oil all parts before assembly.
   b. Use new circlips when possible. Old clips may have lost their tension during removal.
   c. When installing circlips on splined shafts, position them so that their opening falls on top of a spline groove and does not align with a splined washer tooth (**Figure 35**).

2. The assembly sequence for the input shaft (**Figure 36**) is: 4th gear, washer, circlip, 3rd gear, circlip, washer, copper bushing, 5th gear, 2nd gear, copper washer, steel washer, needle bearing, circlip, needle bearing outer race (with O-ring inside).

NOTE
*When installing the input shaft copper bushing, align its oil hole with that of the shaft (**Figure 37**).*

3. The assembly sequence for the output shaft (**Figure 38**) is: 2nd gear, splined washer, circlip, 5th gear, circlip, splined washer, 3rd gear, splined washer, circlip, 4th gear, 1st gear, copper washer, steel washer, needle bearing, circlip, needle bearing outer race.

NOTE
*When installing the output shaft 4th gear, do not grease the steel balls to hold them in place. These balls must be able to move freely. Insert the balls in the holes with a smaller outer diameter (**Figure 39**), not in the large diameter set of holes.*

4. Spin 1st gear on the output shaft (**Figure 38**). If it does not spin freely, a thinner 0.020 in. (0.5

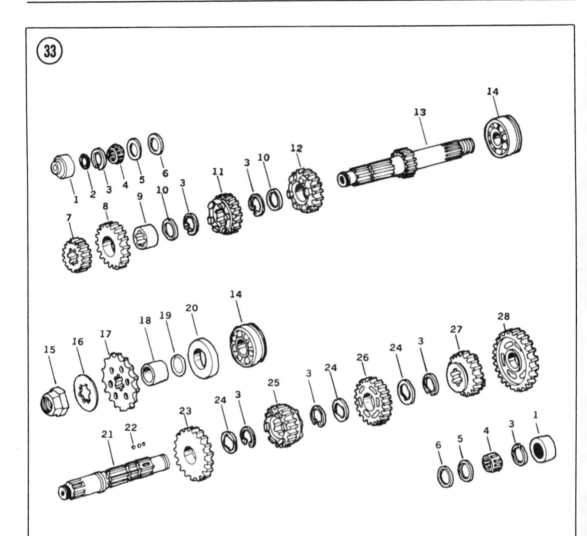

TRANSMISSION GEARSETS

1. Bearing outer race
2. O-ring
3. Circlip
4. Needle bearing
5. Steel washer
6. Copper washer
7. Input 2nd gear
8. Input 5th gear
9. Copper bushing
10. Washer
11. Input 3rd gear
12. Input 4th gear
13. Input shaft
14. Ball bearing
15. Sprocket nut
16. Lockplate
17. Output sprocket
18. Sprocket spacer
19. O-ring
20. Oil seal
21. Output shaft
22. Neutral balls, 5/32 in.
23. Output 2nd gear
24. Splined washer
25. Output 5th gear
26. Output 3rd gear
27. Output 4th gear
28. Output 1st gear

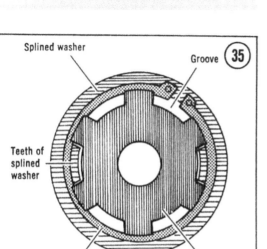

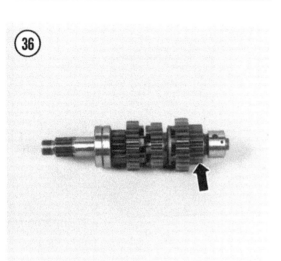

mm) steel washer is available to replace the standard 0.040 in. (1.0 mm) washer outside the bronze washer next to 1st gear. See your Kawasaki dealer.

5. Check the clearance between 2nd gear on the input shaft and the bronze washer outside it (**Figure 36**). If the clearance is not 0.004-0.012 in. (0.1-0.3 mm), change the steel washer (outside the bronze washer) as required. See your Kawasaki dealer for different thickness washers.

6. Install the gear sets in the upper crankcase half, meshing the shift fork fingers with the gear grooves (**Figure 40**). Rotate the bearings until the dowel pins seat and make sure the alignment rings are fully seated in the bearing and case grooves.

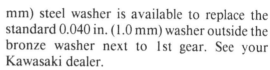

Table 1   KZ750 TRANSMISSION WEAR LIMITS

Item	In.	mm
Gear backlash	0.010	0.25
Gear fork groove width	0.207	5.25
Gear/shaft or bushing clearance	0.006	0.16
Shaft journal OD	0.786	19.96
Shaft bearing race ID	1.025	26.04
Shift fork finger thickness	0.185	4.7
Shift fork pin diameter		
Shift rod forks	0.309	7.85
Shift drum fork	0.312	7.93
Shift drum groove width	0.325	8.25
Shift drum detent spring length	1.21	30.7

NOTE: If you own a 1982 or later model, first check the Supplement at the back of the book for any new service information.

CHAPTER SEVEN

FUEL AND EXHAUST SYSTEMS

This chapter includes removal and repair procedures for the carburetor, fuel tank, fuel tap, air suction system and exhaust system. See Chapter Three for idle speed and idle mixture adjustment. Detailed carburetor specifications are given in **Table 1** at the end of this chapter.

CARBURETOR OPERATION

The following paragraphs explain the basic operation of carburetors, which may be helpful in troubleshooting a problem you suspect is caused by carburetion. If you are disassembling a carburetor, go on to *Carburetor Service* later in this chapter.

The "constant vacuum" carburetor has a rotating butterfly throttle valve, along with an engine vacuum-controlled sliding valve that carries the jet needle (**Figure 1**). This type of carburetor is less susceptible than some carburetors to a stall in acceleration when the throttle is snapped open, because the vacuum slide will not rise until gradually increasing engine vacuum pulls it up. The constant vacuum carburetor is also not as sensitive to changes in altitude as some carburetors are.

Float Mechanism

To assure a steady supply of fuel, the carburetor is equipped with a float valve through which fuel flows by gravity from the gas tank into the float bowl.

Inside the bowl is a pair of floats which move up and down with the fuel level. Resting on the float arm is a float needle, which rides inside the float valve. As the float rises, the float needle rises inside the float valve and blocks it so that when the fuel has reached the required level in the float bowl, no more can enter.

Vacuum Slide

The vacuum slide position is controlled by a diaphragm that has engine vacuum on top and atmospheric pressure on the bottom. The slide moves up and down with engine vacuum; any change in atmospheric pressure (such as a change in altitude) also moderates the slide position slightly to maintain a constant air/fuel ratio. The vacuum slide affects the fuel mixture from 1/4 to 3/4 throttle.

Pilot and Primary
Main Fuel Systems

NOTE
This description describes carburetor operation at a steady speed. When accelerating, the vacuum slide lags behind the throttle valve.

The carburetor's purpose is to supply and atomize fuel and mix it in correct proportions with air that is drawn in through the air intake. At primary throttle openings (from idle to 1/8 throttle) a small amount of fuel is siphoned through the pilot jet by suction from the incoming air. As the throttle is opened further (from 1/4 to 1/2 throttle) the vacuum slide begins to rise and the air stream also begins to siphon fuel through the primary main jet. From 1/2 to 3/4 throttle, the vacuum slide continues to rise and fuel also siphons through the needle jet. The tapered needle allows the needle jet to flow more fuel as the needle rises with the vacuum slide. From 3/4 to full throttle, the vacuum slide is fully open and fuel siphons through the secondary main jet and needle jet. At full throttle the needle is lifted far enough to permit the secondary main jet to flow at full capacity.

Choke System

The choke system consists of a rotating butterfly choke valve and a fast idle cam and linkage. When the choke valve closes the carburetor opening it causes a very high vacuum in the carburetor bore. Fuel flows from all jets and mixes with air coming into the carburetor to provide a rich mixture for cold starting. When the choke is used, the fast idle cam pushes the throttle valve open slightly to keep the engine from stalling.

CARBURETOR TROUBLESHOOTING

If the mixture is too lean at any or all throttle settings, the engine may overheat. It may generate brown exhaust smoke. It may stutter at high rpm. The performance (acceleration and top speed) will fall off. You may be able to confirm this by checking the spark plugs. If the mixture is too lean across the rpm scale, the spark plugs will be white and their electrodes may be rounded. While riding the motorcycle, use the starter plunger to see if the performance improves with what would normally be an overrich mixture.

If the mixture is too rich at any or all throttle settings, the engine may be sluggish and blubbery. It may generate black exhaust smoke. It may perform best while still cold. If the mix-

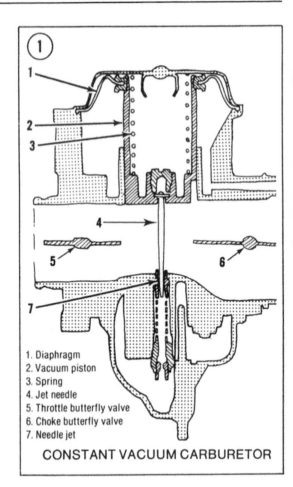

1. Diaphragm
2. Vacuum piston
3. Spring
4. Jet needle
5. Throttle butterfly valve
6. Choke butterfly valve
7. Needle jet

CONSTANT VACUUM CARBURETOR

ture is too rich across the rpm scale, the spark plugs may be black and sooty. Take out the air cleaner element and ride the motorcycle to see if the performance improves with what would normally be too lean a mixture. If it does, the mixture is too rich.

CAUTION
Do not ride the bike without the air cleaner element installed for any longer than is necessary to diagnose the problem.

Diagnosing the Problem

The fact that the mixture being burned is too rich or too lean does not necessarily indicate that the carburetion is at fault. The motorcycle may as easily have an ignition or compression problem.

If the mixture is wrong all up and down the rpm range, check the obvious fuel system components. For example, if the mixture is too rich, check for a clogged air cleaner element or too high a fuel level in the float bowl. If the mixture is too lean, check the fuel tap strainer and the fuel lines for blockage, check for too low a fuel level and check for an air leak at the rubber carburetor holders.

Before taking apart the carburetor, you should first check out the spark plugs and ignition timing and the cylinder compression.

If the engine won't rev to high rpm, check for a hole in the vacuum diaphragm.

Miscellaneous Carburetor Problems

Water in the carburetor float bowl and a sticking slide needle can result from careless washing of the motorcycle. To remedy the problem, remove and clean the carburetor bowl, main jet and any other affected parts. Be sure to cover the air intake when washing the machine.

Be sure that the carburetor mounting clamps are tight.

If gasoline leaks past the float bowl O-ring, high speed fuel starvation may occur. Varnish deposits on the outside of the float bowl are evidence of this condition.

Dirt in the fuel may lodge in the float valve and cause an overrich mixture. As a temporary measure, tap the carburetor lightly to dislodge the dirt. Clean the fuel tank, fuel valve, fuel line and carburetor at the first opportunity should this occur.

REJETTING

NOTE
This book covers U.S. models subject to governmental emission control laws. These laws subject motorcycle dealers and their employees to heavy fines for modifying emission-related components. Although Federal law does not cover modification by the motorcycle owner, some states have laws that prohibit emission-related modifications by owners. Check the laws in your area before you change carburetor parts.

Do not try to solve a carburetion problem if all the following conditions hold true:

a. The engine has held a good tune in the past with standard jetting.

b. The engine has not been modified.

c. The engine is being operated at the same altitude, climate and average speeds as in the past.

Rejetting the carburetors may be required if any of the following conditions hold true:

a. A nonstandard air filter element is being used.

b. A nonstandard exhaust system is being used.

c. Any of the top end parts (piston, camshaft, compression ratio, etc.) have been modified.

d. The motorcycle is in use at considerably higher or lower altitudes or in a markedly hotter, colder, wetter or drier climate than in the past.

e. The motorcycle is being operated at considerably higher speeds than before and changing to colder spark plugs has not solved the problem.

f. Someone has changed the jetting in your carburetor.

g. The motorcycle has never held a satisfactory engine tune.

The original jets and jet needle are listed in **Table 1** at the end of this chapter.

CARBURETOR VARIABLES

The following parts of the carburetor can be changed to alter the fuel mixture. Each part has the most effect over a narrow range of throttle openings, but each also has a lesser effect over a broader range of throttle openings.

Pilot Jet and Screw

The pilot jet and idle mixture setting affect mixture from 0 to about 1/8 throttle. As pilot jet numbers increase, the fuel mixture gets richer. As the idle mixture (pilot air) screw is opened (turned out) the mixture gets leaner.

On U.S. models, the idle mixture screw is sealed under a plug.

Primary Main Jet

The primary main jet affects the mixture from 1/4 to 1/2 throttle. Larger numbers provide a richer mixture, smaller numbers a leaner mixture.

Jet Needle

The jet needle affects the mixture from 1/2 to 3/4 throttle. Constant vacuum type carburetors have only one fixed needle position. The only alteration possible is through substitution of a different needle number or by raising the needle by putting a washer under it; as the needle is raised the mixture gets richer.

Needle Jet

Only one size needle jet is available on KZ750 motorcycles.

Secondary Main Jet

The secondary main jet controls the mixture at full throttle and has some effect at lesser throttle openings. Each jet is stamped with a number. Larger numbers provide a richer mixture, smaller numbers a leaner mixture.

NOTE
Kawasaki uses Reverse type secondary main jets, which have a round head. Do not substitute hex head Mikuni or Amal jets. The numbering systems are not equivalent; a given jet number will flow different amounts of fuel.

CARBURETOR SERVICE

There is no set rule regarding frequency of carburetor overhaul. A motorcycle used strictly for street riding may go 30,000 miles or more without needing a carburetor overhaul. Operation in dusty areas or poor air cleaner maintenance may shorten the useful life of the carburetor. See **Table 1** at the end of this chapter for carburetor specifications.

Removal/Installation

Remove all 4 carburetors as an assembled unit. Replacement of an individual carburetor is described in this chapter under *Separation*.
1. Put the bike up on its centerstand.
2. Remove the side panels from the motorcycle.
3. Remove the fuel tank; see *Fuel Tank Removal* in this chapter.
4. Pull the overflow tubes out of the bottom of the air cleaner housing.

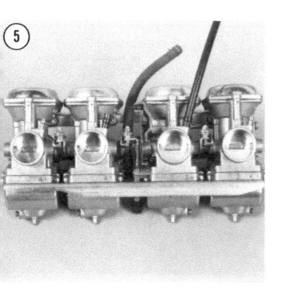

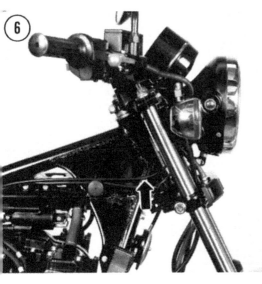

5. Slide the hose clamps up (**Figure 2**) and pull the vacuum hoses up off the carburetors.

6. Loosen the clamps or springs at the front and rear of the carburetor (**Figure 3**) and pull the carburetor assembly free from its mounts.

7. Loosen the lower throttle cable play adjuster and disconnect the throttle cable from the pulley and bracket (**Figure 4**).

8. Remove the carburetor assembly from the motorcycle (**Figure 5**).

9. To install, reverse the removal steps. Note the following:

    a. Install the No. 2 (from the left) carburetor clamp with its opening at the top. Install

the other 3 clamps with their opening at the bottom. Tighten the mounting clamps securely.

> *CAUTION*
> *Make sure there are no air leaks where the carburetors join with the front rubber holders. You should be able to feel the carburetors bottom out in the holders on both sides. Any leakage here will cause a lean fuel mixture and engine damage.*

    b. Route the throttle cable between the steering head and the right fork tube (**Figure 6**). The cable must not be twisted, kinked or pinched.

    c. Adjust the carburetors. Refer to Chapter Three.

Disassembly/Assembly

Refer to **Figure 7**. Most carburetor disassembly can be done without separating an individual carburetor from the assembly.

We recommend disassembling only one carburetor at a time to prevent accidental interchange of parts.

1. Remove the diaphragm cover screws (**Figure 8**) and the cover and spring.

2. Pull out the slide and diaphragm assembly (**Figure 9**). Be careful not to puncture or tear the diaphragm.

3. To remove the needle from the slide, unscrew the retainer (**Figure 10**) and push the needle out.

4. Remove the float bowl screws, lockwashers, float bowl and O-ring (**Figure 11**).

5. Push out the float pivot pin (A, **Figure 12**) and remove the float assembly.

6. Remove the float needle spring clip (B, **Figure 12**) and the float needle.

7. Unscrew the secondary main jet (A, **Figure 13**) and primary main jet (B).

8. Unscrew the needle jet holder under the secondary main jet.

9. Remove the plastic plug and O-ring (C, **Figure 13**) and unscrew the slow jet under it.

10. Turn the carburetor upright so the needle jet (**Figure 14**) falls out or push it out from the top. The primary main jet bleed pipe should also come out.

7

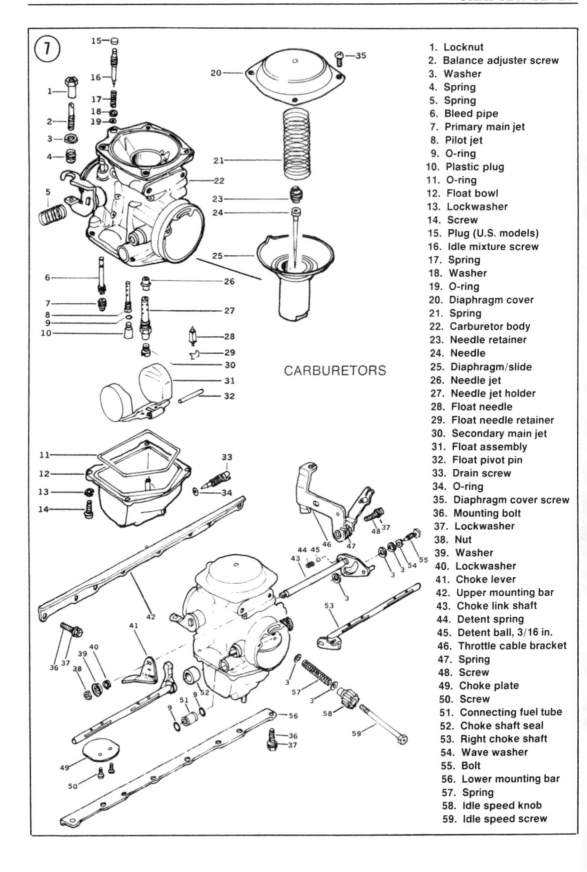

CARBURETORS

1. Locknut
2. Balance adjuster screw
3. Washer
4. Spring
5. Spring
6. Bleed pipe
7. Primary main jet
8. Pilot jet
9. O-ring
10. Plastic plug
11. O-ring
12. Float bowl
13. Lockwasher
14. Screw
15. Plug (U.S. models)
16. Idle mixture screw
17. Spring
18. Washer
19. O-ring
20. Diaphragm cover
21. Spring
22. Carburetor body
23. Needle retainer
24. Needle
25. Diaphragm/slide
26. Needle jet
27. Needle jet holder
28. Float needle
29. Float needle retainer
30. Secondary main jet
31. Float assembly
32. Float pivot pin
33. Drain screw
34. O-ring
35. Diaphragm cover screw
36. Mounting bolt
37. Lockwasher
38. Nut
39. Washer
40. Lockwasher
41. Choke lever
42. Upper mounting bar
43. Choke link shaft
44. Detent spring
45. Detent ball, 3/16 in.
46. Throttle cable bracket
47. Spring
48. Screw
49. Choke plate
50. Screw
51. Connecting fuel tube
52. Choke shaft seal
53. Right choke shaft
54. Wave washer
55. Bolt
56. Lower mounting bar
57. Spring
58. Idle speed knob
59. Idle speed screw

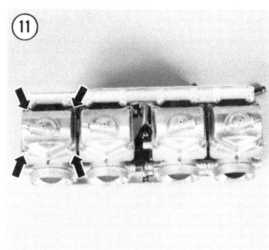

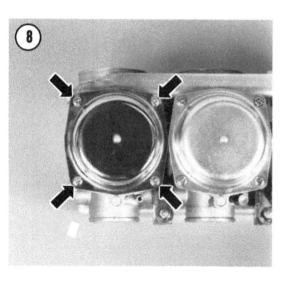

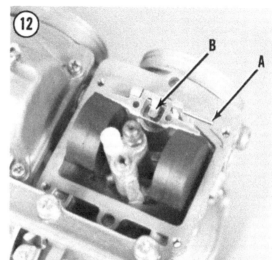

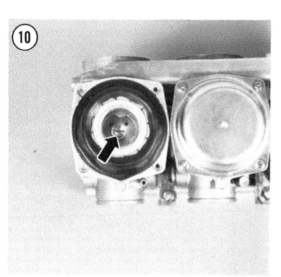

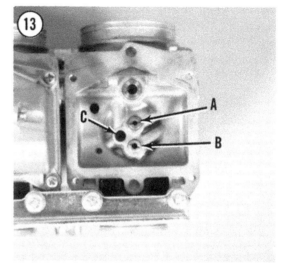

7

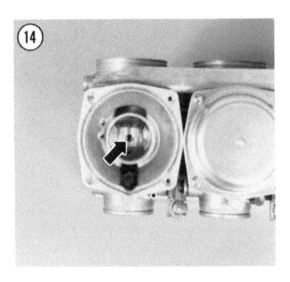

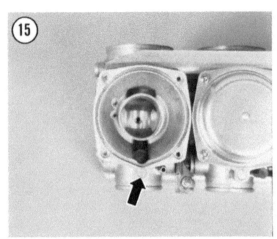

NOTE
Do not try to remove the throttle valve plates; they are precisely matched to the carburetor body and bypass ports. The throttle valve plates are not available separately. Replace the carburetor body if these plates are damaged.

NOTE
*On U.S. models, the idle mixture screw is covered by a plug (**Figure 15**) bonded in place at the factory. When disassembling the carburetor for overhaul, the plug should be removed in order to clean the passage with compressed air. Before removing the idle mixture screw, count the number of turns it takes to seat it lightly.*

11. Remove the idle mixture screw, spring, washer and O-ring.
12. Perform *Cleaning and Inspection.*
13. To assemble, reverse these steps. Note the following:
  a. On U.S. models, install the mixture screw; turn it in until it seats lightly, then back it out the same number of turns noted during removal. Install a new plug and seal the edges of the plug lightly with silicone sealant.
  b. Install the throttle slide diaphragm with the tab in the carburetor body notch (**Figure 16**). Lift the slide a little to take stress off the diaphragm as you tighten the cover screws. Then check that the slide moves up and down freely.

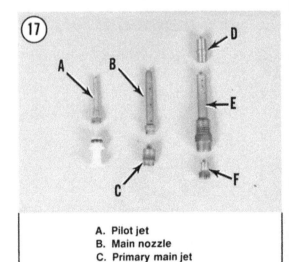

A. Pilot jet
B. Main nozzle
C. Primary main jet
D. Needle jet
E. Needle jet holder
F. Secondary main jet

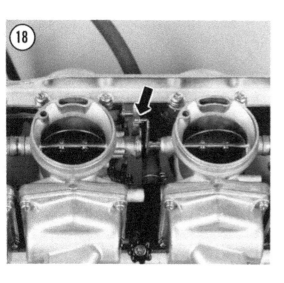

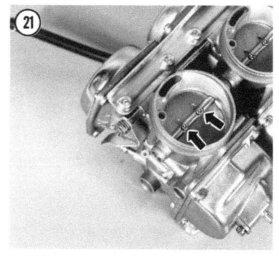

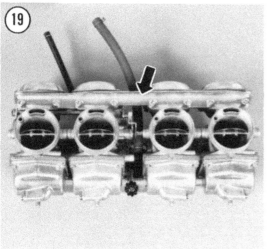

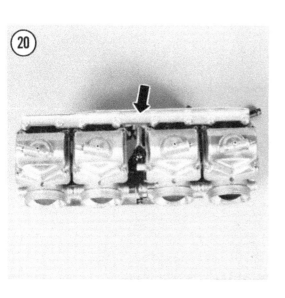

c. Tighten the jets securely, but be careful not to strip their threads (**Figure 17**).

d. After assembly, turn the carburetor upside down and listen to be sure the floats are moving freely.

e. Perform *Fuel Level Inspection* in this chapter and *Carburetor Adjustment* in Chapter Three.

Separation

See **Figure 7**. The carburetors are joined by 2 common choke valve shafts and 2 mounting bars. The left pair of carburetors can be easily separated from the right pair, but a choke shaft repair kit should be obtained from your Kawasaki dealer before the 2 carburetors in a pair are separated.

1. Disconnect the choke return spring from the lever between the middle carburetors (**Figure 18**).

2. Remove the 8 bolts, lockwashers and the upper mounting bar (**Figure 19**).

3. Remove the 8 bolts, lockwashers and the lower mounting bar (**Figure 20**).

4. Separate the left and right carburetor pairs.

5. Remove the choke plate screws (**Figure 21**) and remove the choke plates.

6. Pull out the 2 choke shafts and separate the carburetors.

7. To assemble, reverse these steps. Note the following:

a. Use new choke plate screws furnished with the choke plate repair kit. Crimp the screws in place after installation; use the crimping adapter supplied with the Kawasaki repair kit. A pair of Vise Grips may do the job as well.

> *CAUTION*
> *The choke plate screws must be crimped in place to prevent possible loosening of the screws and entry into the combustion chambers. This would cause severe engine damage.*

b. Make sure each connecting fuel tube has O-rings at both ends.

c. Install the linkage spring as shown between carburetors No. 1 and 2 and between carburetors No. 3 and 4 (A, **Figure 22**).

d. Synchronize the throttle valve plates visually so that they each have an equal gap at the throttle bore when closed (B, **Figure 22**); loosen the adjuster locknuts, turn the adjusters as required and tighten the locknuts (**Figure 23**).

Cleaning and Inspection

1. Thoroughly clean and dry all parts. If a special carburetor cleaning solution is used, all non-metal parts must be removed (gaskets, O-rings, etc.).

2. Blow out all the passages and jets with compressed air. Don't use wire to clean any of the orifices; wire will enlarge them.

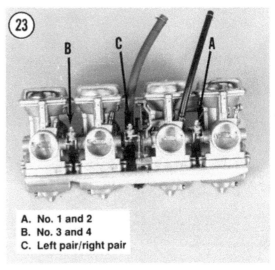

A. No. 1 and 2
B. No. 3 and 4
C. Left pair/right pair

3. Check the cone of the float needle and replace it if it is scored or pitted (**Figure 24**).

4. Examine the end of the idle mixture screw for grooves or roughness. Replace it if damaged. Replace a worn O-ring.

5. Check the O-rings on the float chamber drain plug and on the float bowl. Replace them if damaged.

6. Inspect the slide diaphragm. If there are any pin-holes or other damage, replace the diaphragm.

FUEL LEVEL INSPECTION

The fuel level in the carburetor float bowl is critical to proper performance. The fuel flow rate from the bowl up to the carburetor bore depends not only on the vacuum in the throttle bore and the size of the jets, but also upon the

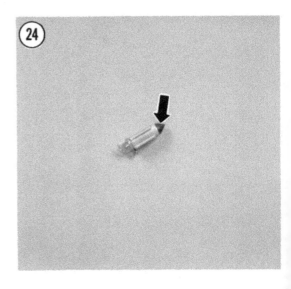

fuel level. Kawasaki gives a specification of actual *fuel level*, measured from the top edge of the float bowl with the carburetor held level (**Figure 25**).

This measurement is more useful than a simple float height measurement because actual fuel level can vary from carburetor to carburetor, even when their floats are set at the same height. However, fuel level inspection requires a special clear tube that attaches to the overflow tube (**Figure 25**).

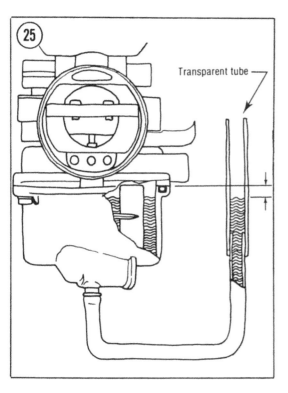

Transparent tube

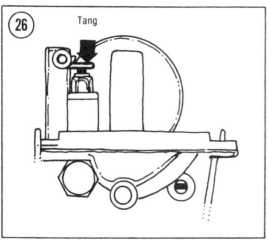

Tang

The fuel level is adjusted by bending the float arm tang.

1. Prop up the front end of the motorcycle so that the carburetors are parallel to the ground. The carburetors must be level.

2. Connect the drain tube to the bottom of the float bowl and attach a transparent hose (**Figure 25**). The hose must have an inside diameter not less than 1/4 in. (6 mm) so that capillary action will not draw the fuel up inside, giving a false reading.

3. Hold the clear tube against the carburetor body and turn the fuel tap to PRI (prime). With the carburetor level to the ground, check the fuel level in the tube. It should be slightly below the bottom edge of the carburetor body. See **Table 1** for your bike's fuel level specification.

NOTE
*Take your reading just after the fuel level has **risen** to its maximum in the tube. If you raise the tube (and the fuel drops in the tube) you'll probably get a faulty level reading. Turn the fuel tap ON, drain the tube and try it again, forcing the fuel level to rise against surface tension within the tube.*

4. If the fuel level is incorrect, adjust the float height; remove the float bowl from the carburetor and bend the float tang (**Figure 26**) as required to get the right fuel level. Install the float bowl and recheck the fuel level.

NOTE
If you want the fuel level lower (a greater distance below the carburetor body) bend the tang so the floats stick up higher when held upside down.

FAST IDLE ADJUSTMENT

Adjustment of the fast idle linkage should rarely be required, but if the engine stalls when the choke is used on a cold engine or if the idle speed is too high when the choke is used, check that there is about 3/16 in. (5 mm) clearance beween the fast idle link pin and the fast idle cam when the choke is fully OFF (**Figure 27**).

If the clearance is incorrect, bend the stop tab on the fast idle link to get the proper clearance.

7

CRANKCASE BREATHER

The crankcase breather separates oil mist droplets from blowby gas and routes the oil back to the crankcase via a drain hole. The vapors are routed to the air cleaner housing. No maintenance is required; but if the oil drain is not kept clear during engine assembly or if the engine is overfilled with oil, oil can be sucked up into the air cleaner housing.

Removal/Installation

1. Remove the breather hose and cover bolt with its O-ring (**Figure 28**).
2. Remove the cover and O-ring.
3. When installing the breather cover, make sure the O-rings are in good condition and make sure the breather cover pin seats in front of the pin cast into the crankcase.

AIR SUCTION SYSTEM (U.S. MODELS)

The air suction system (**Figure 29**) consists of a vacuum switch valve, 2 air suction valves (reed valves) and air and vacuum hoses. This system does not pressurize air, but does introduce fresh air into the exhaust ports when the valves are opened by momentary pressure differentials in exhaust gas pulses.

The vacuum switch (**Figure 30**) normally allows fresh air pulses into the exhaust port but

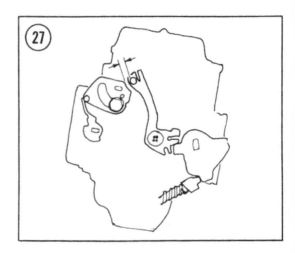

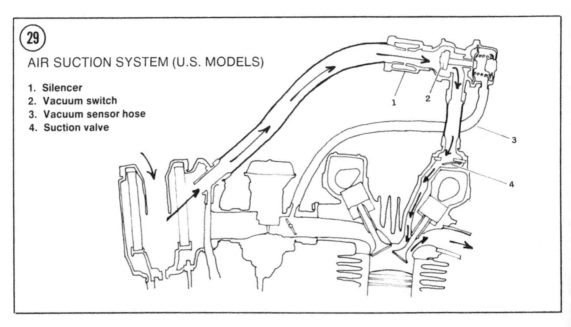

AIR SUCTION SYSTEM (U.S. MODELS)

1. Silencer
2. Vacuum switch
3. Vacuum sensor hose
4. Suction valve

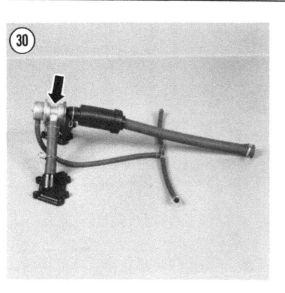

shuts off air flow during engine braking. This helps prevent backfiring in the exhaust system due to the greater amount of unburned fuel in the exhaust gas during deceleration.

The air suction valves, on top of the valve cover, are basically check valves. They allow the fresh air to enter the exhaust port and prevent any air or exhaust from reversing back into the system.

Valve Removal/Installation

If the engine idle is not smooth, if engine power decreases seriously or if there are any abnormal engine noises, remove the air suction valves and inspect them.

1. Check that the ignition switch is OFF.
2. Remove the fuel tank; see *Fuel Tank Removal* in this chapter.
3. Slide up the lower hose clamps and pull the hoses off the air suction valve covers (**Figure 31**).
4. Remove the 8 bolts securing the air suction valve covers (**Figure 31**).
5. Remove the covers and pull the valves up out of the valve cover (**Figure 32**).
6. Check the reed valves for cracks, folds, warpage or any other damage (**Figure 33**).
7. Check the sealing lip coating around the perimeter of the assembly. It must be free of grooves, scratches or signs of separation from the metal holder.

NOTE
The valve assembly cannot be repaired. It must be replaced if damaged.

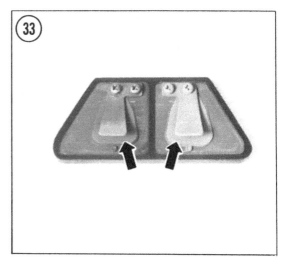

8. Wash off any carbon deposits between the reed and the reed contact area with solvent.

CAUTION
Do not scrape deposits off or the assembly will be damaged.

9. Install by reversing the removal steps.

Vacuum Switch Test

Inspect the vacuum switch if there is backfiring during deceleration or other abnormal engine noise.

1. Run the engine until it is warm.
2. Rev the engine to 4,000 rpm and snap the throttle shut. Note the intensity and frequency of any backfiring for comparison later in this test.
3. Shut the engine off and, at the air cleaner housing, disconnect and plug the hose from the vacuum switch (**Figure 34**).
4. Start the engine, rev it to 4,000 rpm and compare the backfiring to what you heard before. If the backfiring is the same, there is nothing wrong with the vacuum switch.
5. If the backfiring is different, the vacuum switch is faulty. Install a new switch.

FUEL TANK

As water and dirt accumulate in the fuel tank, engine performance will deteriorate. The fuel system should be cleaned when the engine is cold.

WARNING
Some fuel may spill during these procedures. Work in a well-ventilated area at least 50 feet from any sparks or flames, including gas appliance pilot lights. Do not smoke in the area. Keep a BC rated fire extinguisher handy.

Removal/Installation

1. Check that the ignition switch is OFF.
2. Put the bike up on its centerstand.
3. Swing the seat open or remove it.
4. Remove the bolt, washer and collar at the rear of the fuel tank.
5. Turn the fuel tap ON and disconnect the fuel and vacuum lines at the fuel tap. Lift up the rear of the tank slightly, if necessary.

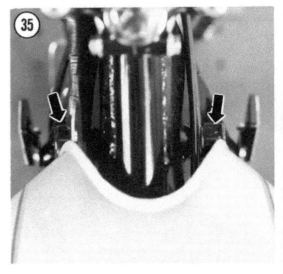

6. On the KZ750-H, disconnect the fuel level sensor wires.

7. Remove the tank; pull it up and to the rear.

8. Discard the fuel in the tank and pour about a pint of clean fuel into the tank, install the cap, slosh the fuel around for about a minute and pour it out in a safe container.

9. To install, reverse the removal steps. Note the following:

    a. Insert the tank brackets carefully over the frame grommets (**Figure 35**). Don't pinch any wires or control cables.

    b. The vacuum hose is smaller than the fuel line (**Figure 36**).

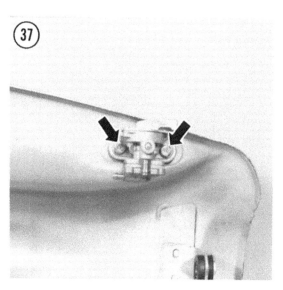

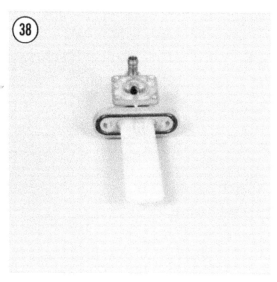

    c. Partially fill the tank with fresh fuel and check for leaks around the tap and at the fuel line connections.

FUEL TAP

Vacuum-operated fuel taps have no OFF position. The tap should pass no fuel in ON or RES until a running engine provides the vacuum required to operate the diaphragm valve. In PRI (prime) the tap will pass fuel whether or not the engine is running.

Removal/Installation

> *WARNING*
> *Some fuel may spill during these procedures. Work in a well-ventilated area at least 50 feet from any sparks or flames, including gas appliance pilot lights. Do not smoke in the area. Keep a BC rated fire extinguisher handy.*

1. Remove the fuel tank.
2. Turn the fuel tap to PRI and drain the fuel into a clean gas can.
3. Remove the 2 fuel tap mounting bolts and the tap and O-ring (**Figure 37**).
4. Inspect the fuel tap mounting O-ring and clean the feed tube screen whenever you remove the tap from the tank (**Figure 38**).
5. To install, reverse the removal steps. Note the following:

    a. Make sure that fuel does not flow in the ON position.

    b. The vacuum hose is smaller than the fuel hose (**Figure 36**).

    c. Check for leakage after you install the fuel tap.

Inspection

See **Figure 39**.

Disassemble the tap and check that the O-rings and diaphragm are clean and undamaged. Look for pin-holes in the diaphragm. Any bit of debris on the valve O-ring (**Figure 40**) will prevent the valve from closing.

The groove in the diaphragm plate must face the O-ring (**Figure 41**). Make sure the dia-

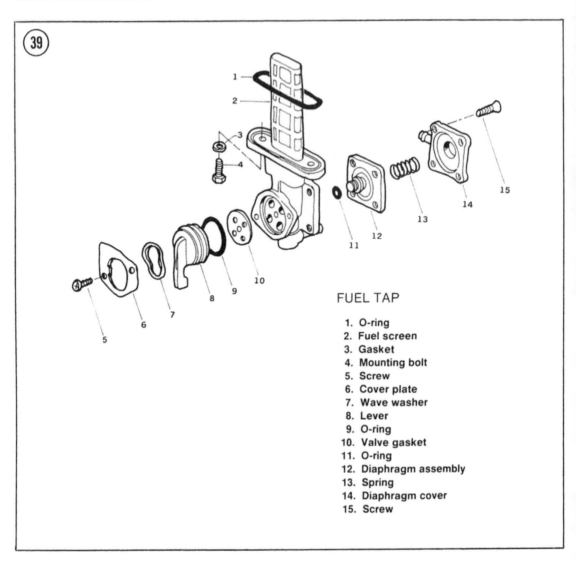

FUEL TAP

1. O-ring
2. Fuel screen
3. Gasket
4. Mounting bolt
5. Screw
6. Cover plate
7. Wave washer
8. Lever
9. O-ring
10. Valve gasket
11. O-ring
12. Diaphragm assembly
13. Spring
14. Diaphragm cover
15. Screw

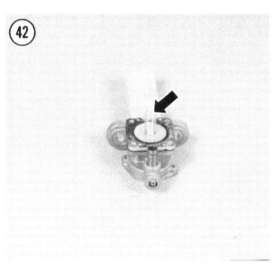

phragm spring is in place (**Figure 42**). Install the diaphragm cover with its tube facing forward on the KZ750-E, L and to the rear on the KZ750-H.

FUEL LEVEL SENDING UNIT

Models with a fuel level gauge or warning light have a sending unit mounted in the bottom of the fuel tank (**Figure 43**). When installing a sending unit, make sure the O-ring is in good condition (**Figure 44**).

EXHAUST SYSTEM

The exhaust system requires no maintenance other than to make sure the connecting clamps are tight.

Removal/Installation

See **Figure 45**.
1. Loosen the crossover pipe clamps (**Figure 46**).
2. Remove the rear footpeg mounting bolts (**Figure 47**).
3. Remove the exhaust pipe holder nuts (**Figure 48**) and pry the holders free from the studs.
4. Pull the exhaust pipe and muffler assemblies out of the cylinder head and remove the split collars (**Figure 49**).
5. To install, reverse the removal steps. Note the following:

    a. Do not mix up the No. 2 (left) and No. 3 (right) exhaust pipes. Each has an identifying number (**Figure 50**).

    b. Use a new gasket in the cylinder head exhaust ports (**Figure 51**).

    c. Make sure the gaskets at the pipe/muffler joint and at the crossover pipe are in good condition.

    d. Tighten the exhaust pipe holder nuts first, gradually and evenly, then the rear footpeg bolts, then the middle clamp and crossover bolts.

    e. After the job is complete, run the motor and check for leakage. Tighten the clamps again after the engine has cooled down.

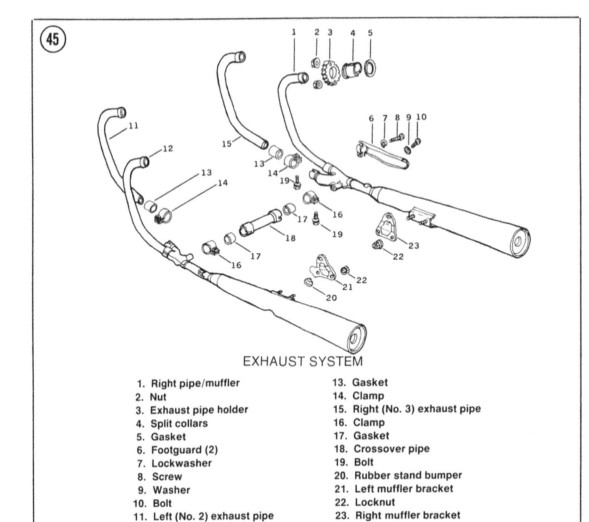

EXHAUST SYSTEM

1. Right pipe/muffler	13. Gasket
2. Nut	14. Clamp
3. Exhaust pipe holder	15. Right (No. 3) exhaust pipe
4. Split collars	16. Clamp
5. Gasket	17. Gasket
6. Footguard (2)	18. Crossover pipe
7. Lockwasher	19. Bolt
8. Screw	20. Rubber stand bumper
9. Washer	21. Left muffler bracket
10. Bolt	22. Locknut
11. Left (No. 2) exhaust pipe	23. Right muffler bracket
12. Left pipe/muffler	

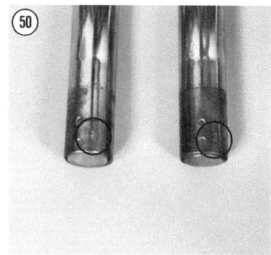

Table 1   KZ750 CARBURETOR SPECIFICATIONS

Item	Specification
Size/type	U.S.: CV34 mm-30
	Others: CV34 mm
Primary main jet	#62
Secondary main jet	#125
Pilot jet	
KZ750-E,H	#38
KZ750-L	#35
Jet needle	N01A
Mixture screw	U.S.: preset
	Others: 2 turns out
Fuel level	0.016 in. (4.0 mm)

NOTE: If you own a 1982 or later model, first check the Supplement at the back of the book for any new service information.

CHAPTER EIGHT

ELECTRICAL SYSTEM

The electrical system includes the battery, ignition system, charging system, electric starter, lighting, fuel level sensor, and horn.

Table 1 and **Table 2** are at the end of the chapter.

WIRING DIAGRAMS

Wiring diagrams are located at the end of this book.

FUSES

There are 3 main fuses in a fuse box under the left side cover. Remove the electrical cover screw and the cover (**Figure 1**). Slide the fuse box out and open it (**Figure 2**). The main fuse is 20A and the headlight and taillight fuses are l0A. Inside the cover are spare fuses; always carry spares.

Some 1981 models have 2 additional 10A fuses for electrical accessories. They are in a separate fuse box under the electrical cover (**Figure 3**).

Whenever a fuse blows, find out the reason for the failure before replacing the fuse. Usually, the trouble is a short circuit in the wiring. This may be caused by worn-through insulation, a disconnected wire shorting to ground or possibly a wire pinched during assembly or installation of parts.

WARNING
Never substitute metal foil or wire for a fuse. Never use a higher amperage fuse than specified. An overload could result in fire and destruction of the bike.

BATTERY

The bike is equipped with a 12 volt, 12 ampere-hour battery with an electrolyte specific gravity of 1.280 at 68° F when fully charged.

NOTE
In very warm climates an electrolyte with a specific gravity of 1.260 is used and you should subtract 0.020 from all specified test readings of specific gravity. If you are uncertain of the electolyte installed in your motorcycle, any local motorcycle dealer should be able to tell from the label of the acid container he uses to initially service batteries.

Battery electrolyte testing and battery charging may be required after long periods (more than a month) of inactivity or when electrical trouble arises.

The battery is the heart of the electrical system. The majority of electrical system troubles can be attributed to neglect of this vital component.

More water evaporates from the battery in warmer climates, but excessive use of water may be an indication that the battery is being overcharged. It is advisable to check the voltage regulator if this situation exists.

WARNING
*Study the **Safety Precautions** before servicing the battery.*

Safety Precautions

While working with batteries, use care to avoid spilling or splashing the electrolyte. The electrolyte is a sulfuric acid solution, which can destroy clothing and cause serious chemical burns. If you get any electrolyte on your clothing, body or any other surface, neutralize it immediately with a solution of baking soda and water, then flush with plenty of clean water.

WARNING
Electrolyte splashed into the eyes is extremely dangerous. Wear safety glasses while working with batteries. If electrolyte is splashed into the eye, call a doctor immediately, force the eye open and flood it with cool water for about 5 minutes.

WARNING
When batteries are being charged, highly explosive hydrogen gas forms in the cells of the battery. Some of this gas escapes through the filler openings and may form an explosive atmosphere around the battery. Sparks, flames or a lighted cigarette can ignite the gas, causing a battery explosion and possible serious personal injury. Follow these precautions to help prevent accidents.

1. Do not smoke or permit any flame near a battery being charged or which has been charged recently. Keep the battery away from gas-operated home appliances.
2. Do not disconnect or connect live circuits at the battery terminals, because a spark will occur when a live circuit is connected or broken. Turn off the ignition switch first or disconnect the circuit away from the battery. When using a battery charger, don't plug the charger in until

the battery clips have been securely attached. Unplug the charger before you remove the clips from the battery.

Removal

Disconnect the negative (-) ground cable first (A, **Figure 4**), then the positive (+) cable. This minimizes the chance of a tool shorting to ground when disconnecting the "hot" positive cable. Remove the battery holder screw and the holder (B, **Figure 4**).

If the motorcycle will not be used for an extended period, remove the battery from the machine, charge it fully and store it in a cool, dry place. Recharge the battery every 2 months while it is in storage and again before it is put back into service.

Installation

Be very careful when installing the battery to connect it properly. If the battery is installed backward, the electrical system may be damaged.

1. Clean the battery terminals, case and tray. Coat the terminals with Vaseline or silicone spray to retard corrosion of the terminals.
2. Connect the positive (+) terminal first, then the negative (-) ground. Don't overtighten the clamps.
3. Check to make sure the cable terminals won't rub against any metal parts (like the seat). Slide the plastic boot over the positive (+) terminal.
4. Connect the battery vent tube and make sure it isn't pinched anywhere. Keep the end away from the mufflers and drive chain. The corrosive gases could cause damage.

Specific Gravity Testing

Hydrometer testing is the best way to check battery condition. Use a hydrometer with numbered graduations from 1.100 to 1.300, rather than one with color-coded bands.

To use the hydrometer, squeeze the rubber ball, insert the tip into the cell and release the ball. Draw enough electrolyte to float the weighted float inside the hydrometer. Note the number in line with surface of the electrolyte

(**Figure 5**); this is the specific gravity for this cell. Return the electrolyte to the cell from which it came.

The specific gravity of the electrolyte in each battery cell is an excellent indication of that cell's condition. A fully charged cell will read 1.260-1.280, while a cell in good condition reads from 1.230-1.250 and anything below 1.140 is discharged.

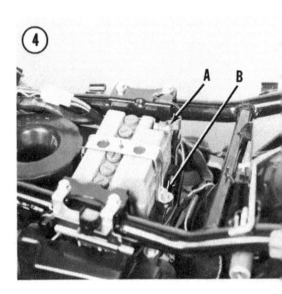

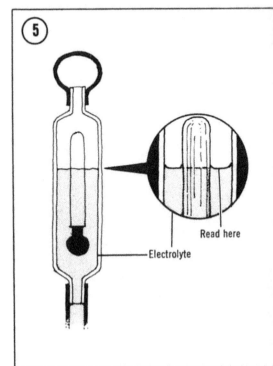

Read here

Electrolyte

Specific gravity varies with temperature. For each 10° that electrolyte temperature exceeds 80° F, add 0.004 to the reading indicated on the hydrometer. Subtract 0.004 for each 10° below 80° F.

NOTE
In very warm climates an electrolyte with a specific gravity of 1.260 is used and you should subtract 0.020 from all specified test readings of specific gravity. If you are uncertain of the electolyte installed in your motorcycle, any local motorcycle dealer should be able to tell from the label of the acid container he uses to initially service batteries.

Repeat this measurement for each battery cell. If there is more than 0.050 difference (50 points) between cells, battery condition is questionable.

If the cells test in the poor range, the battery requires recharging. The hydrometer is useful for checking the progress of the charging operation. **Figure 6** shows the approximate state of charge.

It is most important to keep the battery fully charged during cold weather. A fully charged battery freezes at a much lower temperature than one which is partially discharged. Freezing temperature depends on specific gravity, as shown in **Table 1**.

Charging

WARNING
*Do not smoke or permit any open flame in any area where batteries are being charged or immediately after charging. Highly explosive hydrogen and oxygen gases are formed during the charging process. Be sure to reread **Safety Precautions** at the beginning of this section.*

CAUTION
Always disconnect the battery cables before connecting charging equipment or you may damage part of the bike's charging system. It is preferable to remove the battery from the motorcycle.

Motorcycle batteries are not designed for high charge or discharge rates. A motorcycle battery should be charged at a rate not exceeding 10 percent of its ampere-hour capacity. That is, do not exceed 0.5 ampere charging rate for a 5 ampere-hour battery or 1.2 amperes for a 12 ampere-hour battery. This charge rate should continue for about 10 hours if the battery is completely discharged or until the specific gravity of each cell is up to 1.260-1.280, corrected for temperature.

Some temperature rise is normal as a battery is being charged. Do not allow the electrolyte temperature to exceed 110° F. Should the temperature reach that figure, discontinue charging until the battery cools, then resume charging at a lower rate.

1. Remove the battery from the motorcycle or disconnect the negative (-) ground cable, then the positive (+) cable.
2. Before you switch on or plug in the charger, connect the positive charger lead to the positive battery terminal and the negative charger lead to the negative battery terminal.

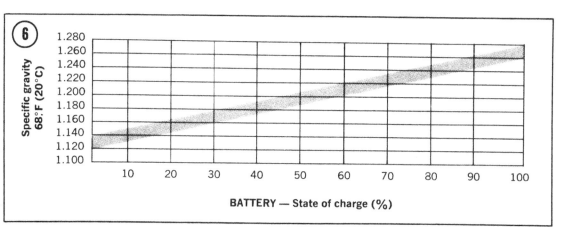

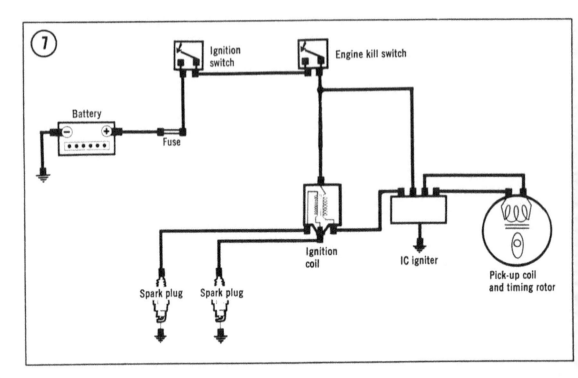

3. Remove all vent caps from the battery, set the charger at 12 volts and switch it on. If the output of the charger is variable, it is best to select a setting that doesn't exceed 10% of the battery's ampere-hour capacity.

4. When you want to check the state of charge, turn the charger off or unplug it, disconnect the leads and check the specific gravity. It should be within the limits specified in **Figure 6**. If it is, and remains stable for one hour, the battery is charged.

5. Install the battery. Connect the positive (+) cable first, then the negative (-) ground.

IGNITION SYSTEM

Operation

The ignition system consists of 4 spark plugs, 2 ignition coils, an IC igniter unit and 2 timing pickup units. **Figure 7** is a diagram of the transistorized ignition circuit for 2 cylinders. The Kawasaki transistorized ignition system is similar to a contact point ignition system. It works much the same, with these differences:

  a. Mechanical contact points are replaced by magnetic triggering pickup coils. The elimination of contact breaker points means that periodic adjustment of point gap and ignition timing is no longer required. Once set properly, initial timing should not require adjustment for the life of the motorcycle.

  b. An intermediate electronic switch, the battery-powered IC igniter, receives the weak signals from the pickup coils and uses them to turn the ignition coil primary current on and off.

  c. The ignition coil has a special low resistance primary winding that helps it produce a powerful spark at high rpm.

  d. The transistorized ignition system's dwell angle *increases* slightly as rpm increases. This is a characteristic of the magnetic pickup coils.

  e. Ignition timing is *not* adjustable.

Ignition takes place every 180° of crankshaft rotation; each pickup unit fires every 360° of crankshaft rotation. The firing order is 1-2-4-3. The left pickup unit fires cylinders No. 1 and 4; the right pickup unit fires cylinders No. 2 and 3. One of the spark plugs in each pair is fired (harmlessly) on the exhaust stroke.

The ignition coil fires through 2 spark plugs wired in series. If one of the plugs fails to fire, so will the other. If one plug develops a weak spark, so will the other.

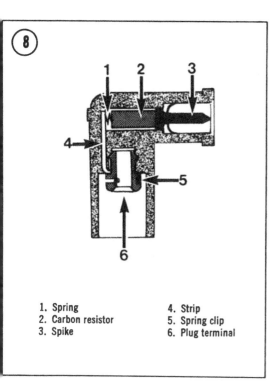

8

1. Spring
2. Carbon resistor
3. Spike
4. Strip
5. Spring clip
6. Plug terminal

The ignition coil primary current is normally off until the ignition timing rotor cam approaches the pickup coil. As the rotor cam approaches the pickup coil, a pickup coil signal builds to a level that turns the IC igniter on, allowing primary current to flow through the ignition coil. As the rotor cam passes the pickup coil, the trigger signal reverses polarity and turns the IC igniter off. The sudden stoppage of current through the ignition coil primary winding causes the magnetic field to collapse. When this occurs, a very high voltage (up to about 20,000 volts) is induced in the secondary winding of the ignition coil. This high voltage is sufficient to jump the gap at the spark plug, causing the plug to fire.

SPARK PLUGS

See *Spark Plugs* in Chapter Three for regular inspection and adjustment of spark plugs.

A resistor-type spark plug cap is used to reduce radio interference (**Figure 8**). If you have a high-speed misfire, check the resistance between the plug terminal and the plug wire spike. Normal is about 10,000 ohms. If resistance is more than 20,000 ohms, the cap is faulty and should be replaced.

Troubleshooting

If you suspect that 2 of the spark plugs are failing to fire or are delivering weak sparks, check them as follows:

1. Remove the spark plugs from the engine and compare them to the spark plug chart in Chapter One.

2. If visual inspection shows that one or both of the spark plugs are defective, discard them and install new ones.

3. Check both spark plugs (whether new or used) by putting on their spark plug wires and taping them to the cylinder head fins, so that metal touches metal.

4. Wheel the motorcycle over to a dark corner. Turn on the ignition, operate the starter and observe the sparks. If the plugs throw sparks that are intermittent, feeble, orange-yellow in color or do not make snapping sounds when they jump, the sparks are weak.

5. If one or both of the spark plugs throwing weak sparks are old, fit new plugs to the spark plug wires and observe the sparks again. Watch also for arcing inside the spark plug cap.

6. If the used spark plugs fail to fire or generate weak sparks and if the new spark plugs throw strong sparks (bluish in color, accompanied by a snapping sound), then one or both of the old spark plugs are defective.

IGNITION ADVANCE

The ignition advance mechanism advances the ignition (fires the spark plugs sooner) as engine speed increases. If it does not advance smoothly, the ignition timing will be incorrect at high engine rpm. The advancer must be lubricated periodically to make certain it operates freely.

The ignition advance mechanism is bolted to the right end of the crankshaft, under the timing cover. It can be removed after removing the pickup coil assembly.

Disassembly/Lubrication/Assembly

1. Hold the larger nut and remove the advancer mounting bolt, large nut and the advancer (**Figure 9**).

2. Remove the cam from the advancer body; hold the base steady, turn the advancer cam

until the arms come out of the cam, then pull the cam off its pivot.

3. Remove the weight C-clips, washers, weights and thrust washers.

4. Install by reversing the removal steps. Note the following:

    a. Grease the groove inside the rotor, the weight pivots and the weight arms that fit into the rotor.

    b. When assembling the advancer unit, align the rotor peak and "TEC" mark (**Figure 10**).

    c. Check for free movement and full weight return by the advancer springs.

    d. When installing the advancer assembly, align the notch in the back of the advancer with the pin in the crankshaft (**Figure 11**).

IGNITION COIL

An ignition coil can fail in any of 3 ways. It can develop an open circuit (broken wire) in the primary windings or the secondary windings, in which case the coil won't function at all, or it can develop a partial short circuit, arcing to bridge some of the secondary windings. If that happens, the coil will generate weak sparks at the electrodes of the spark plug.

Removal/Installation

1. Turn the ignition switch off.

2. Remove the fuel tank; see *Fuel Tank Removal* in Chapter Seven.

3. Disconnect the leads to the ignition coils.

4. Remove the coil mounting bolts and the coils (**Figure 12**).

5. Install by reversing the removal steps. Note the following:

    a. The *red* leads go to the positive (+) coil terminals. The *green* wire goes to the left coil and the *black* wire goes to the right coil.

    b. Be sure to connect the spark plug leads to the correct spark plug, starting with No. 1 on the left.

Testing

1. If the coil condition is doubtful, there are several checks which can be made. Disconnect the coil wires before testing. Measure coil primary resistance, using an ohmmeter set at R x 1. The resistance should be *about* 2 1/4 ohms.

2. Measure coil secondary resistance; remove the resistor-type plug caps from the leads and measure between the secondary leads. The resistance should be *about* 15,000 ohms.

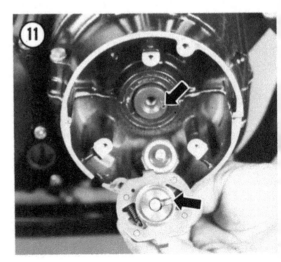

3. Measure the resistance between the coil primary *red* wire and ground (coil core) and between the coil secondary lead and ground (coil core). There should be no continuity (infinite resistance). A reading of any continuity indicates a short circuit.

4. If a coil does not meet these resistance values, it must be replaced. If the coil exhibits visible damage, it should be replaced.

5. If a coil is generating weak sparks, substitute the other coil in the circuit and check the sparks again. If the other coil generates stronger sparks at the electrodes of the spark plugs, there is a short circuit in the secondary windings of the original coil; install a new coil.

IGNITION PICKUP COILS

Inspection

1. Remove the right side cover (or raise the seat) and disconnect the 4-pole pickup coil connector (**Figure 13**).

2. With an ohmmeter set at R x 100, measure the resistance between the 2 pairs of leads: *Black* and *blue* (No. 1 and 4) and *yellow* and *red* (No. 2 and 3). The resistance should be 400-500 ohms.

3. Set the ohmmeter at its highest scale and check the resistance between either lead for each pair of cylinders and chassis ground. There should be no continuity (infinite resistance).

4. If the pickup coils fail either of these tests, check the wiring to the coils and replace the coil(s) if the wiring is okay.

Removal/Installation

The pickup coils are under the timing cover on the right side of the engine.

1. Open the right side cover (or raise the seat) and disconnect the 4-pole pickup coil/IC igniter connector (**Figure 13**).

2. Remove the 2 timing cover screws and the cover and gasket (**Figure 14**).

3. Remove the 3 timing plate screws and the plate (**Figure 15**).

IC IGNITER

The operation of the IC igniter (**Figure 16**) can be checked simply by removing one of each pair of spark plugs (No. 1 and 2 or No. 3 and 4). Ground the plug against the cylinder head while the plug lead is connected, turn the ignition ON and touch a screwdriver to the pickup coil core (**Figure 17**). If the IC igniter is good, the plug will spark.

Remember that the IC igniter is battery-powered and will not function if the battery is dead. The following IC igniter test can be made on the motorcycle.

1. Remove one of each pair of spark plugs and ground it against the cylinder head while its plug wire is connected.

2. Disconnect the 4-pole connector from the pickup coils (**Figure 13**).

3. Turn the ignition ON and connect positive (+) 12 volts to the *black* lead and negative (-) 12 volts to the *blue* lead. As the voltage is connected, the plug should spark.

4. Repeat the test for the other pair of plugs, with the positive (+) to the *yellow* lead from the IC igniter and negative (-) to the *red* lead.

5. If the IC igniter fails these tests, install a new one. If the IC igniter passes these tests but you still have an ignition problem that can't be traced to any other part of the ignition system, substitute an IC igniter that you know is good and see if that solves the problem. Some transistorized ignition troubles just won't show up on your workbench.

CHARGING SYSTEM

The charging system consists of the battery, alternator and voltage regulator/rectifier. **Figure 18** is a schematic diagram of the charging system. The rectifier and regulator are combined in one solid-state unit.

The alternator generates an alternating current (AC) which the rectifier converts to direct current (DC). The regulator maintains the voltage to the battery and load (lights, ignition, etc.) at a constant voltage, regardless of variations in engine speed and electrical power load.

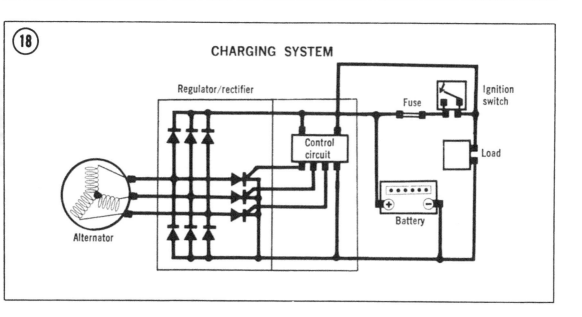

CHARGING SYSTEM

Regulator/rectifier

Ignition switch

Fuse

Control circuit

Load

Battery

Alternator

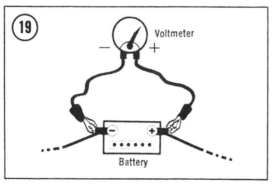

Voltmeter

Battery

Charging System Test

Whenever charging system trouble is suspected, make sure the battery is good before going any further. Check the specific gravity; see *Specific Gravity Testing*. Battery voltage should be above 12 volts. If not, charge the battery. If the battery is okay and all connections are clean and tight, inspect the charging system.

Initial Inspection

1. Start the engine and let it reach normal operating temperature.
2. Connect a 0-20 DC voltmeter to the battery as shown in **Figure 19**. Bring the engine speed from idle to 4,000 rpm, observing the voltage as you go.
3. Turn the headlight on and off, observing the voltage.

NOTE
If your bike doesn't have a headlight switch, disconnect a wire to the headlight or remove and reinstall the headlight fuse.

4. The voltage should be at or near battery voltage at idle and it should increase with engine speed, up to about 14.5 volts. If the reading is much higher (more than about 16.5 volts), the regulator/rectifier is defective and should be replaced. If the reading is less than specified or does not increase with rpm, check the alternator output and resistance and regulator/rectifier resistance.

Alternator Output Test

1. Remove the engine sprocket cover; see *Sprocket Cover Removal/Installation* in Chapter Six.
2. Disconnect the 3 *yellow* leads from the alternator (**Figure 20**).
3. Start the engine and connect an AC voltmeter (0-250 volts) to every pair of *yellow* leads in turn (there are 3 combinations). The meter should read about 50 volts AC. If it is much lower, the alternator is defective. Check stator coil resistance.

Alternator Stator Resistance

1. With the engine OFF and an ohmmeter set on a 1 ohm scale, check resistance between

every pair of *yellow* leads in turn (there are 3 combinations). The resistance should be about 0.5-0.7 ohms.

2. Set the ohmmeter on the highest scale and check resistance between each *yellow* lead and ground. Resistance should be infinite.

3. If the stator coil resistance is okay, but alternator output is low, the rotor has probably been demagnetized. Replace the rotor.

Regulator/Rectifier Resistance

1. Remove the left side cover and open the electrical panel (**Figure 1**).

2. Disconnect the 6-pin connector and the *white/red* lead from the regulator/rectifier, which is beneath the battery box.

> *CAUTION*
> *The **white/red** lead is "hot." Do not short circuit the voltage regulator when connecting the test leads.*

3. With an ohmmeter set at R x 10 or R x 100, measure the resistance between each *yellow* lead and the *white/red* lead and between each *yellow* lead and the *black* lead. Keep the same meter lead on the *white/red* and the *black* leads in turn. Note the readings (there are 6 combinations).

4. Reverse the meter polarity (use the opposite probes to make the connections) and repeat the tests.

There should be at least 10 times as much resistance in one direction as in the other. If any 2 leads show the same resistance in both directions, the regulator/rectifier is faulty and should be replaced.

ALTERNATOR

The alternator rotor is mounted on the left end of the crankshaft. The stator is mounted inside the left engine cover.

Stator Removal/Installation

1. Remove the engine sprocket cover; see *Engine Sprocket Cover Removal/Installation* in Chapter Six.

2. Disconnect the 3 *yellow* leads from the alternator (**Figure 20**).

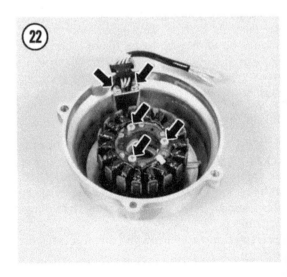

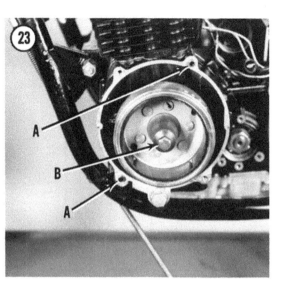

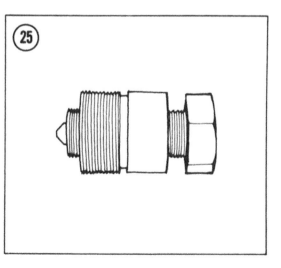

3. Remove the 4 alternator cover screws and the cover and gasket (**Figure 21**). There are 2 locating dowel pins.

4. To remove the stator coils from the cover, remove the 3 coil Allen bolts and the wire guide screws (**Figure 22**).

5. To install, reverse the removal steps. Note the following:

    a. Use a locking agent such as Loctite Lock N' Seal on the stator coil Allen bolts.

    b. Make sure the 2 cover dowel pins are in place (A, **Figure 23**).

Rotor Removal/Installation

Rotor replacement is usually necessary only if the rotor magnets have been damaged by mechanical shock or heat. The rotor can be removed after removing the alternator cover/stator.

CAUTION
Rotor removal requires a puller such as the one illustrated in the procedure below. Don't pry or hammer on the rotor itself. Damage is sure to result and you may destroy the rotor magnetism.

1. Remove the rotor bolt (B, **Figure 23**).

NOTE
To remove the rotor without a special tool, you can lock the engine by shifting the transmission into high gear and stepping on the brake pedal or by removing the clutch cover and stuffing a copper penny or clean rag, folded several times, between the secondary gear and the clutch ring gear.

2. Loosen the rotor. A simple bolt with 15 mm threads to match the large threaded rotor hole will work (**Figure 24**). Universal rotor pullers (**Figure 25**) are also available at most motorcycle shops. Back out the center bolt, screw the outer part all the way into the rotor, then screw in the inner bolt to pull the rotor off. You may have to alternate tapping on the puller bolt sharply with a hammer and tightening the bolt some more, but don't hit the rotor.

3. Remove the puller and the rotor.

4. To install, reverse the removal steps. Note the following:

a. Inspect the inside of the rotor carefully for any bits of metal or small parts that may have been picked up by the rotor magnets. Remove them to prevent damage when the engine starts.

b. Use a solvent to clean any oil from the tapered crankshaft end.

c. Torque the rotor bolt to 50 ft.-lb. (7.0 mkg).

STARTING SYSTEM

The starting system consists of the starter motor, starter clutch, starter solenoid, starter lockout switch and the starter button. **Figure 26** is a schematic diagram of the starting system.

When the clutch lever is pulled and the starter button is pressed, it engages the solenoid switch that closes the circuit. Then electricity flows from the battery to the starter motor.

The starter motor is a 12 volt DC motor geared to the secondary shaft through an idler gear and the clutch gear on the secondary shaft. The starter motor is connected mechanically to the crankshaft and can rotate it when the engine is not running. The starter clutch (between the idler and the clutch gear) uncouples the clutch gear from the secondary shaft when the engine is running.

Removal/Installation

1. Disconnect the battery negative (-) ground lead.

2. Remove the starter cover and the engine sprocket cover; see *Engine Sprocket Cover Removal/Installation* in Chapter Six.

3. Remove the 2 starter motor mounting bolts (**Figure 27**).

4. Pull the starter free, pull back the rubber boot and disconnect the starter cable at the motor (**Figure 28**).

5. To install, reverse the removal steps. Note the following:

a. Make sure the starter case terminal and mounting bosses are clean.

b. Oil the O-ring on the end cap of the motor assembly (**Figure 29**).

c. Make sure the starter terminal is protected by the rubber cover.

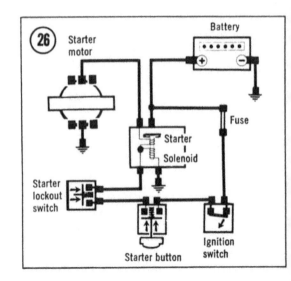

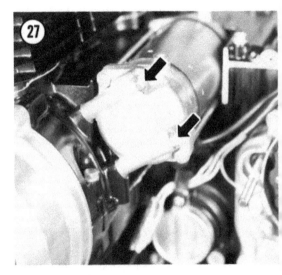

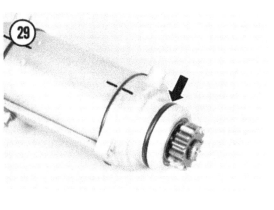

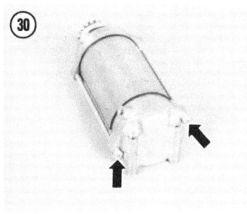

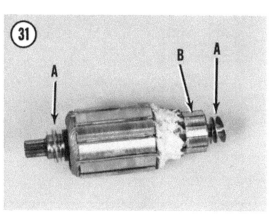

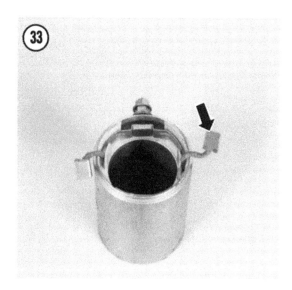

Disassembly/Assembly

1. Remove the starter motor case screws (**Figure 30**) and separate the cases.

> *NOTE*
> *Write down how many thrust washers are used at each end of the armature (A, **Figure 31**) and install the same number when reassembling the starter.*

2. Clean all grease, dirt and carbon dust from the armature, case and end covers.

> *CAUTION*
> *Do not immerse brushes or the wire windings in solvent or the insulation might be damaged. Wipe the windings with a cloth lightly moistened with solvent and dry thoroughly.*

3. Measure the length of the brushes (**Figure 32** and **Figure 33**). If a brush is worn shorter than 1/4 in. (6 mm), it should be replaced.

4. Inspect the condition of the commutator (B, **Figure 31**). The mica in the normal commutator is cut below the copper (**Figure 34**). A worn commutator is also shown where the copper is worn to the level of the mica. A worn commutator can be undercut, but it requires a specialist. Take the job to your Kawasaki dealer or motorcycle repair shop.

5. Inspect the commutator bars for discoloration. If a pair of bars are discolored, that indicates grounded armature coils.

8

6. Check the electrical continuity between pairs of armature bars (**Figure 35**) and between the commutator bars and the shaft (**Figure 36**). There should be continuity between pairs of bars but not between the bars and the shaft. If there is continuity, there is a short circuit and the armature should be replaced.

7. Inspect the field coil by checking continuity from the cable terminal to the brush wire. If there is a short or open, the case should be replaced.

8. Inspect the front and rear cover bearings for damage. Replace the starter if they are worn or damaged.

9. Assemble by reversing the removal steps. Note the following:

   a. Check that the 2 case O-rings are in good condition.

   b. Apply high-temperature grease to the planetary drive gears and shafts (A, **Figure 37**).

   c. Align the case tab with the notch in the planetary gear case (B, **Figure 37**).

   d. Align the brush plate tongue with the end cap groove.

   e. Align the marks on the case and covers (A, **Figure 38**).

   f. Install the starter motor gear with the beveled end of its teeth facing out (B, **Figure 38**).

STARTER CLUTCH

The starter clutch is mounted on the secondary shaft, inside the crankcase (**Figure 39**). To remove the starter clutch, see *Secondary Shaft Removal* in Chapter Four.

The starter motor meshes with an idle gear, which in turn is meshed with the starter clutch on the secondary shaft. The starter clutch gear spins freely in one direction on the secondary shaft.

The starter clutch locks the starter clutch gear to the secondary shaft when the starter motor is turning the engine, by jamming the clutch rollers between the clutch gear and rotor (**Figure 40**).

After the engine starts, the stationary clutch gear hub rolls the rollers back against their springs and frees the crankshaft from the clutch.

Inspection

To check the operation of the starter motor clutch, remove the starter motor. Try pushing the top of the starter idler gear in both directions by hand (**Figure 41**). The gear should not turn at all to the rear. It should turn freely and quietly to the front. If the gear fails to operate in that manner, the clutch assembly is malfunctioning.

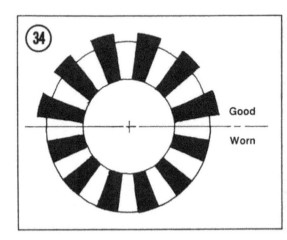

Good

Worn

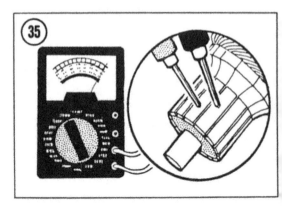

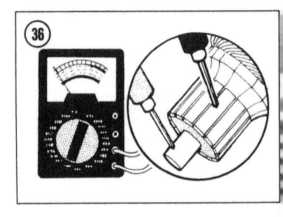

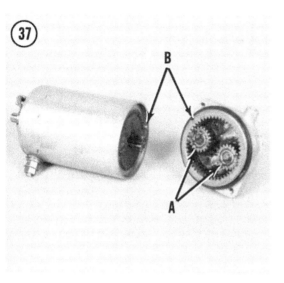

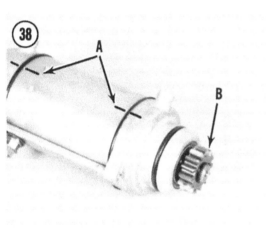

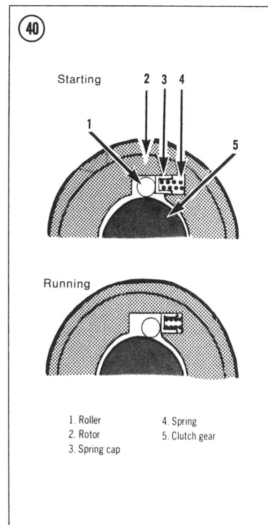

Starting

Running

1. Roller   4. Spring
2. Rotor   5. Clutch gear
3. Spring cap

8

Disassembly/Assembly

The starter clutch can be disassembled after removing the secondary shaft from the engine; see *Secondary Shaft and Starter Clutch* in Chapter Four.

1. Remove the 3 rollers, spring caps and springs from the clutch (**Figure 42**).

2. Remove the 3 Allen screws that mount the clutch body to the rotor.

3. When assembling the clutch, use a locking agent such as Loctite Lock N' Seal on the 3 starter clutch Allen bolts.

STARTER SOLENOID

Testing

Before testing the starter solenoid, make sure the battery is fully charged, with adequate electrolyte, and make sure all the connections between the battery and solenoid are corrosion-free and tight.

Remove the left side cover and electrical cover. The solenoid is to the rear of the fuse box (**Figure 43**).

Turn on the ignition, check that the kill switch is ON, pull in the clutch lever (if your model has a starter lockout) and operate the starter pushbutton. Listen for the loud clicking noise that tells you the solenoid is working.

If the solenoid does not click, check for battery voltage at the *black* wire to the solenoid when the starter button is pushed. If there is battery voltage but the solenoid doesn't click and all connections are clean and tight, the solenoid is bad and should be replaced. If battery voltage is not available at the *black* wire, there is an open somewhere in the circuit.

Starter Lockout Switch Replacement

Push up on the bottom of the switch locking tab (**Figure 44**) and pull out the switch. To install, push the switch in until you feel the tab lock in place.

LIGHTING SYSTEM

Table 2 lists replacement bulbs for the lights and indicators.

If a light stops burning, check first for a burned-out or broken bulb. If bulbs burn out

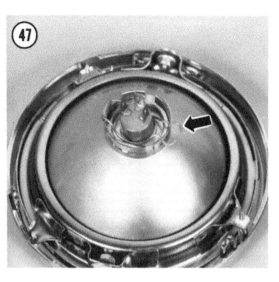

frequently, check for a low level of electrolyte in the battery or for a faulty voltage regulator that could be overcharging the battery.

If the problem is not in the bulb, check the wiring from the socket back to the battery. Measure voltages with the ignition turned ON or resistances with the ignition switch turned OFF and the fuse removed from its holder.

Headlight Adjustment

Adjust the headlight horizontally and vertically according to Department of Motor Vehicle regulations in your area.

To adjust the headlight horizontally, turn the screw on the lower right side of the headlight (A, **Figure 45**). For vertical adjustment, loosen the bottom positioning bolt (**Figure 46**) and tilt the headlight as required. Tighten the positioning bolt.

> *NOTE*
> *If the headlight is too tight to move, remove the headlight assembly and loosen the headlight mounting nuts.*

Headlight Replacement
(Quartz Halogen Models)

1. Remove the 2 mounting screws (B, **Figure 45**) on each side of the headlight housing.
2. Pull the trim bezel and headlight unit out and disconnect the electrical connector from the backside.
3. Remove the dust cover from the back of the light housing, open the spring clip (**Figure 47**) and remove the lighting element. Install a new element.

> *CAUTION*
> *Use a clean cloth to grasp the quartz bulb. Don't handle a quartz bulb with your bare fingers or a dirty rag. They will leave oil on the bulb and cause it to burn out early.*

Headlight Replacement
(Sealed Beam Models)

1. Remove the 2 mounting screws (B, **Figure 45**) on each side of the headlight housing.
2. Pull the trim bezel and headlight unit out and disconnect the electrical connector from the backside.

3. Remove the 2 top and bottom pivot screws and the horizontal adjusting screw (**Figure 48**). Take off the outer rim.

4. Remove the 2 screws from the inner rim and remove the sealed beam unit.

5. When installing a new sealed beam unit, be sure the "TOP" mark faces up.

6. Adjust the headlight as described under *Headlight Adjustment*.

Reserve Lighting Unit

KZ750-H motorcycles imported into the U.S. and Canada are equipped with a reserve lighting system that automatically switches from a burned-out headlight filament to the other filament. At the same time, a warning light will illuminate on the instrument cluster. Always replace the headlight when one filament is burned out.

Replace the reserve lighting unit (**Figure 49**) if it does not function properly.

Front Brakelight Switch Replacement

Push up on the bottom of the switch locking tab (**Figure 50**) and pull out the switch. To install, push the switch in until you feel the tab lock in place.

Turn Signal Flasher

The turn signal flasher is under the left side cover, under the electrical cover (A, **Figure 51**).

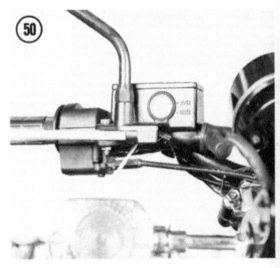

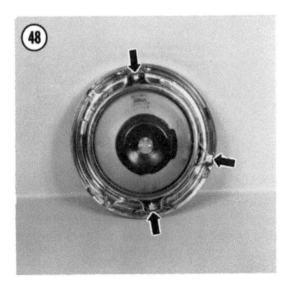

Hazard Flasher

The hazard flasher is under the left side cover, under the electrical cover (B, **Figure 51**).

TURN SIGNAL CANCELLING SYSTEM

Some models are equipped with an automatic turn signal cancelling system that turns off the turn signals after 4 seconds have passed and after the bike has traveled 50 meters. **Figure 52** is a schematic diagram of the system.

The distance sensor is a switch in the speedometer that opens and closes as the front wheel turns. The turn signal control unit counts 4 seconds after you push the turn signal switch to ON, then it electrically counts the number of front wheel revolutions detected by the distance sensor. After both time and distance conditions have been met, the turn signal control unit energizes a solenoid in the turn signal switch that pushes the switch to OFF.

Troubleshooting

1. In case of trouble with the system, check all wiring and connectors first.

2. If the wiring connectors are securely connected, check the distance sensor as follows:
   a. Remove the headlight and disconnect the 4-pin connector from the speedometer.
   b. Connect an ohmmeter to the *red* and *light green* leads from the speedometer.
   c. Disconnect the speedometer cable at its lower end and turn the inner cable by hand. If the distance sensor is working properly, the ohmmeter should show the sensor making and breaking continuity 4 times per revolution. If not, install a new speedometer.
   d. Reconnect the 4-pin connector.

3. If the distance sensor is okay, check the solenoid in the turn signal switch as follows:
   a. Remove the fuel tank (see *Fuel Tank Removal* in Chapter Seven).
   b. Turn the turn signal switch ON.
   c. Disconnect the 9-pin connector from the turn signal switch. Momentarily apply a positive (+) 12 volt signal to the *white/green* lead from the turn signal switch, using a wire connected to the positive battery terminal.

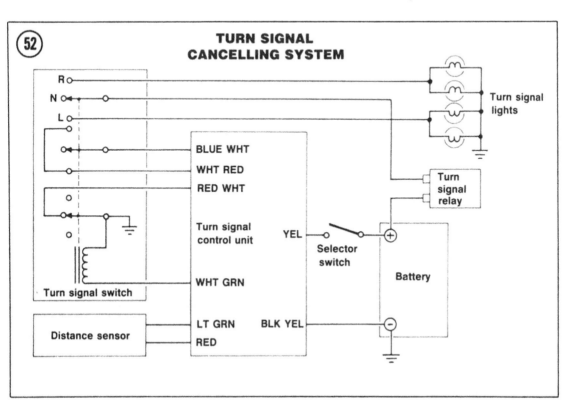

d. If the solenoid is okay, it will push the switch to OFF. If not, install a new turn signal switch.

e. Reconnect the 9-pin connector and install the fuel tank.

4. If the solenoid is okay, open the turn signal switch at the handlebar and clean the switch contacts. No replacement parts are available for the switch; if faulty, it must be replaced.

5. If cleaning the switch contacts doesn't solve the problem, check the turn signal control unit as follows:

a. Open the turn signal switch at the handlebar.

b. Set a voltmeter at 25 volts DC and connect the positive (+) probe to the *white/green* solenoid terminal from the turn signal control unit (not the grounded side of the solenoid). Connect the voltmeter negative (-) probe to the grounded side of the solenoid.

c. With the ignition switch ON and the turn signal switch selector set at "A" (automatic), push the turn signal switch ON.

d. Raise the front wheel and spin it at least 30 revolutions. If the turn signal control unit is okay, the voltmeter will show battery voltage as the control unit tries to energize the solenoid. If not, replace the control unit (the control unit is in front of the battery).

FUEL LEVEL SENSOR

Some models have a low fuel level sensor and a warning light that illuminates when there is less than about 1/5 of a tank of fuel remaining. The system uses the same bulb as the brake light failure indicator. The sensor is mounted in the fuel tank (**Figure 53**). If the warning light does not illuminate correctly, replace the sensor; see *Fuel Level Sensor* in Chapter Seven.

HORN

The horn can not be repaired, but it can be adjusted if prolonged use causes a change in the pitch of the sound. An ammeter (0-5 amps) is required during adjustment to prevent misadjustment and excessive current draw.

Disconnect the *black* lead at the horn and connect the ammeter positive (+) probe to the horn and the negative (-) probe to the *black* lead. Loosen the adjusting screw locknut (**Figure 54**) and turn the adjuster as required, while keeping the horn current draw under 2.0-3.0 amps.

> *CAUTION*
> *Don't turn the adjuster in (clockwise) too far or the battery drain will be excessive and the horn may burn out.*

VOLTMETER

Some models are equipped with a voltmeter. The voltmeter indicates battery voltage when the engine is not running. When the engine is running, the voltmeter indicates charging voltage.

To inspect the meter, remove the headlight and disconnect the 3-pole connector from the tachometer (KZ750-H) or the 6-pole connector from the instrument panel (KZ750-E, L). Use an ohmmeter to measure the resistance between the *brown* and *black/yellow* leads from the meter. If the resistance does not fall between 60-80 ohms, the meter is faulty.

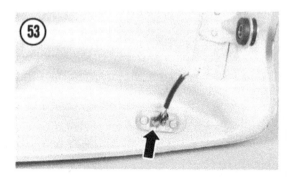

Table 1   BATTERY FREEZING TEMPERATURES

Specific Gravity	Freezing Temperature Degrees F	Specific Gravity	Freezing Temperature Degrees F
1.100	18	1.200	−17
1.120	13	1.220	−31
1.140	8	1.240	−50
1.160	1	1.260	−75
1.180	−6	1.280	−92

Table 2   KZ750 LIGHT BULB REPLACEMENT

Item	Watts (all bulbs are 12 volt)
Headlight	
1980 KZ750-E (U.S., Canada)	Sealed beam 60/50W
1980 KZ750-E (Europe)	Quartz halogen 60/50W
Others	Quartz halogen 60/55W
Tail/brakelights	
1980 KZ750-E (Europe, Australia)	8/21W
All others (Europe, Australia)	5/21W
All others (U.S., Canada)	8/27W
Turn signals	
Europe	23W
Others	21W
Turn signal/running light	23/8W
Meters & indicators	3.4W
City light Europe	4W

8

NOTE: If you own a 1982 or later model, first check the Supplement at the back of the book for any new service information.

CHAPTER NINE

WHEELS, TIRES, AND BRAKES

This chapter describes disassembly and repair of the front and rear wheels, hubs, tires and brakes.

Wear limits and torque specifications are given in **Tables 1-5** at the end of the chapter.

WHEELS

All models are equipped with cast aluminum wheels.

Runout

Wheel rim runout is the amount of "wobble" a wheel shows as it rotates. If the wheel has been subjected to a heavy impact or if you have any cause to suspect the wheel doesn't run "true," inspect the wheel runout.

1. Remove the tire from the wheel rim so that it doesn't distort the rim; see *Tire Removal* in this chapter.
2. Support the wheel on its axle and measure the axial and radial runout with a dial indicator (**Figure 1**).
3. The maximum allowable wheel rim runout on cast wheels is:
   a. Axial (side-to-side): 0.02 in. (0.5 mm)
   b. Radial (up-and-down): 0.03 in. (0.8 mm)

If the runout exceeds these limits, check the condition of the wheel bearings and install new ones if worn.

4. Cast wheels can not be straightened; they must be replaced if warped or damaged.

CAUTION
Do not attempt to straighten a cast wheel. The wheels will crack if a force strong enough to bend them is applied.

Balance

An unbalanced wheel can be unsafe. Balance weights applied to the light side of the wheel will correct imbalance. Before attempting to balance a wheel, detach the drive chain, check to make sure the wheel bearings are in good condition and properly lubricated and that the brake does not drag and keep the wheel from turning freely.

Support the motorcycle with the wheel and tire off the ground. Spin the wheel and allow it to come to rest by itself. Mark the part of the wheel at the bottom and spin it again. If it does not come to rest at the same position, the balance of the wheel may already be acceptable. If you're unsure, spin it and check it again. If the wheel comes to rest at the same position imbalance is indicated. In this case, add a weight to the top of the wheel and spin the wheel to check the effect.

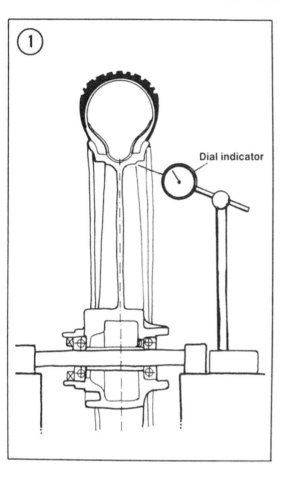

Dial indicator

NOTE

*Kawasaki offers weights that can be crimped on the aluminum rims (**Figure 2**). You may have to let some air out of the tire to install this type of weight. "Tape-A-Weight" or similar adhesive weights are available at motorcycle or auto supply stores. These kits contain test weights and strips of adhesive-backed weights that can be cut to the desired length and attached directly to the rim.*

Begin with the lightest weight and increase the size as necessary. When the wheel comes to rest at a different point each time that it is spun, consider it balanced. Tightly crimp weights applied to the rim lip so they won't be thrown off.

FRONT WHEEL

Removal/Installation

1. Support the motorcycle with the front wheel off the ground.
2. Loosen the speedometer cable nut and disconnect the cable at the wheel (A, **Figure 3**).
3. Remove the brake hose guide bolt from one side of the fender (B, **Figure 3**).
4. Remove the 2 brake caliper mounting bolts (C, **Figure 3**) from the same side of the fender and support the caliper to keep tension off the brake hose.

9

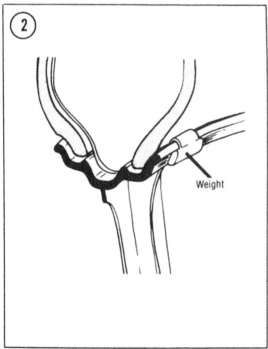

Weight

5A. On the KZ750-E, L, loosen the 4 axle clamp nuts, then loosen the 2 axle nuts (**Figure 4**). Remove the axle clamp nuts, lockwashers and the clamps.

5B. On the KZ750-H, remove the axle nut, then loosen the pinch nut (**Figure 5**) and pull out the front axle.

6. Remove the wheel and speedometer drive assembly. There is a collar on the right side of the wheel.

> *CAUTION*
> *Do not set the wheel down on the brake disc surface. It may get warped or scratched.*

> *NOTE*
> *You may want to insert a piece of wood in the caliper in place of the disc. If the brake lever is accidentally squeezed the caliper piston could be dislocated, which may require caliper disassembly and brake system bleeding.*

7. To disassemble the wheel hub, see **Figure 6**. Refer to *Wheel Bearings and Seals* in this chapter. When installing a brake disc, torque the brake disc mounting bolts as specified in **Table 3**.

8. To install, reverse the removal steps. Note the following:

　a. On the KZ750-H, assemble the wheel and axle assembly with the axle nuts flush with the ends of the axle (**Figure 7** and **Figure 8**).

　b. On the KZ750-E, L, install the wheel and axle assembly, install the axle clamps with lockwashers and tighten the clamp nuts finger-tight. The arrow on the clamp must point to the front, so there will be a gap at the rear of the clamp (**Figure 9**).

　c. Do not tighten the axle nuts until after the speedometer cable is connected to the gearbox. Tighten the cable nut with pliers.

> *NOTE*
> *If the cable won't seat easily, slowly turn the wheel until the cable end aligns with the drive.*

　d. On the KZ750-E, L, tighten the axle nuts, the front clamp nuts and then the rear

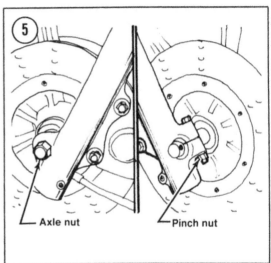

Axle nut　　　　　　Pinch nut

clamp nuts to the proper torque (see **Table 2**).

　e. On the KZ750-H, install the wheel, speedometer drive and collar between the fork legs. The speedometer drive tangs must fit both the hub notches (**Figure 10**) and the notches in the drive gearbox. Torque the axle nut, then the pinch nut as specified in **Table 2**.

> *CAUTION*
> *Make sure the speedometer drive tangs are correctly seated in both the wheel hub and in the speedometer drive gearbox before you tighten the axle nut or you may damage the drive.*

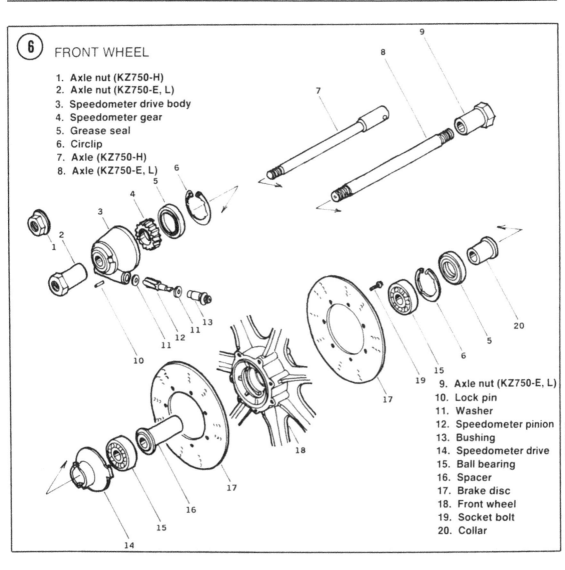

6 FRONT WHEEL

1. Axle nut (KZ750-H)
2. Axle nut (KZ750-E, L)
3. Speedometer drive body
4. Speedometer gear
5. Grease seal
6. Circlip
7. Axle (KZ750-H)
8. Axle (KZ750-E, L)

9. Axle nut (KZ750-E, L)
10. Lock pin
11. Washer
12. Speedometer pinion
13. Bushing
14. Speedometer drive
15. Ball bearing
16. Spacer
17. Brake disc
18. Front wheel
19. Socket bolt
20. Collar

9

7

8

f. Clean any oil or dirt from the brake disc with a non-oily solvent.

g. Install the brake caliper and torque the caliper mounting bolts as specified in **Table 3**.

h. After the wheel is installed, spin it to make sure it turns freely and apply the brake several times to make sure the pads are seated against the disc.

SPEEDOMETER GEAR LUBRICATION

The speedometer drive gears at the front wheel should be lubricated with high-temperature grease according to the maintenance schedule (**Table 4**, Chapter Three).

1. Remove the front wheel from the motorcycle.

2. Clean all old grease from the housing and gear and apply high-temperature grease to the gear.

3. Install the front wheel on the motorcycle.

REAR WHEEL

Removal/Installation

1. Support the motorcycle with the rear wheel off the ground.

2. Loosen the rear brake caliper anchor bolt (A, **Figure 11**) and free the brake line from its guides on the swing arm.

3. Remove the cotter pin and loosen the axle nut (B, **Figure 11**).

4. Loosen the chain play adjuster locknuts (C, **Figure 11**) and the adjusters. Kick the rear wheel forward far enough to free the drive chain from the rear sprocket.

5. Remove the axle nut and pull out the axle while holding the caliper up. There is a thin spacer on the left side of the wheel and a thicker collar on the right side.

6. Remove the rear wheel.

CAUTION
Do not set the wheel down on the brake disc surface. It may get warped or scratched.

NOTE
Insert the axle back through the swing arm and caliper or tie the caliper up to keep stress off the brake hose.

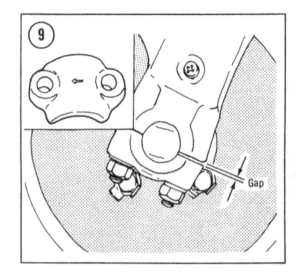

NOTE
You may want to insert a piece of wood in the caliper in place of the disc. If the brake pedal is accidentally pushed the caliper piston could be dislocated, which may require caliper disassembly and brake system bleeding.

7. To disassemble the wheel hub, see **Figure 12**. Refer to *Wheel Bearings and Seals* in this chapter. When installing a brake disc, the beveled side of the holes face the wheel. Torque the brake disc mounting bolts as specified in **Table 3**. When installing a new rear sprocket, the numbered side of the sprocket must face out (A, **Figure 13**). Torque the sprocket nuts as specified in **Table 2** and bend

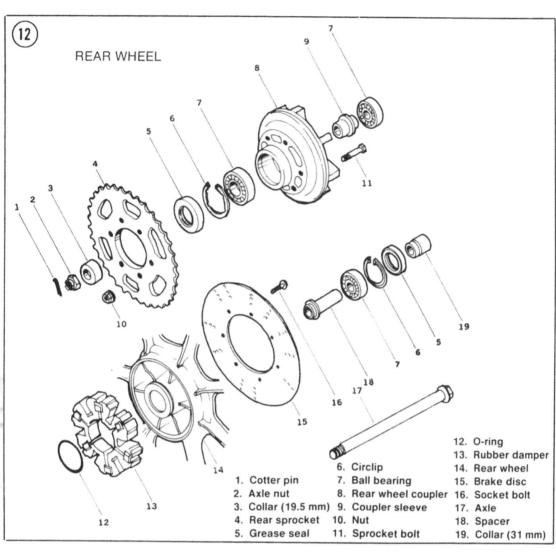

(12) REAR WHEEL

1. Cotter pin
2. Axle nut
3. Collar (19.5 mm)
4. Rear sprocket
5. Grease seal
6. Circlip
7. Ball bearing
8. Rear wheel coupler
9. Coupler sleeve
10. Nut
11. Sprocket bolt
12. O-ring
13. Rubber damper
14. Rear wheel
15. Brake disc
16. Socket bolt
17. Axle
18. Spacer
19. Collar (31 mm)

9

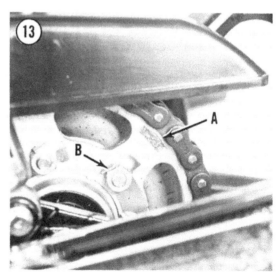

up a tab on the lockplate against each sprocket nut (B, **Figure 13**).

8. To install, reverse the removal steps. Note the following.

    a. Inspect the wheel coupling damper (A, **Figure 14**) and install a new one if worn.

    b. Inspect the hub O-ring and grease it (B, **Figure 14**).

    c. Clean any oil or dirt from the brake disc with a non-oily solvent.

    d. Insert the axle through the right chain adjuster (alignment marks facing up), caliper, thick collar, wheel assembly, thin spacer, left chain adjuster (alignment marks facing up) and the axle nut. Then install the wheel assembly in the swing arm.

    e. Position the drive chain on the sprocket.

    f. Adjust drive chain play. See Chapter Three.

    g. Torque the rear axle nut and torque link nut as specified in **Table 2**.

    h. After the wheel is installed, spin it to make sure it turns freely and apply the brake several times to make sure the pads are seated against the disc.

REAR SPROCKET

Replace the sprocket any time the teeth are worn or undercut as shown in **Figure 15**. A worn sprocket will quickly wear out a new drive chain. The sprocket is removed during the *Rear Wheel Removal/Installation* procedure in this chapter.

Install a new sprocket with the numbered side facing out (A, **Figure 13**). Torque the sprocket nuts as specified in **Table 2** and bend up a tab on the lockplate against each sprocket nut (B, **Figure 13**).

WHEEL BEARINGS AND SEALS

The original wheel bearings are not fully sealed and require periodic lubrication in accordance with the maintenance schedule (**Table 1**) in Chapter Three.

If bearing replacement is necessary, fully sealed bearings are available from any good bearing specialty shop. Be sure you take your

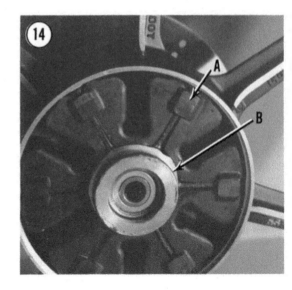

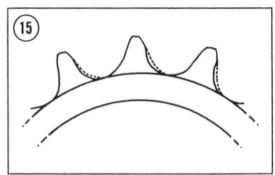

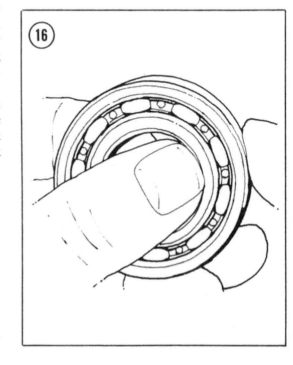

old bearing along to ensure a perfect matchup. Fully sealed bearings provide better protection from dirt and moisture that may get into the hub.

Inspection and Lubrication

1. Clean away all old grease.
2. Check the inner and outer bearing races and balls for cracks, galling or pitting. Rotate the bearings by hand and check for roughness; they should make no noise (**Figure 16**). Replace any worn or damaged bearings.

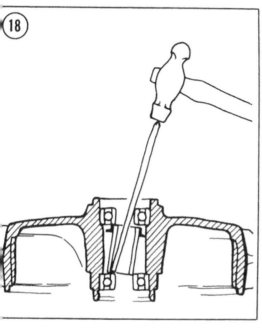

3. Pack the bearings thoroughly with waterproof grease, forcing the grease around the balls.
4. Inspect all grease seals for wear or damage and replace them if necessary.

Replacement

To disassemble the wheel hubs, refer to **Figure 6** and **Figure 12**.
1. Remove any seal retaining circlips (**Figure 17**).
2. Insert a rod into one side of the hub. Move the spacer aside and tap the bearing out the other side of the hub by working around its diameter (**Figure 18**).

> *NOTE*
> *It is not necessary to remove the seal first. The seal will be driven out with the bearing.*

3. Remove the spacer and tap out the opposite bearing.
4. Carefully drive a new bearing into its bore, tapping evenly around the outer race. Invert the hub and set the spacer in place. Install the other bearing on the axle and insert the axle through the spacer and the bearing that has been installed. Carefully tap the other bearing into its bore, tapping evenly around the outer race. Install a new seal into the bore and gently tap it into place.

> *NOTE*
> *When installing semi-sealed wheel bearings, be sure the sealed side faces out.*

AXLE

Support each axle in V-blocks 4 inches apart as shown in **Figure 19** and check the runout with a dial indicator. Replace the axle if it is bent more than 0.028 in. (0.7 mm) or if it can not be straightened to less than 0.008 in. (0.2 mm) runout.

> *NOTE*
> *You can also check axle runout by rolling it on a piece of plate glass and slipping a feeler gauge under any visible gap.*

9

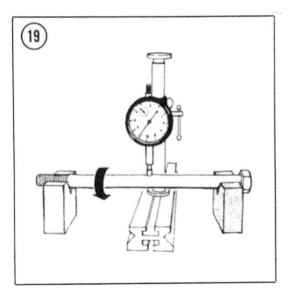

TIRES

The original tire cross-sectional sizes and tread pattern are often the best. See **Table 4** and **Table 5** for the standard sizes and inflation pressures.

> *WARNING*
> *Do not install an inner tube on rims designed for tubeless tires. The tube will cause an abnormal heat buildup in the tire.*

Tubeless tires have the word "TUBELESS" molded in the tire sidewall (**Figure 20**) and the rims have "TUBELESS" marked on a wheel spoke (**Figure 21**).

When a tubeless tire is flat, your best recourse is to take it to a motorcycle dealer for repair. Punctured tubeless tires should be removed from the rim to inspect the inside of the tire and to apply a combination plug/patch from the inside. Don't rely on a plug or cord repair applied from outside the tire. They might be okay on a car, but they're too dangerous on a motorcycle.

After repairing a tubeless tire, don't exceed 50 mph (80 kph) for the first 24 hours. Never race on a repaired tubeless tire. The patch could work loose from tire flexing and heat.

Removal

Removal of tubeless tires from their rims can be very difficult because of the

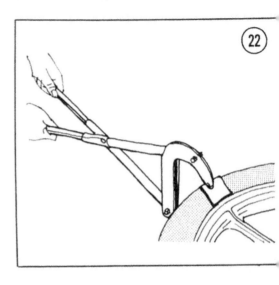

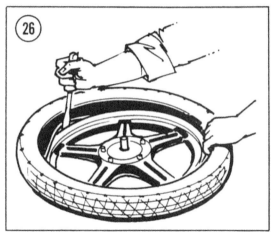

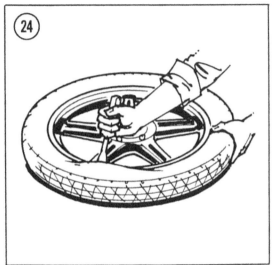

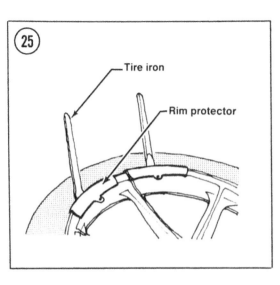

exceptionally tight tire bead/rim seal. Breaking the bead seal may require the use of a special tool such as the one available from Kawasaki (**Figure 22**). If you have trouble breaking the seal, we recommend you take the tire to a motorcycle dealer.

> *CAUTION*
> *The inner rim and tire bead area are sealing surfaces on a tubeless tire. Do not scratch the inside of the rim or damage the tire bead.*

1. Mark the valve stem location on the tire, so the tire can be installed in the same position for easy balancing (**Figure 23**).

2. Unscrew the valve core to deflate the tire.

3. Press the entire bead on both sides of the tire into the center of the rim. Use a rubber mallet or step on the tire with your heels. A special bead seal breaker may be required.

4. Lubricate the beads with soapy water.

5. Insert a tire iron under the bead next to the valve (**Figure 24**). Force the bead on the opposite side of the tire into the center of the rim and pry the bead over the rim with the tire iron.

> *CAUTION*
> *Use rim protectors (**Figure 25**) or insert scraps of leather between the tire irons and the rim to protect the rim from damage.*

6. Insert a second tire iron next to the first to hold the bead over the rim. Then work around the tire with the first tire iron, prying the bead over the rim (**Figure 26**).

7. Stand the tire upright. Insert the tire iron between the second bead and the side of the rim that the first bead was pried over (**Figure 27**). Force the bead on the opposite side from the tire iron into the center of the rim. Pry the second bead off the rim, working around as with the first.

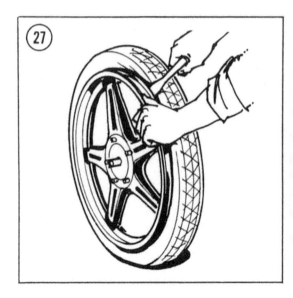

Installation

1. Install a new valve stem whenever you have the tire off the rim (**Figure 28**). Rubber deteriorates with age and valve stem replacement will never be as convenient as now.
2. Carefully inspect the tire for any damage, especially inside.
3. If the tire has a directional arrow, it should point in the normal direction of rotation (**Figure 29**).
4. Position the tire next to the rim with your mark at the valve stem hole.

> *NOTE*
> *On a new tire, a colored spot near the bead indicates a lighter point on the tire. This spot should be placed next to the valve stem (**Figure 23**).*

5. Lubricate both beads of the tire with soapy water.
6. Push the backside of the tire down into the center of the rim and insert the valve stem through the stem hole in the wheel. The lower bead should go into the center of the rim and the upper bead outside. Work around the tire in both directions (**Figure 30**). Use a tire iron for the last few inches of bead (**Figure 31**).
7. Press the upper bead into the center of the rim opposite the valve. Pry the bead into the rim on both sides of the initial point with a tire iron, working around the rim to the valve (**Figure 32**).
8. Inflate the tire slowly to seat the beads in the rim. Bounce the wheel several times while turning the wheel. This helps seat the bead. If an initial air seal is hard to get, your motorcycle or auto repair shop may have a bead seater to make the job easy.
9. After inflating the tire, check to see that the beads are fully seated and that the tire rim lines

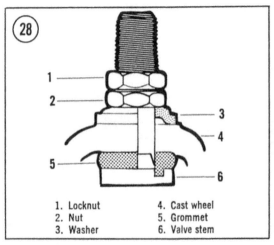

1. Locknut	4. Cast wheel
2. Nut	5. Grommet
3. Washer	6. Valve stem

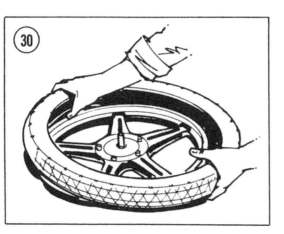

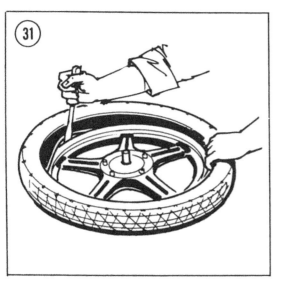

next to the rim are the same distance from the rim all the way around the tire. If the beads won't seat, deflate the tire, relubricate the rim and beads with soapy water and reinflate the tire.

Repair

Do not rely on a plug or cord patch applied from outside the tire. Use a combination plug/patch applied from inside the tire (**Figure 33**).

1. Remove the tire from the rim.

2. Inspect the rim inner flange. Smooth any scratches on the sealing surface with emery cloth. If a scratch is deeper than 0.5 mm (0.020 in.), the wheel should be replaced.

3. Inspect the tire inside and out. Replace a tire if any of the following is found.

    a. A puncture larger than 1/8 in. (3 mm)

    b. A punctured or damaged sidewall

    c. More than 2 punctures in the tire

4. Apply the plug/patch, following the instructions supplied with the patch.

BRAKES

Repair of brake systems is extremely critical work. Don't take shortcuts and don't work with makeshift tools. When a procedure calls for a locking agent such as Loctite Lock N' Seal, use it. If a torque specification is given, use a torque wrench. Recheck your work and, when

9

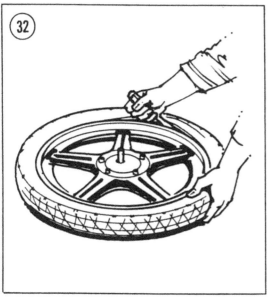

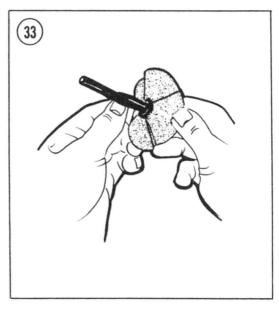

you're finished, test your brakes in a safe area. If you have any doubts about your ability to do the job according to these procedures, have the work done by your Kawasaki dealer or qualified specialist.

See **Table 1** for brake wear limits. See **Table 3** for disc brake system torque specifications.

WARNING
Brake fluid is an irritant. Keep it away from your eyes and off your skin.

The following precautions *must* be observed when servicing brake systems.

1. Never reuse old brake fluid.
2. Never use fluid from a container that has been left open.
3. Use only disc brake fluid clearly marked "DOT 3" or higher.
4. Don't leave the fluid reservoir cap off for too long. The fluid will absorb moisture from the air and will boil at low temperatures.
5. Don't contaminate the brake discs or pads with brake fluid, gasoline or any lubricants (including graphite or pencil lead).
6. Brake fluid can ruin paint and plastic. If you spill any, wipe it up immediately.
7. If you open a bleed valve or loosen a brake line fitting, bleed air from the system.

BRAKE SYSTEM BLEEDING

Bleed the hydraulic brake system whenever the brake lever or pedal action feels spongy or soft, after brake fluid has been changed or whenever a brake line fitting has been loosened.

1. Remove the reservoir cap and add brake fluid. Don't let the fluid level drop below the minimum mark during this procedure or air will enter the system and you'll have to start over.
2. Install the reservoir cap. Connect a clear plastic hose to the bleed valve at the caliper (**Figure 34**). Run the other end of the hose into a clean container. Fill the container with enough brake fluid to keep the end submerged.

NOTE
On dual hydraulic disc models, always bleed the caliper farther away from the

master cylinder first. Otherwise, air bubbles may remain in the brake system.

3. Slowly apply the brake lever or pedal several times. Hold the lever or pedal in the ON position. Open the bleed valve about one-half turn. Allow the lever or pedal to travel to its limit. When this limit is reached, tighten the bleed screw. As the fluid enters the system, the level will drop in the reservoir. Maintain the level at about 3/8 in. (10 mm) from the top of the reservoir to keep air from being drawn into the system.
4. Continue to pump the lever or pedal, open and close the bleed valve and fill the reservoir until the fluid emerging from the hose is completely free of bubbles.

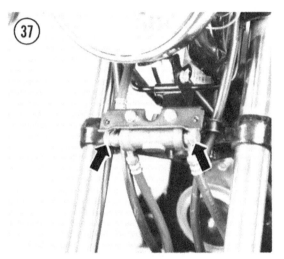

NOTE
Do not allow the reservoir to empty during the bleeding operation or more air will enter the system. If this occurs, the entire procedure must be repeated.

5. Hold the lever or pedal down, tighten the bleed valve, remove the bleed tube and install the bleed valve dust cap.

6. Add fluid to correct the level in the reservoir. Slowly pump the brake lever or pedal several times (reservoir cap still off) until no air bubbles rise up through the fluid from holes at the bottom of the reservoir. (This bleeds air from the master cylinder end of the line.) Install the reservoir cap tightly.

7. Test the feel of the brake lever or pedal. It should be firm and should offer the same resistance each time it's operated. If it feels spongy, it is likely that there is still air in the system and it must be bled again.

WARNING
Do not ride the motorcycle until you are sure the brakes are working with a solid feel.

BRAKE FLUID CHANGE

Change the brake fluid according to the maintenance schedule (**Table 1**, Chapter Three) or whenever the fluid becomes contaminated by water or dirt. Brake fluid changing is the same as bleeding, but continue pumping fluid through the system until the fluid leaving the bleed valve is clean and free of air bubbles.

BRAKE LINE REPLACEMENT

A brake hose should be replaced whenever it shows cracks, bulges or other damage. The deterioration of rubber by ozone and other atmospheric elements may require hose replacement every 5 years.

1. Before replacing a brake hose, inspect the routing of the old hose carefully, noting any guides and grommets the hose may go through (**Figure 35** and **Figure 36**).

2. Disconnect the banjo bolts securing the hose at either end (**Figure 37** and **Figure 38**) and remove the hose with banjo bolt and 2

washers at either end. Plug the open brake lines.

3. Install a new hose with 2 new washers at each banjo bolt. Make sure the hose fitting is correctly positioned in its attachment notch (**Figure 39**).

4. Torque the banjo bolts as specified in **Table 3**.

5. Fill the reservoir and bleed the system as described in this chapter.

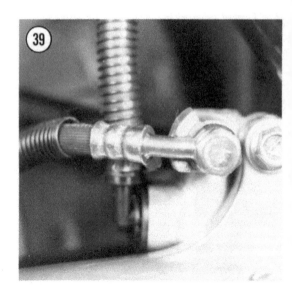

BRAKE PADS

Removal

There is no recommended mileage interval for changing the friction pads in the disc brake. Pad wear depends greatly on riding habits and conditions. See *Brake Pad Inspection* in Chapter Three.

Always replace all pads (2 per disc) on an axle at the same time.

1. Remove the 2 caliper bolts and lift the caliper off the pad holder (**Figure 40**).

2. Remove the brake pads.

Installation

> *CAUTION*
> *Check with your dealer to make sure the friction compound of the new pads is compatible with the disc material.*

1. Remove the cap from the master cylinder and slowly push the piston into the caliper while checking the reservoir to make sure it doesn't overflow (**Figure 41**). The piston should move freely. You may need to use a C-clamp to push the piston back into the caliper. If the piston sticks, remove the caliper and have it rebuilt by your Kawasaki dealer or a qualified specialist.

2. Make sure the brake pad guides are in place (**Figure 42**).

3. Install the pads with the friction material toward the disc (**Figure 43**).

4. Make sure the anti-rattle spring is installed in the caliper (**Figure 44**).

5. Install the caliper. Torque the caliper bolts (**Figure 40**) as specified in **Table 3**.

6. Support the motorcycle with the wheel off the ground. Spin the wheel and pump the brake until the pads are seated against the disc.

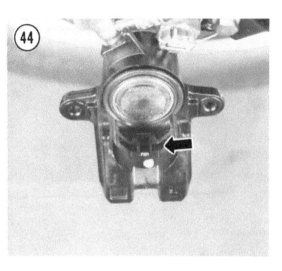

7. Top up the fluid level in the master cylinder if necessary.

> *WARNING*
> *Do not ride the motorcycle until you are sure the brakes are working with a solid feel. If necessary, bleed the brakes to remove any accumulated air from the system.*

BRAKE CALIPERS

Caliper Rebuilding

See **Figure 45** and **Figure 46**.

If the caliper leaks, it should be rebuilt. If the piston sticks in the cylinder, indicating severe wear or galling, the entire unit should be replaced. The factory recommends that the piston fluid seal and dust cover be replaced every other time the pads are replaced.

Rebuilding a leaky caliper requires special tools and experience. We therefore recommend that caliper service be entrusted to your Kawasaki dealer or brake specialist. You will save time, money and possibly your life by removing the caliper yourself and having a professional do the job. Remove the calipers as described in *Brake Pad Removal* and connect the rebuilt calipers as described in *Brake Hose Replacement*.

> *WARNING*
> *Do not ride the motorcycle until you are sure the brakes are operating properly.*

MASTER CYLINDER REBUILDING

If the master cylinder leaks, it should be rebuilt. If the piston sticks in the cylinder, indicating severe wear or galling, the entire unit should be replaced. Rapid darkening of fresh brake fluid is an indication that rubber parts inside the master cylinder have deteriorated.

Rebuilding a master cylinder requires special tools and experience. We therefore recommend that service be entrusted to your Kawasaki dealer or brake specialist. You will save time, money and possibly your life by removing the master cylinder yourself and having a professional do the rebuilding.

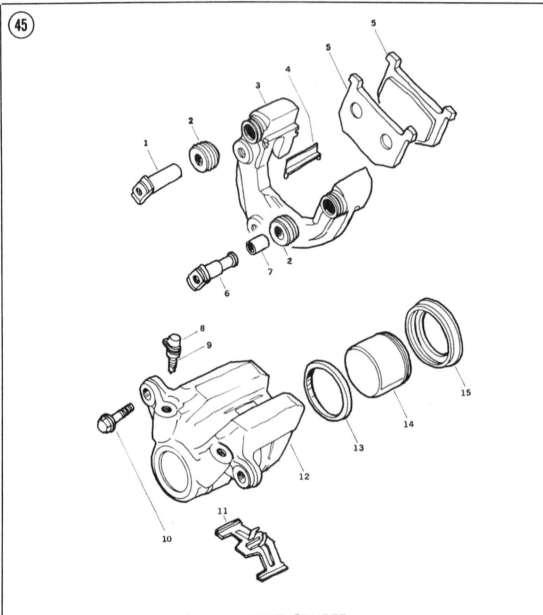

FRONT BRAKE CALIPER

1. Holder shaft
2. Shaft dust cover
3. Caliper holder
4. Pad guide (2)
5. Brake pad
6. Holder shaft
7. Friction boot
8. Rubber cap
9. Bleed valve
10. Caliper shaft bolt (2)
11. Anti-rattle spring
12. Caliper
13. Piston fluid seal
14. Piston
15. Piston dust seal

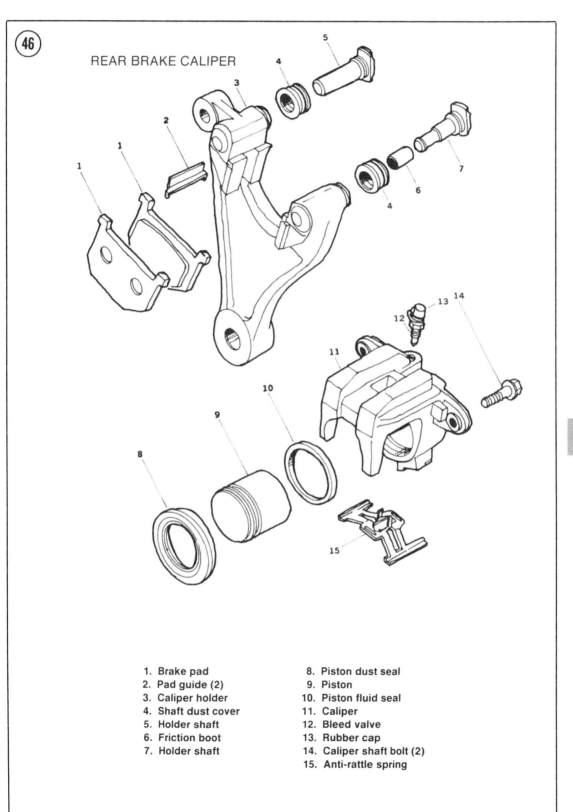

REAR BRAKE CALIPER

1. Brake pad
2. Pad guide (2)
3. Caliper holder
4. Shaft dust cover
5. Holder shaft
6. Friction boot
7. Holder shaft
8. Piston dust seal
9. Piston
10. Piston fluid seal
11. Caliper
12. Bleed valve
13. Rubber cap
14. Caliper shaft bolt (2)
15. Anti-rattle spring

9

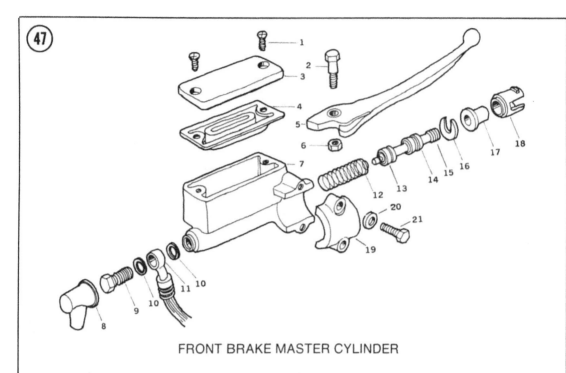

FRONT BRAKE MASTER CYLINDER

1. Cover screw (2)
2. Brake lever pivot bolt
3. Cover
4. Diaphragm
5. Brake lever
6. Locknut
7. Body/reservoir
8. Rubber dust cover
9. Banjo bolt
10. Washer
11. Brake hose

12. Spring
13. Primary cup
14. Secondary cup
15. Piston
16. Piston stop
17. Dust seal
18. Liner
19. Master cylinder clamp
20. Washer (2)
21. Clamp bolt (2)

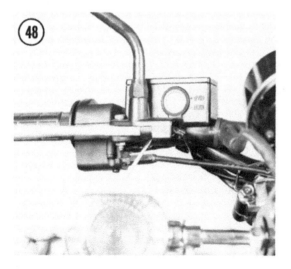

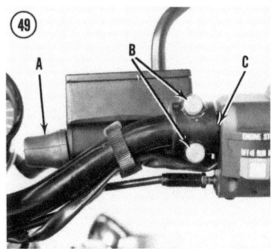

FRONT MASTER CYLINDER

Removal

See **Figure 47**.
1. Remove the right rear view mirror.
2. Push up on the front brake switch locking tab (**Figure 48**) and pull the switch free from the lever bracket.
3. Pull back the rubber boot and remove the banjo bolt (A, **Figure 49**) securing the brake hose to the master cylinder.

CAUTION
Brake fluid ruins paint and plastic surfaces. If you spill any, wipe it up immediately.

4. Remove the 2 clamp bolts and the clamp (B, **Figure 49**). Take off the master cylinder.

Installation

1. Install the master cylinder with the projection on the clamp toward the throttle grip (C, **Figure 49**).
2. Install the clamp bolts and washers. Tighten the upper bolt first, then the lower one.
3. Connect the brake line. Use new washers on either side of the banjo bolt and torque it as specified in **Table 3**.
4. Bleed the brake system as described in this chapter.

REAR MASTER CYLINDER

Removal

See **Figure 50**.
1. Remove the cotter pin from the pushrod link pivot (**Figure 51**). Remove the pin.
2. Remove the banjo bolt and 2 washers (A, **Figure 52**).

CAUTION
Brake fluid ruins paint and plastic surfaces. If you spill any, wipe it up immediately.

3. Remove the 2 rear master cylinder mounting bolts (B, **Figure 52**). Pull the master cylinder free of the brake pushrod and dust cover.

4. Loosen the reservoir hose clamp and pull the hose free from the master cylinder. Tie the hose up high to prevent fluid loss.

Installation

1. Inspect the pushrod boot on the bottom of the master cylinder. Install a new one if cracked or worn.
2. Torque the 2 master cylinder mounting bolts as specified in **Table 3**.
3. Pull the rubber boot up into place on the cylinder.
4. Insert the pushrod pivot pin, install a new cotter pin and spread its ends.
5. Connect the brake line. Use new washers on either side of the banjo bolt and torque it as specified in **Table 3**.
6. Connect the reservoir hose and tighten its clamp.
7. Bleed the brake system as described in this chapter.

BRAKE DISCS

Inspection

It is not necessary to remove the disc from the wheel to inspect it. Small marks on the disc are not important, but radial scratches deep enough to snag a fingernail reduce braking effectiveness and increase pad wear.
1. Measure the thickness at several points around the disc with vernier calipers or a micrometer (**Figure 53**). The disc must be replaced if the thickness, at any point, is less than 0.24 in. (6 mm) on single discs or 0.18 in. (4.5 mm) on dual discs.
2. Check the disc runout with a dial indicator. Raise the wheel being checked, set the arm of the dial indicator against the surface of the disc (**Figure 54**) and slowly rotate the wheel while watching the indicator. If the runout is greater than 0.012 in. (0.3 mm), the disc must be replaced.

Removal/Installation

See *Front Wheel Disassembly* or *Rear Wheel Disassembly* in this chapter for brake disc replacement.

9

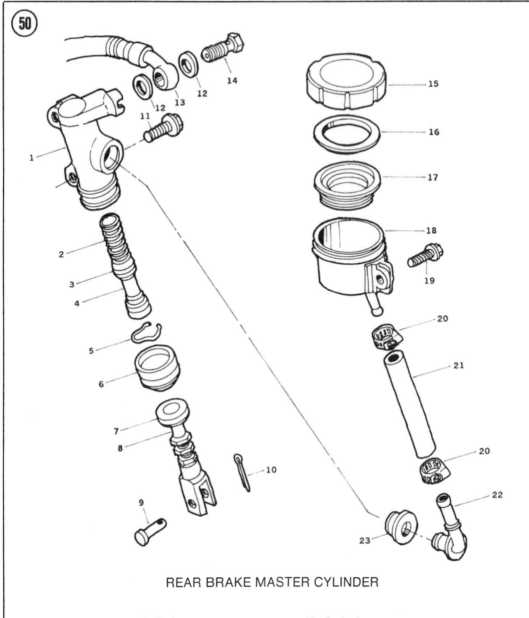

REAR BRAKE MASTER CYLINDER

1. Body
2. Return spring
3. Primary cup
4. Piston
5. Retainer
6. Boot
7. Piston stop
8. Pushrod
9. Pushrod pivot pin
10. Cotter pin
11. Mounting bolt (2)
12. Washer
13. Brake hose
14. Banjo bolt
15. Reservoir cap
16. Washer
17. Diaphragm
18. Reservoir
19. Bolt
20. Hose clamp
21. Reservoir hose
22. Elbow
23. Grommet

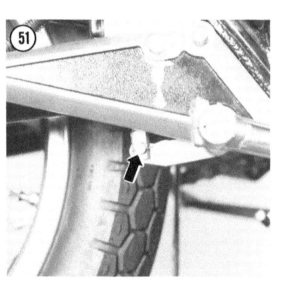

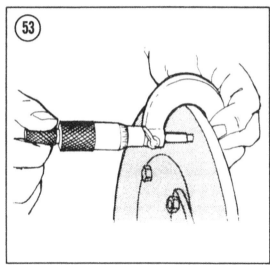

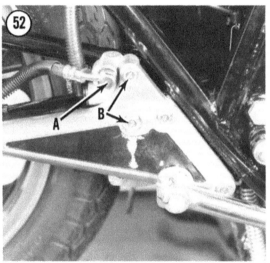

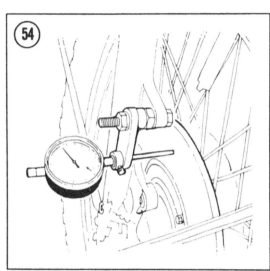

9

Tables are on the following pages.

Table 1   KZ750 DISC BRAKE WEAR LIMITS

Item	In.	mm
Caliper/Disc		
Caliper cylinder ID	1.690	42.92
Caliper piston OD	1.683	42.75
Disc runout	0.012	0.3
Disc thickness		
Front	0.177	4.5
Rear	0.236	6.0
Pad thickness	0.040	1.0
Front Master Cylinder		
Cylinder ID	0.628	15.95
Piston OD	0.622	15.80
Primary cup OD	0.630	16.0
Secondary cup OD	0.646	16.4
Spring free length	1.37	34.7
Rear Master Cylinder		
Cylinder ID	0.554	14.08
Piston OD	0.542	13.77
Primary cup OD	0.555	14.1
Secondary cup OD	0.571	14.5
Spring free length	1.46	37.2

Table 2   KZ750 CHASSIS TORQUES

Item	Ft.-lb.	mkg
Front axle clamp nuts (KZ750-E)	13	1.8
Front axle nut(s)	60	8.0
Front axle pinch bolt (KZ750-H)	14.5	2.0
Fork triple clamp bolts		
Upper	14.5	2.0
Lower	27	3.8
Fork air valves	8.5	1.2
Fork bottom Allen bolt	16.5	2.3
Fork top plugs	16.5	2.3
Handlebar clamp bolts	13	1.8
Rear axle nut	90	12.0
Rear sprocket nuts	30	4.0
Shock absorber mounts	22	3.0
Steering head clamp nut	13	1.8
Steering head top bolt	35	4.5
Steering head adjuster locknut	22	3.0
Swing arm pivot nut	70	10.0
Torque link nuts	22	3.0

Table 3   KZ750 DISC BRAKE TORQUES

Item	Ft.-lb.	mkg
Banjo bolts	22	3.0
Bleed valve	70 inch.-lb.	0.80
Brake lever pivot bolt	25 inch.-lb.	0.30
Brake lever pivot locknut	50 inch.-lb.	0.60
Caliper holder shaft bolts	13	1.8
Caliper mounting bolts	30	4.0
Disc mounting bolts	16.5	2.3
Master cylinder clamp bolts	80 inch.-lb.	0.90

Table 4 KZ750 TIRES AND TIRE PRESSURE (U.S. AND CANADA)

Model/Tire Size	Pressure 0-215 lb. (0-97.5 kg)	@ Load 215-364 lb. (97.5-165 kg)
KZ750-E (tubeless) Front 3.25H-19 4PR	28 psi (200 kPa)	28 psi (200 kPa)
Rear 4.00H-18 4PR	32 psi (225 kPa)	36 psi (250 kPa)
KZ750-H (tubeless) Front 3.25H-19 4PR	25 psi (175 kPa)	25 (175 kPa)
Rear 130/90-16 67H 4PR	22 psi (150 kPa)	25 (175 kPa)

Table 5 KZ750 TIRES AND TIRE PRESSURES (EUROPE)

Model/Tire Size	Pressure @ Load		
	0-210 lb. (0-95 kg)	210-300 lb. (95-136 kg)	300-397 lb. (136-180 kg)
Front KZ750-H,L 3.25H-19 4PR			
Up to 110 mph (180 kph)	25 psi (175 kPa)	25 psi (175 kPa)	25 psi (175 kPa)
Over 110 mph (180 kph)	28 psi (200 kPa)	28 psi (200 kPa)	25 psi (175 kPa)
Rear KZ750-H 130/90-16 67H 4PR KZ750-L 4.00H-18 4PR			
Up to 110 mph (180 kph)	25 psi (175 kPa)	28 psi (200 kPa)	32 psi (225 kPa)
Over 110 mph (180 kph)	28 psi (200 kPa)	32 psi (225 kPa)	36 psi (250 kPa)

9

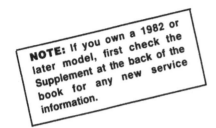

CHAPTER TEN

CHASSIS

This chapter covers the front forks, steering head, rear shock absorbers, drive chain and swing arm. Chassis wear limits and torque specifications are given in **Table 1** and **Table 2** at the end of the chapter.

FRONT FORKS

The Kawasaki front suspension consists of a spring-controlled, hydraulically damped telescopic fork. The damping rate is determined by the viscosity (weight) of the oil used and the spring rate can be altered by varying the *amount* of oil used and the air pressurization of the forks; refer to Chapter Three for these adjustments. Before suspecting major trouble, drain the fork oil and refill with the proper type and quantity. If you still have trouble, such as poor damping, tendency to bottom out or top out or leakage around the fork seals, then follow the service procedures in this section.

To simplify fork service and to prevent the mixing of parts, the fork legs should be removed, serviced and reinstalled individually.

Each front fork leg consists of the fork leg (inner tube), slider (outer tube), fork spring and damper rod with its damper components.

Removal/Installation

1. Raise the front wheel off the ground; support the motorcycle securely under the engine.

2. Remove the front wheel; see Chapter Nine.
3. Remove the front fender.
4. Release all air pressure from the fork leg by removing the rubber cap and pushing the valve core.
5. Loosen the upper fork clamp bolt (A, **Figure 1**).
6. If you are going to disassemble the fork after removal, loosen the fork cap now (B, **Figure 1**).

> *WARNING*
> *The fork is assembled with spring preload. Keep your face away from the fork end. The plug may spring out.*

> *NOTE*
> *On some models, the handlebars may need to be detached for access to the fork plug bolts.*

7. Loosen the lower fork clamp bolt (**Figure 2**).
8. Work the fork leg down and out of the clamps with a twisting motion.
9. Install by reversing the removal steps. Note the following.

   a. On the KZ750-E, L, the top of the fork tube should be even with the top of the clamp (**Figure 3**). The cap will be above the clamp.
   b. On the KZ750-H, the top of the cap should be even with the top of the clamp (**Figure 4**).

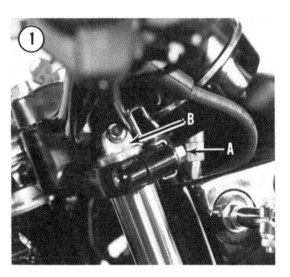

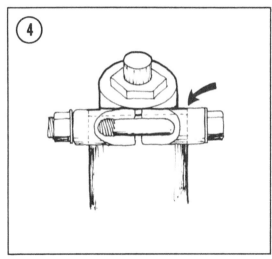

CAUTION
Both fork assemblies must be installed at
exactly the same height to prevent axle
and suspension damage when the sus-
pension is bottomed or fully extended.

c. Torque the fork mounting bolts as specified in **Table 2**.

Disassembly

See **Figure 5**. Disassembly of the fork leg
requires special tools and patience. If you have
trouble taking the fork leg apart, take it to your
Kawasaki dealer to keep from damaging it or
hurting yourself.
1. Release any air pressure from the fork.
2. Hold the upper fork tube in a vise with a
rubber sheet to grip the tube (**Figure 6**). Remove the top cap and spring seat or spacer.

WARNING
The fork is assembled with spring pre-
load. Keep your face away from the fork
end. The plug may spring out.

4. Remove the fork spring.
5. Remove the fork from the vise and pour the
oil out and discard it. Pump the fork several
times by hand to expel the remaining oil.
6. Clamp the fork slider in a vise with a rubber
sheet to grip the slider.
7. Remove the Allen bolt and gasket from the
bottom of the slider (**Figure 7**). The Allen

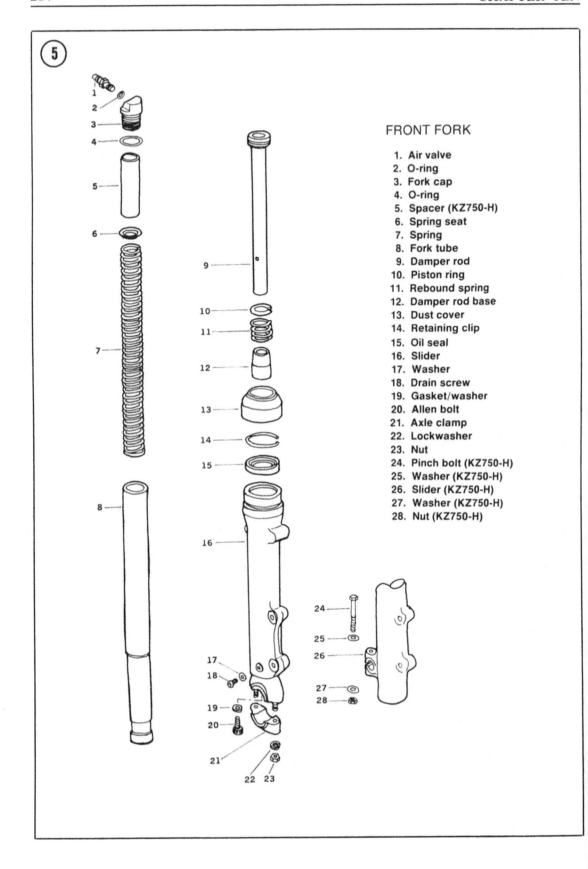

FRONT FORK

1. Air valve
2. O-ring
3. Fork cap
4. O-ring
5. Spacer (KZ750-H)
6. Spring seat
7. Spring
8. Fork tube
9. Damper rod
10. Piston ring
11. Rebound spring
12. Damper rod base
13. Dust cover
14. Retaining clip
15. Oil seal
16. Slider
17. Washer
18. Drain screw
19. Gasket/washer
20. Allen bolt
21. Axle clamp
22. Lockwasher
23. Nut
24. Pinch bolt (KZ750-H)
25. Washer (KZ750-H)
26. Slider (KZ750-H)
27. Washer (KZ750-H)
28. Nut (KZ750-H)

bolt is secured with a locking agent and can be hard to remove. After the Allen bolt is loosened, it may not come out because the damper rod base it screws into may turn with the bolt.   Try one of the following methods:

a. If a heavy-duty air-powered impact wrench is available, try that first.

b. If necessary, you may be able to keep the damper rod base from turning by temporarily installing the fork spring and cap and having an assistant compress the fork while you remove the bottom bolt.

c. If the methods described in Step a and Step b are not sucessful, you will have to keep the damper rod from turning with a special tool on the end of several socket extensions (**Figure 8** or **Figure 9**). Your Kawasaki dealer can tell you which tool to use; a bolt with a 19 mm head (welded to a long rod) may fit the end of the damper rod.

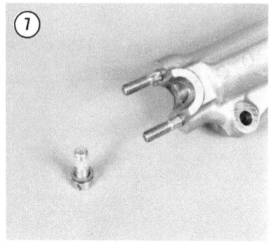

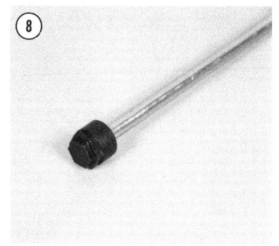

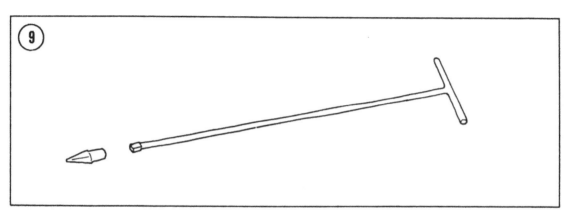

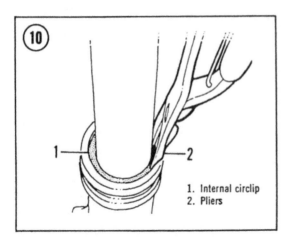

1. Internal circlip
2. Pliers

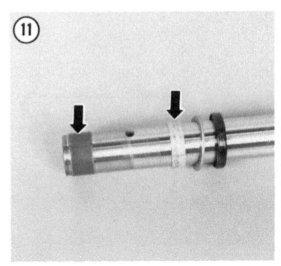

8. Slide the dust cap up off of the fork slider and remove the oil seal retaining clip (**Figure 10**).

9. Pull the fork tube out of the slider.

> *NOTE*
> *These forks have 2 anti-friction rings and a washer (**Figure 11**) on the end of the fork tube. You will probably have to slam the tube apart repeatedly, removing the slider's anti-friction ring, washer and oil seal from the slider as the tube comes out.*

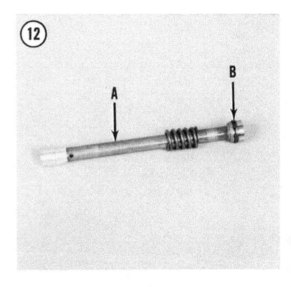

10. Remove the damper rod and rebound spring. The damper rod base will probably be at the bottom of the fork slider.

11. If the oil seal did not come out with the fork tube, it can be pried out now with a screwdriver. Make sure the retaining circlip has been removed.

> *NOTE*
> *It may be necessary to slightly heat the area on the slider around the oil seal prior to removal. Be careful not to damage the top of the slider.*

Inspection

1. See **Figure 5**. Thoroughly clean all parts in solvent and dry. Check the fork tube for signs of wear or scratches that would damage the oil seal.

2. Inspect the fork tube and slider guide bushings (**Figure 11**). The bushings are not available separately and the fork tube and slider must be replaced if the bushings are damaged.

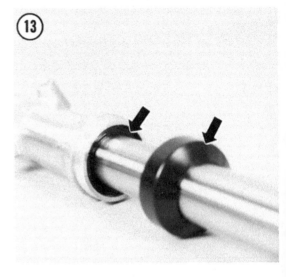

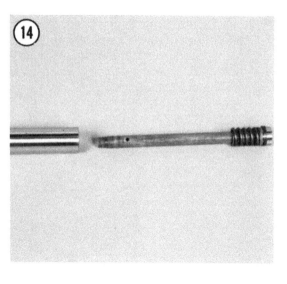

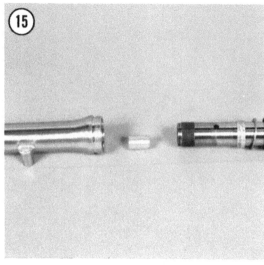

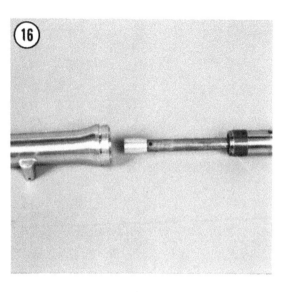

3. Check the damper rod for straightness (A, **Figure 12**). If bent, install a new rod.

4. Carefully check the piston ring for wear or damage (B, **Figure 12**).

5. Inspect the oil seal and dust cap (**Figure 13**) for scoring and nicks and loss of resiliency. Replace if the condition is questionable.

6. Check the upper fork tube for straightness. If bent or severely scratched, it should be replaced.

7. Check the lower slider for dents or exterior damage that may cause the upper fork tube to drag. Replace it if necessary.

8. Measure fork spring free length. Replace the spring if shorter than specified in **Table 1**. Compare the left and right fork springs. Replace them both if one is much shorter than the other.

Assembly

1. See **Figure 5**. Put the damper rod spring onto the damper rod and insert them into the top of the fork tube (**Figure 14**).

2. If the slider's anti-friction ring came out during disassembly, make sure that it, the washer and the oil seal are in place on the fork tube (**Figure 15**). Install the seal with the numbered or marked face up.

3. Insert the damper rod base onto the end of the damper rod (**Figure 16**).

4. Insert the fork tube into the slider. Apply a locking agent such as Loctite Lock N' Seal to the threads of the bottom Allen bolt before installing it with its washer and torque the bolt as specified in **Table 2**. To keep the damper rod base from turning, use the same method you used during disassembly.

5. Tap the oil seal into place until it seats.

NOTE
You may find that a piece of pipe just large enough to slide over the fork tube will seat the oil seal squarely without damaging it. Make sure the seal is not cocked in its hole.

6. Install the oil seal circlip (**Figure 10**) and install the dust cap.

7. Add fresh oil to each fork tube. See *Fork Oil Change* in Chapter Three.

8. Install the fork spring and seat or spacer and install the top plug (**Figure 17**).

10

9. Pressurize the forks. See *Fork Air Pressure* in Chapter Three.

STEERING

If your periodic checks detect excessive steering play, adjust the steering as described in this section. The steering head should also be disassembled, cleaned, inspected for wear on the balls and races and lubricated with a waterproof grease according to the maintenance schedule (**Table 4**, Chapter Three).

Adjustment

1. Raise the front wheel off the ground; support the motorcycle securely under the engine.
2. Remove the fuel tank to protect its finish. See *Fuel Tank Removal* in Chapter Seven.
3. Loosen the steering stem clamp bolt at the rear of the upper triple clamp (A, **Figure 18**) and the head bolt (B). You may have to detach the handlebars to loosen the head bolt.
4. Loosen both upper fork clamp bolts (**Figure 19**).
5. Back the steering stem adjuster (C, **Figure 18**) out one or two turns until it feels free, then turn the adjuster back in with a spanner wrench until you just feel the steering play taken up. There is a spanner wrench in the motorcycle's tool kit.

> *NOTE*
> *Don't back the adjuster out too far or the steering bearing balls may drop down into the steering head. Complete disassembly of the steering would be required to put the balls back in place.*

6. Tighten the adjuster another 1/16 turn.

> *CAUTION*
> *Don't turn the adjuster so tight that you indent the bearing balls into their races. If you do, the steering will be "notchy" and you'll have to replace the races.*

7. Torque the steering stem head bolt as specified in **Table 2** and tighten the rear clamp bolt.
8. Torque the upper triple clamp bolts as specified in **Table 2**.

9. Recheck the steering play as directed in Chapter Three. If the play is still incorrect, disassemble and inspect the steering head.
10. Install the fuel tank. See *Fuel Tank Installation* in Chapter Seven.

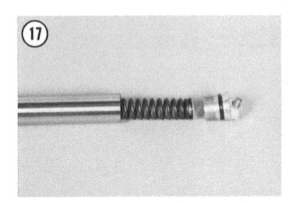

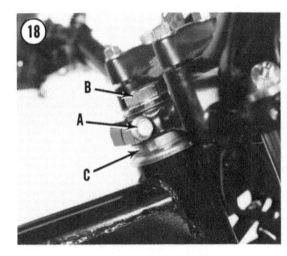

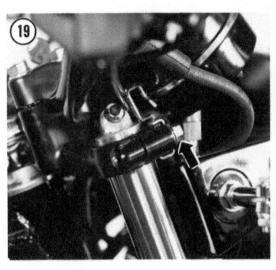

Head Disassembly/Assembly

Refer to **Figure 20**. As you remove parts, note the location of all cable and wiring guides and straps for proper installation during reassembly. Watch what you're doing and make a note of anything you might not remember a week from now if you have to order parts from your dealer. Pay particular attention to control, wiring and instrument cable routing.

1. Raise the front wheel off the ground; support the motorcycle securely under the engine.

2. Remove the fuel tank to protect its finish. See *Fuel Tank Removal* in Chapter Seven.

3. Remove the headlight from its shell and disconnect the wiring connectors. Push the wiring out the back of the shell.

4. Disconnect the speedometer and tachometer cables at the meters (**Figure 21**).

5. Remove the trim plate mounting screws and the trim plate from the front of the lower triple clamp.

6. Remove the 2 brake junction mounting bolts (**Figure 22**).

STEERING HEAD

1. Head bolt
2. Washer
3. Lockwasher
4. Upper triple clamp
5. Adjuster
6. Cap
7. Upper inner race
8. Bearing balls, 1/4 in. (19)
9. Upper outer race
10. Frame head
11. Bearing balls, 1/4 in. (20)
12. Lower inner race
13. Grease seal
14. Washer
15. Lower triple clamp
16. Lower outer race

10

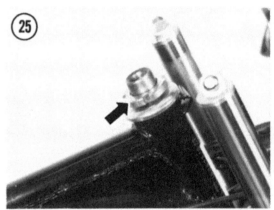

7. Remove the two upper and one lower headlight bracket bolts, lockwashers and washers (**Figure 23**).

8. Pull the headlight bracket and shell assembly free from the steering head.

9. Detach the complete front brake system (master cylinder, junction block and calipers) without disconnecting any brake lines.

10. Remove the handlebar clamp bolts (**Figure 24**) and detach the handlebar assembly. Let it hang down in front of the motorcycle, being careful not to damage any cables or wires.

11. Remove the front wheel. See Chapter Nine.

12. Remove the front fender.

13. Remove the front fork legs. See *Front Forks Removal* in this chapter.

14. If there are any other parts attached to the fork clamps, remove them now.

15. Loosen the upper triple clamp rear bolt, then remove the head bolt, washer and lockwasher (B, **Figure 18**).

16. Remove the upper triple clamp. If necessary, tap it up from the bottom with a soft mallet.

17. Hold the lower triple clamp up to keep it from falling and remove the notched steering adjuster nut with a spanner wrench (**Figure 25**).

18. Remove the lower triple clamp and steering stem assembly from the bottom of the steering head (**Figure 26**).

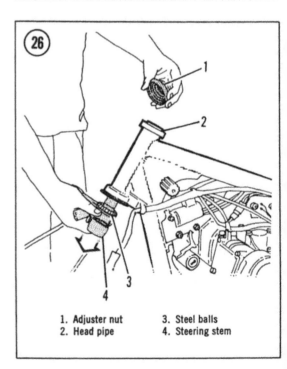

| 1. Adjuster nut | 3. Steel balls |
| 2. Head pipe | 4. Steering stem |

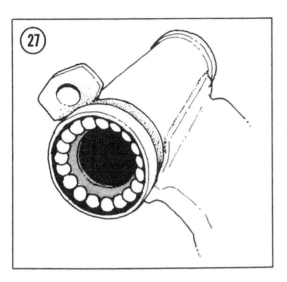

NOTE
Catch the lower bearing balls as they fall
out. Remove any balls stuck in the bot-
*tom bearing race (**Figure 27**).*

19. Remove the steering stem cap and inner race from the top of the steering head (**Figure 28**).

20. Remove the bearing balls from the steering head, catching any that fall out (**Figure 29**).

21. To assemble, reverse the disassembly procedure. Note the following:

a. Stick the bearing balls in place with grease while you install the lower triple clamp and steering stem. There should be 20 balls at the bottom, and 19 balls at the top. All balls are 1/4 in. diameter.

b. Adjust the steering play after installing the upper race, cap and adjuster.

c. Leave the upper triple clamp head bolt loose until after you insert and align the fork legs.

d. Torque the triple clamp bolts and the steering head top bolt as specified in **Table 2**.

e. If you replaced any bearings or races, recheck the steering play after a short ride. If the steering is loose, the bearings weren't fully seated during installation; readjust the steering play.

10

Inspection and Lubrication

1. Clean the bearing races and bearings with solvent.

2. Check for broken welds on the frame around the steering head.

3. Check each of the balls for pitting, scratches or discoloration indicating wear or corrosion. Replace them in upper or lower sets if any are bad.

4. Check the upper and lower races in the steering head. See *Bearing Race Replacement* in this chapter if the races are pitted, scratched or worn.

5. Check the upper and lower *inner* ball races for pitting, scratches or wear.

6. Check the steering stem for cracks.

7. Grease the bearings and races.

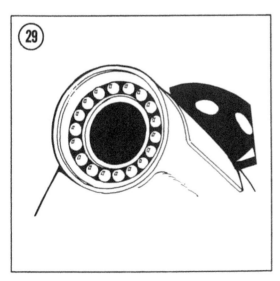

Bearing Race Replacement

The head and steering stem bearing races are pressed into place. Do not remove them unless they are worn and require replacement; ball races are easily bent. Take old races to your dealer to ensure exact replacement.

Steering head

To remove the steering head races, insert a hardwood stick into the head tube and carefully tap the race out from the inside (**Figure 30**). Tap all around the race so that neither the race nor the head tube are bent. To install a race, grease it and fit it into the end of the head tube. Tap it slowly and squarely with a block of wood (**Figure 31**). Make sure it is fully seated. You will notice a distinct change in the hammering sound as the race "bottoms-out."

Steering stem

To remove the steering stem race, try twisting and pulling it up by hand. If it will not come off, carefully pry it up with a screwdriver, while working around in a circle, prying a little at a time. Be careful not to damage the grease seal under the lower race.

Remove the grease seal and washer from the steering stem and install a new seal before reassembling the steering head.

To install the lower stem race, slide it over the steering stem and tap it down with a piece of hardwood or a pipe of the proper size. Work around in a circle so that the race will not be bent. Make sure it is seated squarely and all the way down.

REAR SHOCK ABSORBERS

Service to the original equipment rear spring/gas shocks is limited to inspection for damage to the damper rod, checking for damping and replacing worn mounting bushings and sagged springs.

The KZ750-E, L and H models have adjustable damping at the rear shocks. The relative damping effect for different adjuster settings (**Figure 32**) is described in Chapter Three.

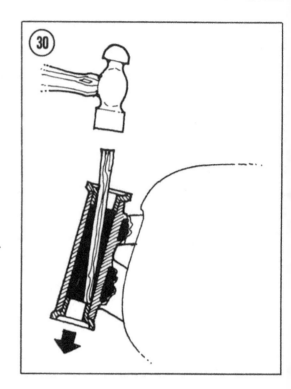

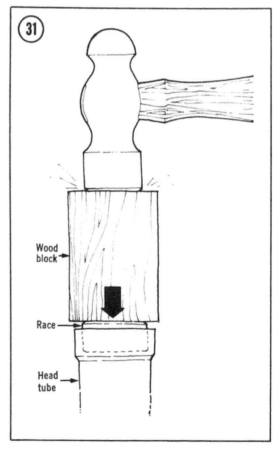

Removal/Installation

Removal and installation of one shock absorber at a time makes the task easier. The unit that remains will maintain the correct distance between the swing arm and the frame.

1. Put the motorcycle up on its centerstand.
2. Loosen the upper shock mounting nuts (A, **Figure 33**).
3A. On the KZ750-H, remove the grab rail mounting bolts and lockwashers and remove the grab rail.
3B. On the KZ750-E, L, remove the grab rail mounting bolt, nut and flat washer.
4. Remove the nut, lockwasher and flat washers from the upper shock absorber mount.
5. Remove the bottom bolt and lockwasher (B, **Figure 33**), then remove the shock absorber.
6. To install, reverse the removal steps. Note the following:

a. Check the rubber shock mounting eye bushings for damage or deterioration and replace them if necessary.
b. On the KZ750-H, put a flat washer on either side of the grab rail end.
c. Torque the mounting nuts/bolts as specified in **Table 2**.
d. Check to see that the spring preload cam adjuster on each shock is turned to the same notch (**Figure 34**).
e. Check to see that the damping adjuster on each shock is turned to the same setting (**Figure 32**).

10

Inspection

Check the shocks by removing them from the motorcycle, compressing them by hand and observing the extension stroke and comparing their "feel." If one has a different feel or if their extension damping differs, replace both with new shocks. Grasp the upper mounting eye and repeatedly compress and extend the damper rod to check for damping resistance. Resistance during extension of the rod should be noticeably greater than during compression. Also, the resistance in both directions should be smooth throughout the stroke. If the shock absorber fails on either of these points, it is unsatisfactory and should be replaced. If damping is noticeably different between the 2 shocks, they should both be replaced.

Visually check the damper rod for bending. If bending is apparent, the unit is unserviceable and should be replaced.

Inspect for shock fluid leakage. If any is present, the shocks should be replaced.

SWING ARM

The swing arm has a sleeve at the pivot, riding inside 2 needle bearings pressed into both ends of the swing arm. The sleeve must be inspected for wear and the needle bearings must be lubricated periodically for long life and good handling. Kawasaki uses an endless drive chain (no master link) on these models for high strength and reliability. The swing arm must be removed to remove the drive chain.

Inspection
(On The Bike)

A worn swing arm pivot will cause poor handling that may be indicated by wheel hop or pulling to one side during acceleration or braking.

1. Place the motorcycle on its centerstand.

2. Disconnect the shock absorbers at the bottom and swing them up out of the way.
3. Grasp the top of the rear wheel and the frame and try to rock the wheel back and forth. If you feel any more than a very slight movement of the swing arm, and the pivot bolt is correctly tightened, the swing arm should be removed and inspected.

NOTE
Make sure any motion you feel is not caused by loose or worn wheel bearings.

Removal/Lubrication/Installation

See **Figure 35**.
1. Remove the rear wheel. See Chapter Nine.
2. Pull the rear brake hose from its swing arm guides and tie the caliper up to keep stress off the brake hose.
3. Remove the drive chain guard (**Figure 36**). The KZ750-H has 3 mounting screws.
4. Remove the lower shock absorber mounting bolt and lockwasher on both sides.
5. Move the swing arm up and down to check for abnormal friction.

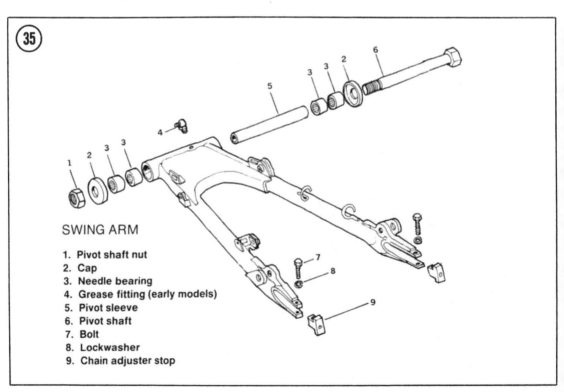

SWING ARM

1. Pivot shaft nut
2. Cap
3. Needle bearing
4. Grease fitting (early models)
5. Pivot sleeve
6. Pivot shaft
7. Bolt
8. Lockwasher
9. Chain adjuster stop

6. Remove the swing arm pivot nut (**Figure 37**).

7. Push out the pivot shaft and move the swing arm to the rear. There is a cap on either end of the swing arm.

8. Push out the swing arm sleeve with a soft metal rod. Take care not to damage the bearings.

9. Clean the old grease from the needle bearings and sleeve and apply a liberal amount of waterproof grease.

10. To install, reverse the removal steps. Note the following:

    a. Be sure to loop the drive chain around the swing arm when installing the swing arm.

    b. Torque the swing arm pivot nut and shock absorber bolts as specified in **Table 2**.

    c. On the KZ750-E, L, when installing the drive chain guard, position it carefully on

the swing arm brackets or the guard rub on the drive chain.

Inspection (Off The Bike)

1. Inspect the needle bearings for discoloration or cracking. Replace them if damaged.

2. To replace the bearings, drive them out with a rod inserted from the opposite end of the swing arm and install new ones carefully to prevent cocking or damage.

> *CAUTION*
> *Do not try to remove the bearings unless you intend to replace them. They will be damaged during removal.*

3. Measure the outside diameter (OD) of the swing arm sleeve that rides in the bearings. If the OD is smaller than the limit in **Table 1**, install a new sleeve.

4. Measure the pivot shaft runout. Roll the shaft across a piece of plate glass and measure any gap with a feeler gauge. If the runout is bigger than the limit in **Table 1**, install a new pivot shaft.

DRIVE CHAIN

Because the drive chain is endless (has no master link), the swing arm must be removed to remove the drive chain from the motorcycle.

> *WARNING*
> *Kawasaki uses an endless chain on this model because an endless chain is stronger than a chain with a master link. Do not cut the chain with a chain cutter or install chain with a master link. The chain may fail and rear wheel lockup and an accident could result.*

Removal/Installation

1. Remove the swing arm as described in this chapter.

2. Remove the engine sprocket cover; see *Engine Sprocket Cover Removal/Installation* in Chapter Six.

3. Remove the drive chain.

4. To install a chain, reverse these steps; check the chain for wear and lubricate and adjust it as described in Chapter Three.

Table 1   KZ750 CHASSIS WEAR LIMITS

Item	In.	mm
Fork spring minimum free length		
KZ750-E,L	19.6	49
KZ750-H	19.0	483
Swing arm sleeve OD	0.865	21.96
Swing arm pivot runout	0.006	0.14
Repair limit	0.028	0.7

Table 2   KZ750 CHASSIS TORQUES

Item	Ft.-lb.	mkg
Front axle clamp nuts (KZ750-E)	13	1.8
Front axle nut(s)	60	8.0
Front axle pinch bolt (KZ750-H)	14.5	2.0
Fork triple clamp bolts		
Upper	14.5	2.0
Lower	27	3.8
Fork air valves	8.5	1.2
Fork bottom Allen bolt	16.5	2.3
Fork top plugs	16.5	2.3
Handlebar clamp bolts	13	1.8
Rear axle nut	90	12.0
Rear sprocket nuts	30	4.0
Shock absorber mounts	22	3.0
Steering head clamp nut	13	1.8
Steering head top bolt	35	4.5
Steering head adjuster locknut	22	3.0
Swing arm pivot nut	70	10.0
Torque link nuts	22	3.0

1982 AND LATER SERVICE INFORMATION

> This supplement provides service information unique to 1982 and later models. All other service information remains unchanged.
>
> The chapter headings in this supplement correspond to those in the main portion of this book. Read the material in this supplement and then read the procedures in the main book before beginning any work.

CHAPTER THREE

LUBRICATION, MAINTENANCE AND TUNE-UP

Refer to **Table 1** for 1982 and later model designations and **Tables 2-4** for general model specifications. Refer to **Table 5** for the 1983 and later maintenance schedule. New torque specifications for 1982 and later models are found in **Tables 6-8**.

ENGINE OIL AND FILTER (ZX750E)

Oil and Filter Change

The efficiency of the turbocharger unit depends upon the engine lubricating system. As with other engine systems, the turbocharger is lubricated with engine oil. The engine oil should be changed every 3,000 miles (5,000 km). Kawasaki recommends oil filter replacement at 6,000 miles (10,000 km) intervals. An oil screen has been installed on turbo models to provide additional oil system filtering and cleaning. The screen is located behind a banjo bolt at the bottom of the engine near the engine drain bolt (**Figure 1**). Remove the banjo bolt (after the engine oil has been drained) and remove the screen. Clean the screen in solvent and allow to air dry before installation. If the screen is damaged, replace it with a new one. Use the following tightening torques when changing the engine oil and filter:

a. Engine drain plug: 3.0 mkg (22 ft.-lb.).
b. Oil filter bolt and oil screen banjo bolt: 2.0 mkg (14.5 ft.-lb.).

CLUTCH ADJUSTMENT

The clutch release mechanism was revised on 1982 models and changed completely on 1983 ZX750A models.

Clutch Release Adjustment (1982)

1. Refer to Chapter Three in the main book for preliminary clutch adjustment steps.
2. Loosen the locknut securing the adjusting screw and turn the screw *out* until some resistance is felt. This indicates the clutch mechanism is just starting to release. Turn the adjuster screw *in* 1/4 turn. Hold the adjuster screw and tighten the locknut to secure the adjustment.
3. Complete the adjustment procedure as outlined in Chapter Three.

11

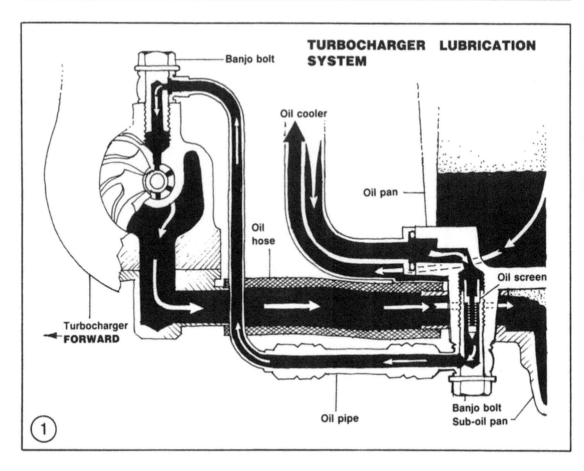

Clutch Release Adjustment (1983-on ZX750)

1. Loosen the large knurled locknut and screw in the cable adjuster on the clutch handlebar lever (**Figure 2**).

2. Loosen the locknuts securing the engine-mounted cable adjuster (**Figure 3**).

3. Turn in the cable adjuster on the engine until 2-3 mm (1/16-1/8 in.) of free play exists in the cable at the handlebar lever (**Figure 4**). Tighten the locknuts to secure the cable adjustment.

4. Future clutch adjustments can be made at the handlebar adjuster until the cable stretches and the proper free play cannot be attained at the handlebar. At that point readjust the cable at the engine-mounted adjuster.

SUSPENSION TUNING

Proper suspension adjustment and "tuning" is necessary to provide optimum handling and comfort.

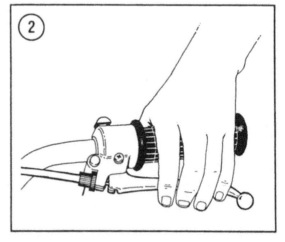

The rear shock absorbers on 1982 KZ750R models have 5 damping positions and 7 spring pre-load positions. The Uni-Trak rear suspension on 1983 ZX750A models has 4 damping adjustment positions as well as an air valve. Air or nitrogen pressure can be added to the shock absorber to alter the

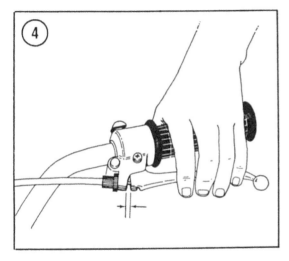

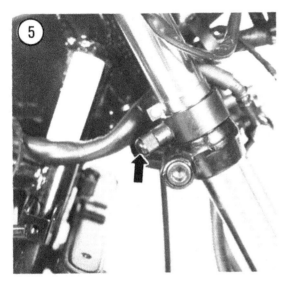

handling characteristics of the rear suspension.

WARNING
All suspension components must be adjusted the same on each side of the motorcycle; for example, the preload for each rear shock absorber spring must be set the same. Unequal adjustments of the forks or rear suspension components may cause handling instabilities.

Pressurizing Front Forks

An equalizing air valve and manifold is fitted to 1982 KZ750R, 1983-on ZX750A and 1984-on ZX750E models (**Figure 5**). The manifold connects both fork tubes so the air pressure is always equal in each tube. Proper air pressure in the forks will ensure the best possible performance. Air pressure in the forks can be used to alter the handling characteristics of the motorcycle and help support the weight of any touring gear or accessories.

1. Refer to *Fork Air Pressure* in Chapter Three of the main book to pressurize the forks. Keep the following points in mind during the procedure:

  a. Make sure motorcycle is placed on a jack or stand so that all weight is removed from the front forks.

  b. Pressurizing the forks is best performed with a bicycle tire pump. If the S & W gauge described in Chapter Three is not available, use a tire gauge. Make sure you use a low-pressure gauge with a scale capable of accurately reading the low air pressure required for the front forks.

  c. When using the gauge make sure it is placed squarely on the air valve and removed quickly to prevent as much air loss as possible. Normal air loss when the gauge is removed is 0.7-1.4 psi (0.05-0.10 kg/cm²).

CAUTION
Never use a high-pressure air supply (such as a service station air hose) to pressurize the fork tubes or the fork seals will be damaged. Use only a hand-operated tire pump for the best results.

2. Slowly bleed air from the fork tubes until the desired air pressure is achieved. Refer to **Table 9** for the normal range of pressure used. Keep in mind that approximately 1 psi is lost every time the gauge is removed from the air valve.

3. Install the air valve protection cap.

Rear Spring Pre-load Adjustment (1982 KZ750R)

Refer to *Rear Shock Absorbers* in Chapter Three of the main book to adjust the pre-load on the rear shock absorbers. Seven pre-load positions are available on 1982 KZ750R models.

Increase the spring pre-load as necessary to compensate for the additional weight of a passenger or touring equipment. If the pre-load is changed it may be necessary to adjust the damping rate on models equipped with external adjustment.

Rear Shock Absorber Damping Adjustment (1982 KZ750R)

Adjust the rear shock damping as outlined under *Shock Absorber* in Chapter Three of the main book. On 1982 KZ750R models, the damping or rebound rate of each rear shock absorber can be adjusted to any one of 5 rates, instead of 4 as on other models. To perform the adjustment, rotate the external adjuster wheel (**Figure 6**) until a definite "click" is heard or felt. The adjuster must be positioned on a specific number (No. 1 is softest; No. 5 is stiffest). If the adjuster is placed between any 2 numbers the damping rate is automatically 5. Make sure both rear shock absorber damping rates are the same. The standard damping setting is 1.

Rear Shock Absorber Damping Adjustment (Uni-Trak)

On 1983 and later ZX750A models with Uni-Trak rear suspension, the damping rate of the rear shock can be adjusted to any one of 4 rates. To adjust the damping, remove the right-hand side cover and push the adjuster plunger all the way in as far as possible (A, **Figure 7**). This position is 1, the softest (B, **Figure 7**). Pull out on the plunger until a click

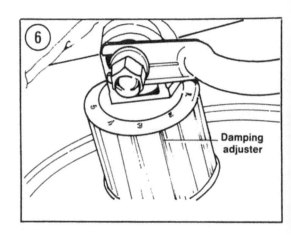

Damping adjuster

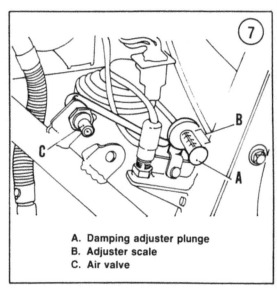

A. Damping adjuster plunge
B. Adjuster scale
C. Air valve

is felt. Each click provides progressively stiffer damping. No. 4 position (stiffest) is reached with the plunger all the way out. For solo riding, damping position 1 or 2 is normal. An increase in damping is necessary if riding with a passenger or if riding in an aggressive manner.

Pressurizing Rear Shock (Uni-Trak)

The single rear shock absorber on models with Uni-Trak rear suspension is equipped with an air valve so that air or nitrogen pressure can be injected into the shock absorber. It is necessary to maintain proper air pressure in the shock absorber to obtain the best possible performance of the rear suspension.

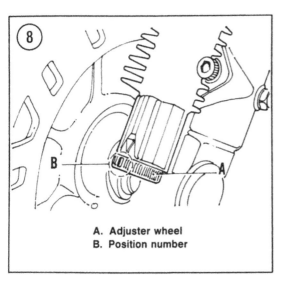

A. Adjuster wheel
B. Position number

Pressurizing the shock absorber is best performed with a special air suspension gauge (Kawasaki part No. 52005-1003). If such a gauge is not available, use a tire gauge. Make sure you use a gauge with a scale capable of accurately reading the low to high range of pressures required for the shock absorber. When using the gauge make sure it is placed squarely on the air valve and removed quickly to prevent as much air loss as possible. Normal air loss when the gauge is removed is 0.7-1.4 psi (0.05-0.10 kg/cm²).

1. Place the motorcycle on the center stand to raise the rear wheel clear of the ground. The rear shock absorber must be cool (room temperature) or the pressure readings may be inaccurate.

2. Remove the right-hand motorcycle side cover. Remove the protection cap (C, **Figure 7**) from the air valve and bleed off any remaining pressure.

3. Connect a bicycle tire pump or regulated nitrogen supply to the air valve.

CAUTION
Never use a high-pressure air supply (such as a service station air hose) to pressurize the shock absorber or the internal seals will be damaged. Use only a hand-operated tire pump for the best results.

4. Pump up pressure in the shock absorber to a pressure slightly higher than desired. The

normal setting for a solo rider weighing approximately 150 lb. (no accessories on the motorcycle) is 0.5 kg/cm² (2-7.1 psi) for the ZX750A and 2.0 kg/cm² (2-28 psi) for the ZX750E. The maximum recommended pressure (within the usable range) is 3.0 kg/cm² (43 psi). More than the normal pressure stated will be required whenever the motorcycle load is higher than normal or the road surface and riding style require a stiffer suspension. As in all suspension adjustments, the optimum air pressure for the rear shock absorber is largely a matter of personal choice based on riding style, weight, desired comfort, etc.

CAUTION
Do not exceed 5.0 kg/cm² (71 psi) or the shock absorber seals may be damaged.

5. Slowly bleed pressure from the shock absorber until the desired pressure is achieved. Keep in mind that approximately 1 psi is lost every time a tire gauge is removed from the air valve.

6. Install the air valve protection cap.

Anti-dive Fork Adjustment
(1983-on ZX750A and 1984-on ZX750E)

The anti-dive mechanism installed on the front forks has a 3-position adjustment. This adjustment allows for more or less anti-dive in the forks to compensate for the load on the motorcycle as well as road conditions. To adjust the rate of anti-dive, rotate the adjuster wheel (A, **Figure 8**) until the desired number is aligned with the arrow on the fork tube (B, **Figure 8**). Position No. 1 is the weakest and position No. 3 is the strongest. Ensure that both anti-dive adjusters are set on the same number or the handling may be unstable.

AIR FILTER

Some 1982 and later models are equipped with an oiled foam air filter element. The filter element should be cleaned and reoiled with every tune-up or at least every 3,000 miles (5,000 km). If the motorcycle is subjected to abnormally dirty conditions, the filter element should be checked and cleaned more frequently.

11

1A. *Non-turbo models:* Remove the air filter element as outlined under *Air Filter* in Chapter Three of the main book.

1B. *Turbo models:* Remove the air filter Allen bolt on the left-hand side of the engine and remove the air filter cover (**Figure 9**). Remove the air filter assembly (**Figure 10**).

2. Carefully separate the foam element from the filter frame.

3. Wash the foam filter element in solvent, then in hot, soapy water. Rinse the element in clean water and squeeze it between your palms to remove as much water as possible. The element can be pressed between several layers of paper towels to speed up the drying process. Allow the element to dry completely.

> *CAUTION*
> *Never wring or twist the foam element during the cleaning or reoiling process, as the foam can easily be damaged.*

4. Carefully examine the foam element for any splits or tears. Replace the element if it is damaged in any way.

5. Apply SAE 30W engine oil or special air filter oil to the element and gently work the foam in your hands until the element is completely saturated with oil. Squeeze the element between your palms to remove all the excess oil.

> *NOTE*
> *Place the filter element and filter oil in a plastic bag to eliminate most of the mess of the job.*

> *NOTE*
> *A good grade of special air filter oil provides better protection against dirt and moisture than plain engine oil.*

6. Thoroughly clean the inside of the air box and the filter sealing flange.

7. Install the filter frame inside the foam element. Lightly grease the top and bottom edges of the element. Install the air filter in the motorcycle by reversing the removal steps.

FRONT FORKS

Fork Oil Change

1982-on non-turbo models

Refer to **Table 9** for fork oil capacities and levels for 1982 and 1983 models. Refer to

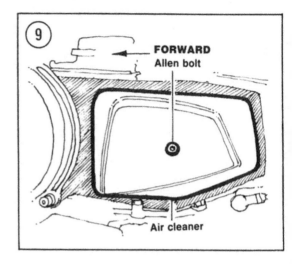

FORWARD
Allen bolt

Air cleaner

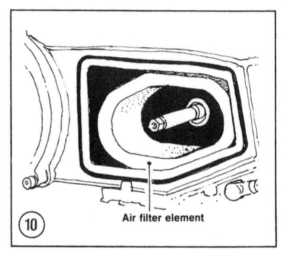

Air filter element

Fork Oil Change in Chapter Three of the main book to change fork oil and set the oil level.

1984-on turbo models

Procedures used to change the fork oil remain the same as for 1983 and earlier models, except when measuring the fork oil level. On these models, it is necessary to remove the fork springs and compress the forks when measuring the fork oil level. Refer to **Table 9** for fork oil capacities and levels.

> *CAUTION*
> *Because both fork tubes must be compressed at the same time to measure the fork oil level, it is advisable to have an assistant on hand when performing this procedure.*

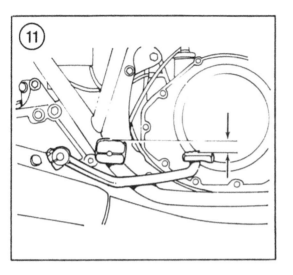

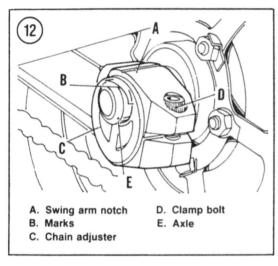

A. Swing arm notch   D. Clamp bolt
B. Marks   E. Axle
C. Chain adjuster

BRAKES

Brake Pedal Height

Adjust the brake pedal height to the specified distance below the top of the foot rest (**Figure 11**). See **Table 10**. Refer to *Brake Pedal Height* in Chapter Three of the main book to perform the adjustment.

AIR SUCTION VALVES

On 1982 models, a gasket was installed under each air suction valve. On 1983 models, the air suction valves were modified and the gaskets are no longer used. On 1982 models, replace the gaskets if the air valves are removed and the gasket is damaged in any way.

VALVE CLEARANCE (ZX750E)

Valve Clearance Inspection

The procedures used to check the valve clearance on the ZX750E models is the same as that described in Chapter Three of the main book, except for the valve measurement position. The correct positions are as follows:

   a. When the No. 1 and No. 4 pistons are at top dead center, adjust the exhaust valve clearances.

   b. When the No. 2 and No. 3 pistons are at top dead center, adjust the intake valve clearances.

DRIVE CHAIN

Drive Chain Adjustment (1983-on ZX750 models)

The drive chain adjustment for ZX750 models is essentially the same as other models; however, the chain adjusters are different. Refer to *Drive Chain* in Chapter Three of the main book for chain inspection, lubrication and adjustment procedures. Refer to **Figure 12** and perform the following additional adjustment steps.

1. Remove the cotter pin securing the rear axle nut and loosen the nut.
2. Loosen the clamp bolts securing each chain adjuster (D, **Figure 12**).
3. Rotate the chain adjusters equally on each side of the motorcycle until 35-40 mm (1 3/8-1 5/8 in.) of slack exists in the drive chain. Turn the rear wheel and check the chain slack in several places. A drive chain will rarely stretch evenly.
4. Ensure that the mark on each adjuster is aligned with the same mark on each side of the swing arm. Tighten the clamp bolts securing the adjusters to 3.3 mkg (24 ft.-lb.).
5. Tighten the axle nut to the following specifications. Secure the nut with a new cotter pin:
   a. ZX750A: 12.0 mkg (87 ft.-lb.).
   b. ZX750E: 9.5 mkg (69 ft.-lb.).

CARBURETOR (ZX750E)

Idle Speed

Procedures used to adjust the idle speed remain the same; the idle speed specification

11

and the location of the idle speed screw have changed. Turn the idle speed screw (**Figure 13**) to set the idle speed as follows:

a. U.S. models: 1,150-1,250 rpm.

b. Non-U.S. models: 1,000-1,100 rpm.

Throttle Synchronization

The procedure for synchronizing the throttle valves is the same as for synchronizing carburetors, as described in Chapter Three of the main book. Refer to **Figure 14** for the adjust screw positions when adjusting the throttle valves.

> *NOTE*
> *Kawasaki recommends the use of the balance adjust tool (part no. 57001-351) to turn the adjust screws.*

FUEL SYSTEM (ZX750E)

Fuel Hose Replacement

Kawasaki recommends that the following pressure fuel hoses (**Figure 15**) be replaced at specific time intervals.

> *WARNING*
> *All ZX750E models are equipped with a fuel pump. When any fuel hose is disconnected, do not turn on the ignition switch. This will activate the fuel pump and cause fuel to pour from the disconnected hose(s). This could present a severe fire hazard.*

The following high-pressure hoses should be replaced every 2 years:

a. Hose between the fuel pump and the fuel distributing pipe.

b. Hose between the fuel distributing pipe and pressure regulator.

The following low-pressure hoses should be replaced every 4 years:

a. Hose between the fuel valve and the fuel filter.

b. Hose between the fuel filter and the fuel pump.

c. Hose between the pressure regulator and the check valve.

CYLINDER COMPRESSION (ZX750E)

Before measuring the cylinder compression on any ZX750E model, disconnect the white/red lead connector at the positive battery terminal. This will stop the fuel injector from operating when using the electric starter.

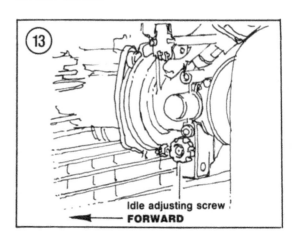
Idle adjusting screw
FORWARD

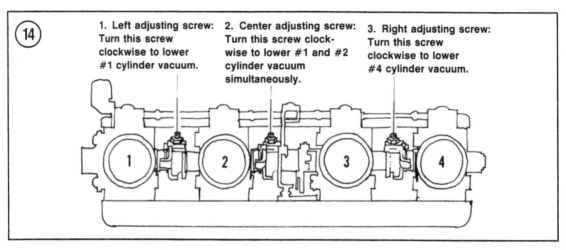
1. Left adjusting screw: Turn this screw clockwise to lower #1 cylinder vacuum.
2. Center adjusting screw: Turn this screw clockwise to lower #1 and #2 cylinder vacuum simultaneously.
3. Right adjusting screw: Turn this screw clockwise to lower #4 cylinder vacuum.

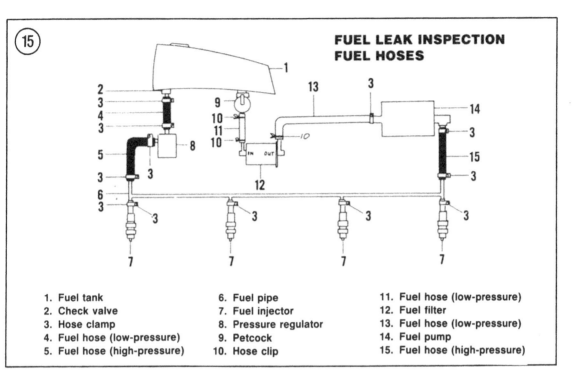

(15) **FUEL LEAK INSPECTION**
FUEL HOSES

1. Fuel tank
2. Check valve
3. Hose clamp
4. Fuel hose (low-pressure)
5. Fuel hose (high-pressure)
6. Fuel pipe
7. Fuel injector
8. Pressure regulator
9. Petcock
10. Hose clip
11. Fuel hose (low-pressure)
12. Fuel filter
13. Fuel hose (low-pressure)
14. Fuel pump
15. Fuel hose (high-pressure)

Table 1 MODEL DESIGNATION

1982	KZ750E; Z750L; KZ750H; KZ750R
1983	KZ750H; KZ/Z750L; ZX750A
1984-1985	ZX750E; KZ750L; ZX750A

Table 2 GENERAL MODEL SPECIFICATIONS (1982)

DIMENSIONS	
Overall length	
KZ750E/Z750L	2130 mm (83.9 in.)
KZ750E/Z750L (see notes 1 & 3)	2190 mm (86.2 in.)
KZ750H	2195 mm (86.4 in.)
KZ750H (see notes 1 & 3)	2210 mm (87.0 in.)
KZ750R	2170 mm (85.4 in.)
KZ750R (see notes 1 & 3)	2215 mm (87.2 in.)
Overall width	
KZ750E/Z750L	835 mm (32.9 in.)
KZ750E/Z750L (see notes 1, 3 & 5)	780 mm (30.7 in.)
KZ750H	810 mm (31.9 in.)
KZ750R	780 mm (30.7 in.)
Overall height	
KZ750E/Z750L	1135 mm (44.7 in.)
KZ750H	1235 mm (48.6 in.)
KZ750R	1220 mm (48.0 in.)
Ground clearance	
KZ750E/Z750L	150 mm (5.9 in.)
KZ750H	155 mm (6.1 in.)
KZ750R	140 mm (5.5 in.)
(continued)	

11

Table 2 GENERAL MODEL SPECIFICATIONS (1982)

DIMENSIONS (cont.)	
Dry weight	
KZ750E	210 kg (463 lb.)
Z750L, KZ750H	211 kg (465 lb.)
KZ750R	217 kg (478 lb.)

ENGINE	
Type	4-cylinder, 4-stroke, air-cooled, dual overhead camshafts
Bore	66.0 mm (2.598 in.)
Stroke	54.0 mm (2.126 in.)
Displacement	738 cc (45.0 cu. in.)
Compression ratio	
KZ750R	9.5:1
All others	9.0:1
Horsepower	
KZ750E/Z750L; KZ750H	74 @ 9,000 rpm
KZ750E/Z750L (see note 6)	77 @ 9,500 rpm
KZ750R	80 @ 9,500 rpm
Valve timing	
Intake open	30° BTDC
Intake close	60° ATDC
Duration (both)	270°
Exhaust open	60° BBDC
Exhaust close	30° ATDC
Carburetors	
KZ750E/Z750L; KZ750H	4 Keihin CV34
KZ750R	4 Mikuni BS34SS

DRIVE TRAIN	
Clutch	Wet multi-plate
Transmission	Constant mesh 5-speed
Primary reduction	2.550 (27/23×63/29)
Final reduction	
KZ750E/Z750L; KZ750R	2.538 (33/13)
KZ750H	2.461 (32/13)
Gear ratios	
1st	2.333 (35/15)
2nd	1.631 (31/19)
3rd	1.272 (28/22)
4th	1.040 (26/25)
5th	0.875 (21/24)

CHASSIS	
Caster	
KZ750E/Z750L; KZ750R	27°
KZ750H	30°
Trail	
KZ750E/Z750L	108 mm (4.25 in.)
KZ750H	121 mm (4.76 in.)
KZ750R	107 mm (4.21 in.)
Front tire	
KZ750E/Z750L; KZ750H	3.25H-19 4PR
KZ750R	100/90V-19
KZ750R (see notes 2 & 4)	100/90-19 57H

(continued)

Table 2 GENERAL MODEL SPECIFICATIONS (1982) (continued)

CHASSIS (cont.)	
Rear tire	
KZ750E/Z750L	4.00H-18 4PR
KZ750H	130/90-16 67H
KZ750R	120/90V-18
KZ750R (see notes 2 & 4)	120/90-18 65H
ELECTRICAL	
Ignition	Transistorized
Battery	12 volt, 12 amp/hr.

1. Australian models.
2. Canadian models.
3. European models.
4. U.S. models.
5. South African models.
6. West German models.

Table 3 GENERAL MODEL SPECIFICATIONS (1983-ON NON-TURBO)

DIMENSIONS	
Overall length	
KZ/Z750L	2215 mm (87.2 in.)
KZ/Z750L (see notes 1, 3, & 4)	2170 mm (85.4 in.)
KZ750H	2195 mm (86.4 in.)
ZX750A	2220 mm (87.4 in.)
ZX750A (see notes 1, 3, & 4)	2190 mm (86.2 in.)
Overall width	
KZ/Z750L	770 mm (30.3 in.)
KZ750H	840 mm (33.1 in.)
ZX750A	760 mm (29.9 in.)
Overall height	
KZ/Z750L	1135 mm (44.7 in.)
KZ/Z750L (see notes 1 & 3)	1130 mm (44.5 in.)
KZ750H	1240 mm (48.8 in.)
ZX750A	1260 mm (49.6 in.)
Ground clearance	
KZ/Z750L	140 mm (5.5 in.)
KZ750H	155 mm (6.1 in.)
ZX750A	150 mm (5.9 in.)
Dry weight	
KZ/Z750L	215 kg (474 lb.)
KZ/Z750L (see notes 1 & 3)	216 kg (476 lb.)
KZ750H	211 kg (465 lb.)
ZX750A	220 kg (485 lb.)
ZX750A (see notes 1 & 3)	219 kg (482 lb.)
ZX750A (see note 4)	221 kg (487 lb.)
ENGINE	
Type	4-cylinder, 4-stroke, air-cooled, dual overhead camshafts
Bore	66.0 mm (2.598 in.)
Stroke	54.0 mm (2.126 in.)

(continued)

11

Table 3 GENERAL MODEL SPECIFICATIONS (1983-ON NON-TURBO)(cont.)

ENGINE (cont.)	
Displacement	738 cc (45.0 cu. in.)
Compression ratio	9.5:1
Horsepower	
KZ/Z750L	80 @ 9,000 rpm
KZ750H	75 @ 9,500 rpm
ZX750A	86 @ 9,500 rpm
ZX750A (see notes 5 & 6)	87 @ 9,500 rpm
ZX750A (see note 3)	85 @ 9,500 rpm
Valve timing (KZ/Z750L; KZ750H models)	
Intake open	30° BTDC
Intake close	60° ATDC
Duration (both camshafts)	270°
Exhaust open	60° BBDC
Exhaust close	30° ATDC
Valve timing (ZX750A models)	
Intake open	38° BTDC
Intake close	68° ATDC
Duration (both camshafts)	286°
Exhaust open	68° BBDC
Exhaust close	38° ATDC
Carburetors	
KZ750H	4 Keihin CV34
KZ/Z750L; ZX750A	4 Mikuni BS34SS

DRIVE TRAIN	
Clutch	Wet multi-plate
Transmission	Constant mesh 5-speed
Primary reduction	2.550 (27/23×63/29)
Final reduction	
KZ/Z750L	2.538 (33/13)
KZ750H	2.461 (32/13)
ZX750A	2.533 (38/15)
Gear ratios	
1st	2.333 (35/15)
2nd	1.631 (31/19)
3rd	1.272 (28/22)
4th	1.040 (26/25)
5th	0.875 (21/24)

CHASSIS	
Caster	
KZ/Z750L	27°
KZ750H	30°
ZX750A	26.5°
Trail	
KZ/Z750L	107 mm (4.21 in.)
KZ750H	121 mm (4.76 in.)
ZX750A	103 mm (4.05 in.)
Front tire	
KZ750H	3.25H-19 4PR
KZ/Z750L	100/90-19 57H
ZX750A	110/90V-18
ZX750A (see notes 1 & 3)	110/90-18 66H

(continued)

Table 3 GENERAL MODEL SPECIFICATIONS (1983-ON NON-TURBO) (continued)

CHASSIS (cont.)	
Rear tire	
KZ750H	130/90-16 67H
KZ/Z750L	120/90-18 65H
ZX750A	130/80V-18
ZX750A (see notes 1 & 3)	130/80-18 66H
ELECTRICAL	
Ignition	Transistorized
Battery	
KZ/Z750L; KZ750H	12 volt, 12 amp/hr.
ZX750A	12 volt, 14 amp/hr.

1. Canadian models.
2. European models.
3. U.S. models.
4. South African models.
5. West German models.
6. Swedish models.

Table 4 GENERAL MODEL SPECIFICATIONS (1984-ON TURBO)

DIMENSIONS	
Overall length	2220 mm (87.4 in.)
Overall width	740 mm (29.13 in.)
Overall height	1,260 mm (49.6 in.)
Ground clearance	155 mm (6.1 in.)
Dry weight	233 kg (514 lb.)
ENGINE	
Type	4-cylinder, 4-stroke air-cooled, dual overhead camshafts
Bore	66.0 mm (2.598 in.)
Stroke	54.0 mm (1.126 in.)
Displacement	738 cc (45.0 cu. in.)
Compression ratio	7.8:1
Valve timing	
Intake open	22° BTDC
Intake close	52° ABDC
Duration (both camshafts)	254°
Exhaust open	60° BBDC
Exhaust close	20° ATDC
Duration (both camshafts)	260°
Carburetion	Digital fuel injection (DFI)
DRIVE TRAIN	
Clutch	Wet multi-plate
Transmission	Constant mesh 5-speed
Primary reduction	1.935 (60/31)
Final reduction	3.066 (46/15)
(continued)	

11

Table 4 GENERAL MODEL SPECIFICATIONS 1984-ON TURBO (cont.)

DRIVE TRAIN (cont.)	
Gear ratios	
1st	2.285 (32/14)
2nd	1.647 (28/17)
3rd	1.272 (28/22)
4th	1.045 (23/22)
5th	0.833 (20/24)
CHASSIS	
Caster	28°
Trail	117 mm (4.61 in.)
Front tire	110/90V-18
Rear tire	130/80V-18
ELECTRICAL	
Ignition	Transistorized
Battery	12 volt, 14 amp hour

Table 5 1983-ON MAINTENANCE SCHEDULE

Every 200 miles (500 km)	• Lubricate the drive chain
Every 500 miles (800 km)	• Check and adjust the drive chain free play
Initial 500 miles (800 km); then every 3,000 miles (5,000 km)	• Check, clean and regap spark plugs • Check and adjust valve clearance • Check throttle grip play; adjust if necessary • Check carburetor idle speed; adjust if necessary • Check carburetor synchronization; adjust if necessary • Check the evaporative emission control system[1] • Change the engine oil and filter • Adjust the clutch • Check the master cylinder brake fluid level; refill as required • Check the front and rear brake light switch operation; adjust and/or replace the switch as required • Check the front steering free play; adjust if necessary
Initial 3,000 miles (5,000 km); then every 6,000 miles (10,000 km)	• Clean the air cleaner element; replace element every 5 cleanings • Check the cylinder head bolt and nut tightness; retighten if necessary • Clean the oil screen[2] • Replace the fuel filter[2]
Every 3,000 miles (5,000 km)	• Check drive chain for wear; replace if necessary • Check the air suction valve; replace if necessary[3] • Check the brake pads for wear; replace if necessary • Check the tires for wear and damage; replace if necessary

(continued)

Table 5 1983-ON MAINTENANCE SCHEDULE (cont.)

Every 6,000 miles **(10,000 km) or 1 year**	• Change the brake fluid • Change the front fork oil • Lubricate the swing arm pivot shaft and the Uni-trak linkage
Every 2 years	• Replace the high-pressure fuel hoses[2] • Replace the anti-dive brake plunger assembly • Replace the master cylinder cup and dust seal • Replace the brake caliper piston seal and dust seal • Lubricate the steering stem bearings • Lubricate the wheel bearings • Lubricate the speedometer drive gear
Every 4 years	• Replace the low-pressure fuel hoses[2] • Replace the brake hose and pipe or nipple

1. 1984 and later California models.
2. Turbocharged models.
3. Non-turbocharged U.S. models.

Table 6 TORQUE SPECIFICATIONS (1982 KZ750R)

	ft.-lb.	mkg
Carburetor mounting Allen bolts	10	1.4
Engine mounting bolts	29	4.0
Engine mounting bracket bolts	17.5	2.4
Oil cooler tube connecting nuts	16	2.2
Front fork air valve	5.7 (69 in.-lb.)	0.8
Handlebar clamp bolts	22	3.0
Steering stem head bolt	22	3.0
Turn signal mounting nuts	9.4 (113 in.-lb.)	1.3
Front brake caliper mounting bolts	22	2.3

Table 7 TORQUE SPECIFICATIONS (1983-ON NON-TURBO)*

	ft.-lb.	mkg
Alternator rotor bolt (all models)	94	13.0
Engine sprocket nut	72	10.0
Clutch spring bolts	6.5 (78 in.-lb.)	0.9
Chain adjuster clamp bolts	24	3.3
Brake caliper mounting bolts (front and rear)	24	3.3
Brake caliper bleed valve	5.8 (69 in.-lb.)	0.8
Metal brake line nipples	16.5	2.3
Brake hose banjo bolts	22	3.0
Handlebar clamp bolts (except ZX750A models)	13.5	1.9
Handlebar clamp bolts	7.2 (87 in.-lb.)	1.0
Handlebar holder bolts	54	7.5

(continued)

11

Table 7 TORQUE SPECIFICATIONS (1983-ON NON-TURBO)* (cont.)

	ft.-lb.	mkg
Steering head clamp nut		
(except ZX750A models)	15	2.1
Steering stem head bolt	31	4.3
Front fork axle clamp nuts		
(except ZX750A models)	10	1.4
Axle clamp nut (ZX750A)	14.5	2.0
Front fork air valves (all models)	5.8 (69 in.-lb.)	0.8
Front fork anti-dive assembly		
mounting bolts	5.1 (61 in.-lb.)	0.7
Front fork anti-dive brake plunger		
mounting bolts	3.2 (39 in.-lb.)	0.45
Rear shock absorber air hose pipe fitting	8.7 (104 in.-lb.)	1.2
Rear shock absorber air valve	5.8 (69 in.-lb.)	0.8
Rear shock absorber mounting nuts		
Upper	27	3.8
Lower	51	7.0
Uni-Trak rocker arm pivot shaft nut	51	7.0
Uni-Trak tie rod nuts		
Upper	27	3.8
Lower	51	7.0

* The specifications in this table are only those that differ from the torque specifications listed in the main book. Unless otherwise noted, the torque specifications in this Table apply *only* to 1983-on ZX750A models.

Table 8 CHASSIS TORQUE SPECIFICATIONS (1984-ON TURBO)

	ft.-lb.	mkg
Front axle nut	43	6.0
Front axle clamp nut	14.5	2.0
Rear axle nut	69	9.5
Handlebar		
Clamp bolts	87 in.-lb.	1.0
Holder bolts	54	7.5
Steering stem head bolt	31	4.3
Anti-dive		
Brake plunger mounting bolts	39 in.-lb.	0.45
Assembly mounting bolts	61 in.-lb.	0.70
Chain adjuster clamp bolts	24	3.3
Front fork bottom bolts	16.5	2.3
Front fork clamp bolts		
Upper	14.5	2.0
Lower	29	4.0
Front fork cap bolt	16.5	2.3
Rear shock absorber nuts		
Upper	27	3.8
Lower	51	7.0
Rear sprocket nuts	29	4.0
Swing arm pivot shaft nut	72	10.0

(continued)

Table 8 CHASSIS TORQUE SPECIFICATIONS (1984-ON TURBO) (cont.)

Uni-trak links		
Rocker arm pivot shaft nut	51	7.0
Tie-rod nuts		
Upper	27	3.8
Lower	51	7.0
Brake hose banjo bolts	22	3.0
Brake pipe nipples	16.5	2.3
Caliper mounting bolts		
Front and rear	24	3.3
Brake disc mounting bolts	16.5	2.3
Front master cylinder clamp bolts	78 in.-lb.	0.90
Torque link nuts	22	3.0

Table 9 FRONT FORK SPECIFICATIONS*

1982 KZ/750H		
Dry capacity	304-312 cc	10.3-10.5.5 oz.
Wet capacity	290 cc	9.8 oz.
All other 1982 models		
Dry capacity	251-259 cc	8.5-8.8 oz.
Wet capacity	240 cc	8.2 oz.
1983-on		
KZ750H	308-316 cc	10.4-10.7 oz.
KZ/Z750L	293-301 cc	9.9-10.2 oz.
ZX750A and L	245-253 cc	8.3-8.6 oz.
ZX750E	267-275 cc	9.0-9.3 oz.
Oil level (fork tube		
extended without spring)		
1982 KZ/Z750H	434-438 mm	17.1-17.2 in.
All other 1982 models	380-384 mm	15.0-15.1 in.
1983 KZ750H	436-440 mm	17.2-17.3 in
Oil level (fork tube		
compressed without spring)		
1983-on KZ/750L	101-105 mm	4.0-4.1 in.
1983-on ZX750A	183-187 mm	7.2-7.4 in.
1984-on ZX750E	174-178 mm	6.8-7.0 in.
Air pressure		
1982 KZ750R		
Standard	0.7 kg/cm²	10 psi
Range	0.6-0.9 kg/cm²	8.5-13 psi
1983 KZ750H	0.5-1.0 kg/cm²	7-14 psi
1983-on KZ/750L	0.6-0.9 kg/cm²	8.5-13 psi
1983-on ZX750A	0.4-0.6 kg/cm²	5.7-8.5 psi
1984-on ZX750E	0.4-0.6 kg/cm²	5.7-8.5 psi

* Fork oil capacities will vary slightly from model to model. Always measure fork oil level to obtain the most accurate oil capacity in each fork tube.

11

Table 10 BRAKE PEDAL HEIGHT

1982	
KZ750E/Z750L	
U.S.	8-12 mm (3/8-1/2 in.)
Except U.S.	13-17 mm (1/2-5/8 in.)
KZ750H	4-8 mm (1/16-1/8 in.)
KZ750R	14-18 mm (9/16-3/4 in.)
1983-on	
KZ750H	4-8 mm (1/16-1/8 in.)
KZ/Z750L	14-18 mm (9/16-3/4 in.)
ZX750A, ZX750E	50.5-54.5 mm (2-2 1/16 in.)

CHAPTER FOUR

ENGINE

Refer to **Table 11** for new engine wear limits for 1982 and later non-turbo models. **Table 12** and **Table 13** list complete engine wear limit and tightening torque specifications for ZX750E models. **Table 14** lists piston ring specifications applicable to 1982 and later models.

OIL COOLER

On 1982 KZ750R and 1983-on ZX750 models, an oil cooler is mounted on the forward frame just below the fuel tank.

Removal/Installation

Refer to **Figure 16** for this procedure.
1. Place a drain pan beneath the oil cooler lines.
2. Loosen both nuts connecting the oil lines to the oil cooler (A, **Figure 17**).

CAUTION
Always use a second wrench to hold the oil fitting on the oil cooler while loosening the oil line nuts or damage to the cooler may result (B, Figure 17).

3. If oil line removal is desired, remove the bolts securing each oil line to the engine and remove the lines. Make sure the O-rings in each oil line mounting flange are not lost.
4. Remove the bolts securing the oil cooler to the motorcycle. Note the location of the metal collars and rubber grommets on each mounting lug.
5. On 1983-on ZX750 models, disengage the center mounting lug from the motorcycle frame (**Figure 18**).
6. Carefully remove the oil cooler.
7. Installation is the reverse of these steps. Keep the following points in mind.
8. Check the O-rings in the lower oil line flanges, if the lines were removed. Replace the O-rings if distorted or damaged.
9. Lightly oil the threads on the oil cooler fittings before installing the oil lines.
10. Install the oil cooler and oil lines. Leave all mounting bolts finger-tight at this time.
11. Tighten the mounting bolts securing the lower oil line mounting flanges.
12. Tighten the bolts securing the oil cooler to the frame.

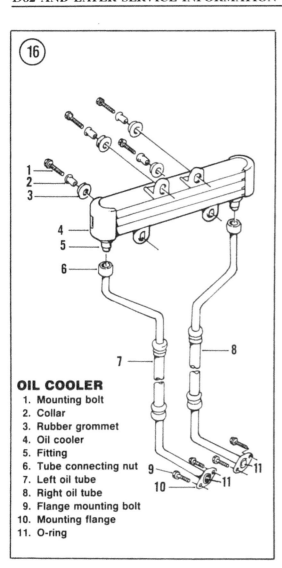

OIL COOLER
1. Mounting bolt
2. Collar
3. Rubber grommet
4. Oil cooler
5. Fitting
6. Tube connecting nut
7. Left oil tube
8. Right oil tube
9. Flange mounting bolt
10. Mounting flange
11. O-ring

13. Use a second wrench to hold the oil cooler fittings and tighten the oil line connecting nuts to 2.2 mkg (16.0 ft.-lb.).

14. Start the engine and check for oil leaks. Check the engine oil level and top off if necessary.

CAMSHAFTS (ZX750E)

Installation

Procedures used to install the camshaft sprocket and time the camshafts remain the same as for 1983 and earlier models, except for the following:

a. The intake and exhaust sprockets are identical. If removed, install the sprocket on the camshaft using the proper holes for each cam (**Figure 19**). The marked side of the sprocket faces the end of the cam with a notch in it. Use a locking agent such as Loctite Lock N' Seal on the sprocket bolts and tighten them as specified in **Table 13**.

b. When installing the cam chain, locate the cam chain pin on the exhaust sprocket in line with the "Z6EX" line (**Figure 20**). Beginning with this pin as zero, count off 45 pins toward the intake cam. The "IN" line on the intake sprocket must lie between the 45th and 46th pins. If it does not align, recheck your pin count and reposition the intake camshaft if required.

11

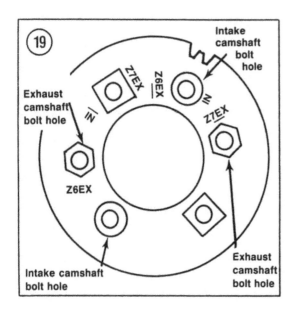

⑲ Intake camshaft bolt hole

Exhaust camshaft bolt hole

Z6EX

Intake camshaft bolt hole

Exhaust camshaft bolt hole

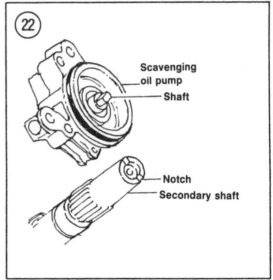

㉒ Scavenging oil pump

Shaft

Notch

Secondary shaft

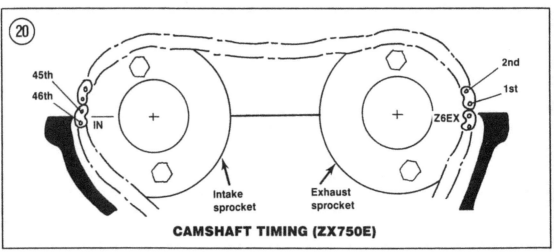

⑳

45th
46th

IN

2nd
1st

Z6EX

Intake sprocket

Exhaust sprocket

CAMSHAFT TIMING (ZX750E)

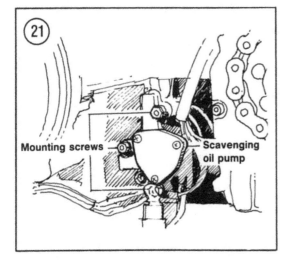

㉑ Mounting screws — Scavenging oil pump

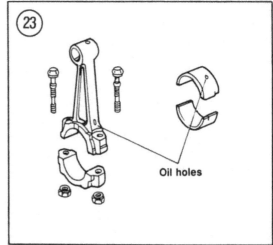

㉓

Oil holes

SCAVENGING OIL PUMP
(ZX750E)

Removal/Installation

1. Remove the external shift mechanism assembly.
2. Loosen the starter cover mounting bolts.
3. Loosen the air filter intake pipe-to-turbocharger seal clamps.
4. Remove the air filter intake pipe mounting screw.
5. Disconnect the breather tube.
6. Remove the air filter housing.
7. Referring to **Figure 21**, remove the scavenging oil pump mounting bolts and remove the oil pump.
8. Inspect the oil pump O-ring; replace it if necessary.
9. Installation is the reverse of these steps; note the following:
   a. Align the scavenging oil pump shaft with the notch in the end of the secondary shaft when installing the oil pump (**Figure 22**).

CRANKSHAFT AND
CONNECTING RODS

On 1982 and later models (except ZX750 models), an oil hole is added to the connecting rods to improve engine cooling and lubrication. When installing the connecting rods ensure that the oil hole is positioned toward the rear of the engine (**Figure 23**). Install the bearing insert with the oil holes in the upper half of the big rod. Ensure that the oil hole in the bearing insert and connecting rod are aligned. The lower bearing insert installed in the bearing cap does not have an oil hole.

Refer to **Table 15** to select connecting rod bearings and **Table 16** to select crankshaft main bearings for all 1982 and later models.

ALTERNATOR ROTOR

A 12 mm bolt is used to secure the alternator rotor on 1983 and later models. Torque the bolt to 13.0 mkg (94 ft.-lb.).

Table 11 ENGINE WEAR LIMITS* (1983-ON NON-TURBO)

	in.	mm
Cam lobe height		
(1983 KZ750H; KZ/Z750L models)	1.423	36.15
Valve head thickness (1983 models)		
Exhaust valve	0.028	0.7
Valve stem OD (1983 models)	0.273	6.94
Valve guide clearance (rocking method)		
All valves	0.013	.33
Valve installed height (1983 ZX750A models)		
Exhaust valve	1.497	38.03

* The specifications in this table are only those that differ from the specifications listed in Chapter Four of the main book.

Table 12 ENGINE WEAR LIMITS (1984-ON TURBO)

	in.	mm
Cam lobe height	1.403	35.65
Camshaft bearing oil clearance	0.0090	0.23
Camshaft journal diameter	0.8625	21.91
Camshaft bearing inside diameter	0.8716	22.14
Camshaft runout	0.0039	0.1
Cam chain 20-link length	5.075	128.9

(continued)

11

Table 12 ENGINE WEAR LIMITS (1984-ON TURBO) (continued)

	in.	mm
Cylinder head warp	0.0019	0.05
Valve head thickness		
Intake	0.019	0.5
Exhaust	0.027	0.7
Valve stem bend	0.0019	0.05
Valve stem diameter	0.273	6.94
Valve guide inside diameter	0.279	7.08
Valve guide clearance (rocking method)	0.0129	0.33
Valve spring free length		
Intake	1.389	35.3
Exhaust	1.586	40.3
Cylinder diameter	2.602	66.10
Piston diameter	2.591	65.81
Connecting rod bearing crankpin clearance	0.0039	0.10
Crankpin diameter	1.3767	34.97
Connecting rod big end side clearance	0.0196	0.50
Crankshaft runout	0.0019	0.05
Crankshaft main bearing clearance	0.0031	0.08
Crankshaft main journal diameter	1.4157	35.96

Table 13 ENGINE TORQUE SPECIFICATIONS (1984-ON TURBO)

	ft.-lb.	mkg
Digital fuel injection system		
Engine temperature sensor	9.5	1.3
Fuel injector mounting bolts	43 in.-lb.	0.50
Oil cooler hoses		
Bolts	87 in.-lb.	1.0
Nuts	69 in.-lb.	2.2
Engine mounting bolts		
Bolts	29	4.0
Bracket bolts	17.5	2.4
Camshaft bearing cap bolts	104 in.-lb.	1.2
Camshaft chain tensioner cap	18	2.5
Camshaft sprocket bolts	11	1.5
Cylinder head		
Bolts	22	3.0
Nuts	29	4.0
Spark plugs	20	2.8
Engine sprocket nut	72	10
Alternator rotor bolt	94	13
Neutral switch	11	1.5
Shift pedal return spring pin	18	2.5
Starter clutch bolts	25	3.5
Secondary shaft nut	43	6.0
Clutch hub nut	98	13.5
Oil pressure switch	11	1.5
Timing rotor mounting bolt	18	2.5
Engine drain plug	27	3.7
Oil filter mounting bolt	14.5	2.0
Oil pipe banjo bolts	14.5	2.0
Connecting rod cap nuts	27	3.7
Exhaust manifold nuts	14.5	20
Connecting pipe bolts	14.5	20

Table 14 PISTON RING SPECIFICATIONS
(1982 KZ750R AND ALL 1983 AND LATER MODELS)

	Standard	Service limit
Piston ring thickness		
Top ring	0.970-0.990 mm	0.90 mm
	(0.038-0.039 in.)	(0.035 in.)
Second ring	1.179-1.190 mm	1.10 mm
	(0.046-0.047 in.)	(0.043 in.)
Ring groove width		
Top ring	1.02-1.04 mm	1.10 mm
	(0.040-0.041 in.)	(0.043 in.)
Second ring	1.21-1.23 mm	1.30 mm
	(0.047-0.048 in.)	(0.051 in.)

Table 15 CONNECTING ROD BEARING SELECTION

	Crank pin code	
	"O"	No mark
Big end code		
"O"	Black	Brown
No mark	Green	Black
Bearing color code/part No. (Upper half, 1982 and 1983 models, except 1983 ZX750A)		
Color	**Kawasaki part No.**	
Green	92028-1156	
Black	92028-1157	
Brown	92028-1158	
Bearing color code/part No. (Lower half, 1982 and 1983 models, except 1983 ZX750A)		
Color	**Kawasaki part No.**	
Green	13034-050	
Black	13034-051	
Brown	13034-052	
Bearing color code/part No. (1983 ZX750A models)		
Color	**Kawasaki part No.**	
Green	92028-1203	
Black	92028-1204	
Brown	92028-1205	

11

Table 16 CRANKSHAFT MAIN BEARING SELECTION

	Crankshaft journal code	
	"1"	No mark
Big end code		
"O"	Brown	Black
No mark	Black	Blue
Bearing color code/part No.		
Color	**Kawasaki part No.**	
Blue	92028-1100	
Black	92028-1101	
Brown	92028-1102	

CHAPTER FIVE

CLUTCH

CLUTCH RELEASE

Disassembly/Assembly (1982)

The clutch release mechanism was slightly modified for 1982 models (**Figure 24**). Refer to *Clutch Release Disassembly/Assembly* in Chapter Five of the main book to service the release mechanism. When installing the mechanism in the engine sprocket cover, ensure that the notch in the ball ramp plate engages the boss in the sprocket cover (**Figure 25**).

Disassembly/Assembly (1983-on ZX750 models)

The release mechanism on 1983 ZX750A models is a rack and pinion type (**Figure 26**). The release lever assembly is mounted in the clutch cover and should not need servicing unless the bearings or seal are defective. When installing the clutch cover, ensure that the arm on the release lever is positioned approximately 30° below horizontal as shown in **Figure 27**.

CLUTCH

Disassembly

Disassemble the clutch mechanism as outlined in Chapter Five of the main book. The clutch release mechanism on 1983 and later ZX750 models is changed as shown in **Figure 26**. When removing the pressure plate take care not to drop or damage the thrust washer and needle thrust bearing. Refer to **Table 17** for clutch wear limits.

Assembly

Assemble the clutch as outlined in Chapter Five of the main book. Take note of the following steps for 1983 and later ZX750 models.

1. Make sure the release mechanism is installed as shown in **Figure 26**.
2. Install the clutch cover so that the arm on the release lever is positioned approximately 30° below horizontal as shown in **Figure 27**.
3. On ZX750E models, install the last friction plate so that its tangs fit in the clutch housing grooves as shown in **Figure 28**.

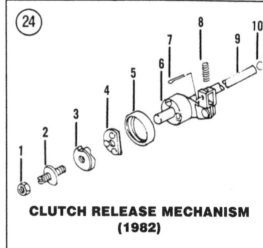

CLUTCH RELEASE MECHANISM (1982)

1. Locknut
2. Adjuster screw
3. Ball ramp plate
4. Ramp ball assembly
5. Seal
6. Release lever
7. Cotter pin
8. Spring
9. Pushrod
10. Steel ball

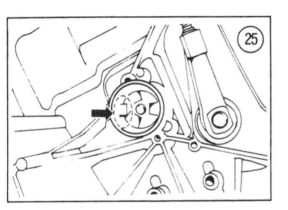

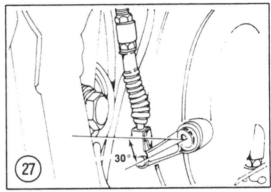

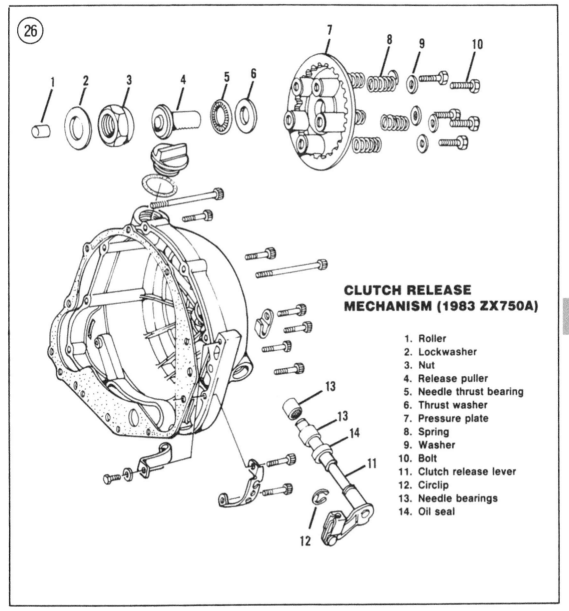

CLUTCH RELEASE MECHANISM (1983 ZX750A)

1. Roller
2. Lockwasher
3. Nut
4. Release puller
5. Needle thrust bearing
6. Thrust washer
7. Pressure plate
8. Spring
9. Washer
10. Bolt
11. Clutch release lever
12. Circlip
13. Needle bearings
14. Oil seal

11

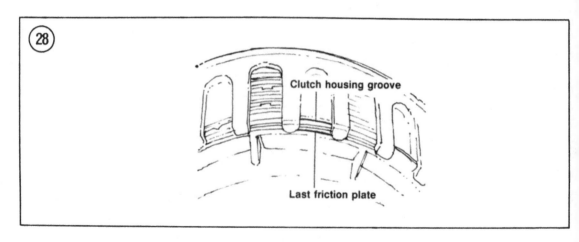

Table 17 CLUTCH WEAR LIMITS (1983-ON)

	mm	in.
Friction plate thickness		
ZX750A	3.4	0.13
ZX750E	2.8	0.11
Disc/plate warp	0.3	0.012

CHAPTER SIX

TRANSMISSION

TRANSMISSION GEARS

Removal/Disassembly

Refer to **Figure 29** for this procedure.

On 1982 and later models, both copper washers (item 6, **Figure 29**) were deleted from the input and output shafts.

On 1983 and later models, the O-ring (item 29, **Figure 29**) is deleted from the output shaft. The sprocket spacer (item 18, **Figure 29**) is press fitted to the output shaft.

Inspection

On 1982 and later models, the inside diameter of the needle bearings on the input and output shafts was modified. The service limit for the shaft bearing race inside diameter (ID) is changed to 27.04 mm (1.064 in.).

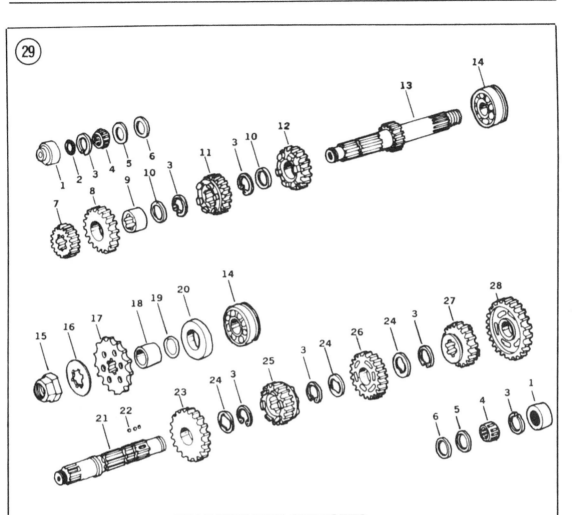

TRANSMISSION GEARSETS

1. Bearing outer race
2. O-ring
3. Circlip
4. Needle bearing
5. Steel washer
6. Copper washer
   (deleted on 1982 models)
7. Input 2nd gear
8. Input 5th gear
9. Copper bushing
10. Washer
11. Input 3rd gear
12. Input 4th gear
13. Input shaft
14. Ball bearing

15. Sprocket nut
16. Lockplate
17. Output sprocket
18. Sprocket spacer
    (press fit to shaft on 1983 models)
19. O-ring (deleted on 1983 models)
20. Oil seal
21. Output shaft
22. Neutral ball, 5/32 in.
23. Output 2nd gear
24. Splined washer
25. Output 5th gear
26. Output 3rd gear
27. Output 4th gear
28. Output 1st gear

11

CHAPTER SEVEN

FUEL AND EXHAUST SYSTEMS

CARBURETOR SERVICE

The carburetors fitted to 1982 KZ750R, 1983-on KZ/Z750L and 1983 ZX750A models are Mikuni BS34 constant velocity (CV) units. The operation and service of the Mikuni carburetors are very similar to the Keihin units used on other models. Refer to **Table 18** for Mikuni carburetor specifications.

Disassembly/Assembly (Mikuni)

Refer to **Figure 30** for this procedure.
1. If it is necessary to separate all 4 carburetors for cleaning or repair, perform the following:
   a. Remove the circlips securing the choke shaft and carefully pull the shaft out of the carburetor brackets. Take care not to lose any of the springs, spring seats or balls in the choke linkage.
   b. Remove the screws securing the upper and lower brackets (**Figure 31** and **Figure 32**).
   c. Carefully separate the carburetors. Note how the throttle linkage is fitted. Ensure that the fuel hoses connecting the carburetors are not damaged.
2. Remove the 4 screws securing the diaphragm cover and remove the cover (**Figure 33**). Lift out the diaphragm spring (**Figure 34**).
3. Carefully lift out the diaphragm assembly (**Figure 35**).
4. Remove the 4 screws securing the float chamber and remove the chamber (**Figure 36**).
5. Carefully remove the main jet and washer (**Figure 37**).

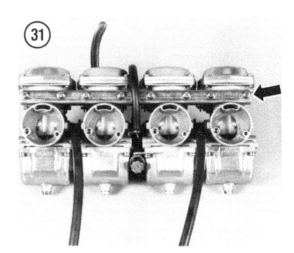

6. Slide the needle jet out of the carburetor body.
7. Push out the hinge pin securing the float assembly and lift out the float (**Figure 38**). The hinge pin is a slight press fit. It may be necessary to use a small punch to start the pin out of the carburetor body.
8. Use a socket and remove the needle valve assembly.

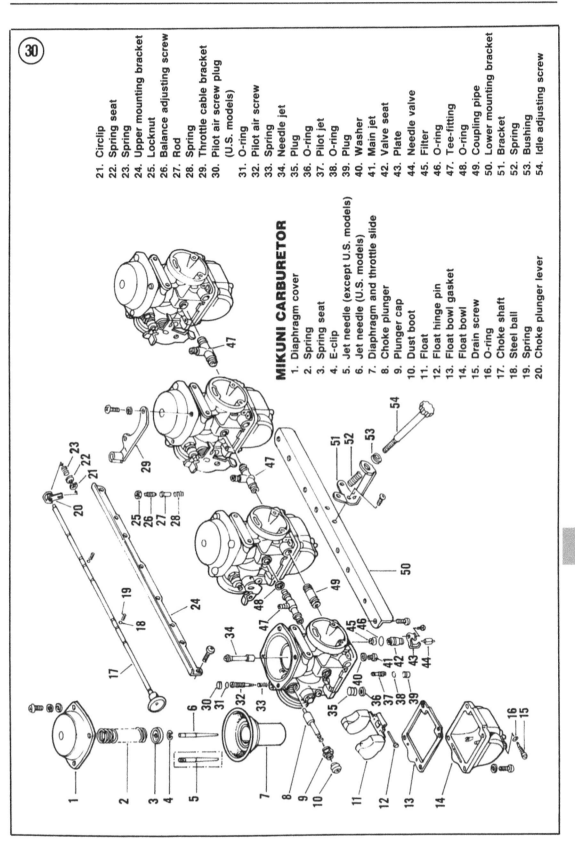

MIKUNI CARBURETOR

1. Diaphragm cover
2. Spring
3. Spring seat
4. E-clip
5. Jet needle (except U.S. models)
6. Jet needle (U.S. models)
7. Diaphragm and throttle slide
8. Choke plunger
9. Plunger cap
10. Dust boot
11. Float
12. Float hinge pin
13. Float bowl gasket
14. Float bowl
15. Drain screw
16. O-ring
17. Choke shaft
18. Steel ball
19. Spring
20. Choke plunger lever
21. Circlip
22. Spring seat
23. Spring
24. Upper mounting bracket
25. Locknut
26. Balance adjusting screw
27. Rod
28. Spring
29. Throttle cable bracket
30. Pilot air screw plug (U.S. models)
31. O-ring
32. Pilot air screw
33. Spring
34. Needle jet
35. Plug
36. O-ring
37. Pilot jet
38. O-ring
39. Plug
40. Washer
41. Main jet
42. Valve seat
43. Plate
44. Needle valve
45. Filter
46. O-ring
47. Tee-fitting
48. O-ring
49. Coupling pipe
50. Lower mounting bracket
51. Bracket
52. Spring
53. Bushing
54. Idle adjusting screw

11

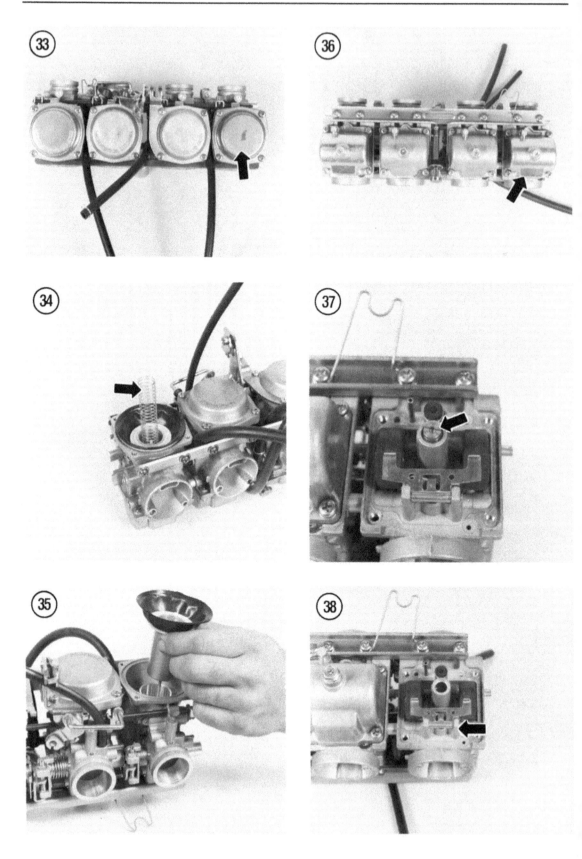

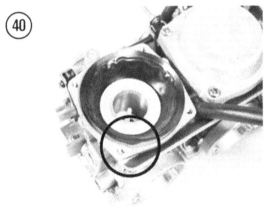

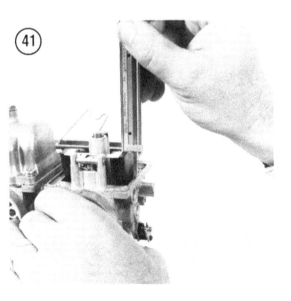

9. Remove the plug and O-ring covering the pilot jet and remove the jet.

10. Carefully turn the pilot air screw in and count the turns until the screw seats *lightly*. On U.S. models, it is necessary to first remove the plug and O-ring covering the air screw.

11. Perform *Cleaning and Inspection* as outlined in Chapter Seven of the main book.

12. Assembly is the reverse of these steps. Keep the following points in mind.

13. Install the needle jet so that the groove engages the locating pin as shown in **Figure 39**.

14. When installing the diaphragm assembly, ensure that the locating tab on the diaphragm is positioned in the locating notch (**Figure 40**).

15. Apply a small amount of blue Loctite (Lock N' Seal No. 2114) to the set screws securing the choke links.

16. Install the pilot air screw until it seats *lightly*, then turn it out the exact number of turns noted during removal. On U.S. models, install a new plug over the screw and seal the plug around the edges with a *small* amount of sealant.

17. Use a caliper and set the float level on each carburetor to 18.6 mm (0.73 in.) as shown in **Figure 41** without the float chamber gasket. Reduce the measurement approximately 1 mm if the gasket is still installed. If the float level is not as specified, carefully bend the tang on the float arm until the correct level is achieved.

18. If the choke shaft was removed, keep the following points in mind during installation:

a. Position the steel balls and springs in place in No. 2 and No. 3 carburetors.

b. Install the spring and spring seat in each choke plunger lever.

c. Lightly grease the shaft and install the shaft through each plunger lever, spring seat and spring.

d. Install the circlips to hold the shaft in position. Make sure the circlips are on the left side of each spring seat.

e. Make sure the choke shaft works smoothly and operates the choke plunger on each carburetor an equal amount.

11

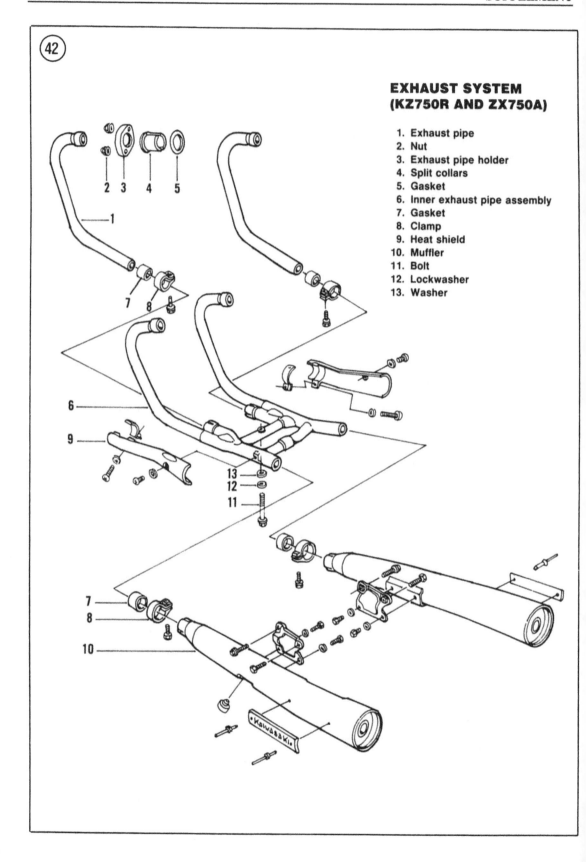

42

**EXHAUST SYSTEM
(KZ750R AND ZX750A)**

1. Exhaust pipe
2. Nut
3. Exhaust pipe holder
4. Split collars
5. Gasket
6. Inner exhaust pipe assembly
7. Gasket
8. Clamp
9. Heat shield
10. Muffler
11. Bolt
12. Lockwasher
13. Washer

AIR SUCTION VALVES

On 1982 models, a gasket was installed under each air suction valve. On 1983 models, the air suction valves were modified and the gaskets are no longer used. On 1982 models, replace the gaskets if the air valves are removed and the gasket is damaged in any way.

EXHAUST SYSTEM

On 1982 KZ750R and 1983-on ZX750A models, the mufflers can be separated from the exhaust pipes. Refer to *Exhaust System* in Chapter Seven of the main book to remove the exhaust system. To remove the mufflers, loosen the clamp bolt securing each muffler to the welded exhaust pipe assembly (**Figure 42**). Turn the mufflers slightly to help break the joint loose and pull the mufflers from the pipes.

FUEL INJECTION SYSTEM

Service Notes

When performing service to the turbocharger and its related systems, observe the following information.

1. An arrow mark is stamped on some parts. The arrow will show either the installed orientation or the direction of rotation of the part. If the arrow is used to show orientation, the part should be installed so that the part faces to the front of the motorcycle. If the arrow represents direction of rotation, install the part so that the arrow points in the direction of rotation.

2. Care should be taken when servicing the digital fuel injection system (DFI). Failure to observe the following notes can result in expensive DFI system damage:

   a. Do not disconnect the battery leads or any other electrical connector when the ignition switch is turned on or while the engine is running.

   b. The DFI system is designed to operate with a 12-volt battery source. Anything other than a 12-volt battery should not be used.

   c. Always remove the battery from the motorcycle before charging to prevent the DFI system from damage caused by excessive peak voltage.

   d. When washing the bike, do not direct a hard water spray on the electrical components, wiring, connectors or wiring harness connected to the DFI system.

   e. Do not reroute the DFI wiring harness. Locating the wires too close to other system leads could cause malfunctioning from external electrical noises.

   f. If a citizens band (CB) radio is mounted on the motorcycle, mount the antenna as far as possible from the DFT control box.

NOTE
After mounting a CB onto the motorcycle, check the engine operation at idle. It is possible that the fuel injection system could be affected by antenna operation.

3. Note the following throttle sensor information when servicing your motorcycle:

   a. The throttle sensor does not require any periodic maintenance.

   b. Do not attempt to alter the throttle sensor position. Special equipment and procedures are required for its adjustment.

   c. Refer all service or troubleshooting related to the throttle sensor to a qualified Kawasaki dealer.

Fuel Hoses

Refer to **Figure 15**.

WARNING
All turbo models are equipped with a fuel pump. When any fuel hose is disconnected, do not turn on the ignition switch. This will activate the fuel pump and cause fuel to pour from the disconnected hoses. This would present a severe fire hazard.

WARNING
When the fuel hoses are disconnected, a small amount of fuel may spill out because of residual pressure in the fuel line. Before disconnecting any hose, cover the end of the hose with a clean shop cloth.

11

When installing fuel hoses, note the following:

  a. Always install *new* high-pressure hose clamps once the old clamps are loosened.

  b. Install the fuel hoses with a minimum of bending to prevent fuel flow obstructions.

  c. Replace any hose that is bent or kinked.

  d. The inner surfaces of all high-pressure hoses are coated with a special material. If this material is damaged, the hose must be replaced.

  e. When installing clamps onto high-pressure hoses, observe the specifications in **Figure 43**.

Fuel Filter
Replacement

Refer to **Figure 44**.

1. Remove the right-hand side cover.

2. Disconnect the filter clamp around the filter.

3. Disconnect the fuel hoses at the filter and remove the filter.

4. Installation is the reverse of these steps; note the following:

  a. Install the fuel filter so that the "IN" mark is facing up.

  b. The hose from the fuel tank connects at the top of the filter and the hose to the fuel pump connects at the bottom of the filter.

Fuel Pump
Removal/Installation

1. Remove the battery case.

2. Disconnect the electrical connector at the fuel pump.

3. Disconnect the 2 fuel hoses at the pump (**Figure 15**). Discard the hose clamps.

4. Installation is the reverse of these steps, noting the following.

5. Refer to *Fuel Hoses* in this section of the supplement when reconnecting the fuel hoses to the fuel pump.

6. If a new fuel pump is being installed, perform the following:

  a. Reconnect the battery.

  b. Install the fuel pump onto the motorcycle. Connect the fuel hoses.

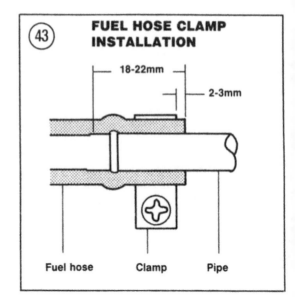

(43) **FUEL HOSE CLAMP INSTALLATION**

18-22mm

2-3mm

Fuel hose   Clamp   Pipe

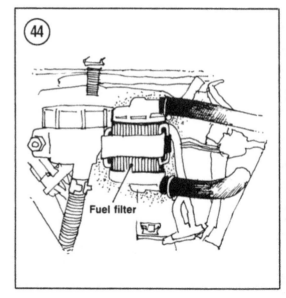

(44)

Fuel filter

  c. Reconnect the fuel pump electrical connector.

  d. Turn on the ignition switch to operate the fuel pump. Repeat this step a few times to bleed the air in the fuel line.

Surge Tank/
Air Filter Housing
Removal/Installation

Refer to **Figure 45** when removing and installing the surge tank and/or air filter housing.

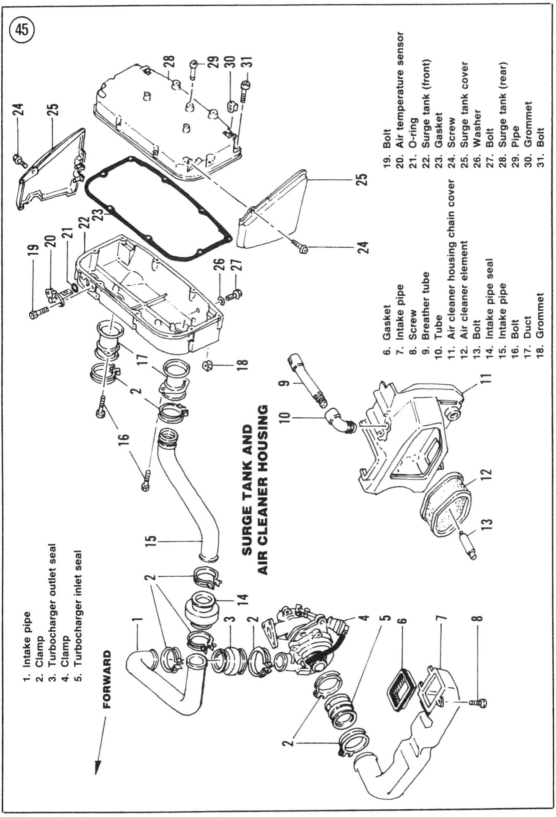

(45)

**SURGE TANK AND
AIR CLEANER HOUSING**

FORWARD

1. Intake pipe
2. Clamp
3. Turbocharger outlet seal
4. Clamp
5. Turbocharger inlet seal
6. Gasket
7. Intake pipe
8. Screw
9. Breather tube
10. Tube
11. Air cleaner housing chain cover
12. Air cleaner element
13. Bolt
14. Intake pipe seal
15. Intake pipe
16. Bolt
17. Duct
18. Grommet
19. Bolt
20. Air temperature sensor
21. O-ring
22. Surge tank (front)
23. Gasket
24. Screw
25. Surge tank cover
26. Washer
27. Bolt
28. Surge tank (rear)
29. Pipe
30. Grommet
31. Bolt

11

Digital Fuel Injection
Control Unit
Removal/Installation

1. Turn the ignition switch off.
2. Remove the seat.
3. Disconnect the white/red fuel injection system connector from the positive battery terminal.
4. The digital fuel injecion control unit is located on top of the rear fender (**Figure 46**). Carefully pull the dust cover away from the DFI connector.
5. Disconnect the 21-pin connector at the DFI control unit.
6. Remove the fasteners securing the DFI control unit and remove it.
7. Installation is the reverse of these steps.

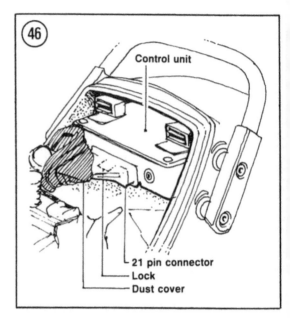

Turbocharger
Removal/Installation

Refer to **Figure 47** during this procedure.
1. Remove the fairing.
2. Drain the engine oil as described in this supplement and in Chapter Three of the main book.
3. Remove the external shift mechanism.
4. Disconnect and remove the air filter intake pipe at the turbocharger.
5. Remove the air cleaner housing.
6. Remove all intake pipes at the turbocharger assembly.
7. Loosen the clamps and remove the left- and right-hand muffler bolts. Then remove the mufflers.

NOTE
Store the mufflers so that they will not be damaged.

8. Remove the banjo bolt that connects the oil pipe to the oil pan. See **Figure 1**.
9. Remove the clamp that connects the oil hose to the sub-oil pan. See **Figure 1**.
10. Remove the exhaust pipe holder nuts at the cylinder heads. Repeat for each cylinder.
11. Remove the exhaust system connecting pipe bolt at the bottom of the pipe.
12. Remove the exhaust manifold (with front pipes attached), turbocharger and connecting pipe as an assembly.

13. Disconnect and remove the pipes at the turbocharger.
14. Remove the oil pipe at the turbocharger, if necessary.
15. Check that the turbocharger wheel turns freely when turned by hand. If not, replace the turbocharger.
16. Check all connecting pipes, hoses and clamps for wear or damage. Replace worn or damaged parts.
17. Installation is the reverse of these steps; note the following.
18. Install the oil pipe, elbow, oil hose and damper onto the turbocharger.
19. Install the exhaust manifold onto the turbocharger, using a new gasket.
20. Install the connecting pipe and a new gasket onto the turbocharger. Install the attaching bolts finger-tight.
21. Place the turbocharger/connecting pipe assembly into position in the motorcycle. Install the attaching bolts finger-tight.
22. Attach the oil pipe to the engine using the banjo pipe and 2 sealing washers. Install the screen in the banjo bolt in the correct direction before installing the bolt.
23. Tighten the installed components in the following order:
   a. Exhaust pipe holders and clamps.
   b. Install the connecting pipe on the oil pan.

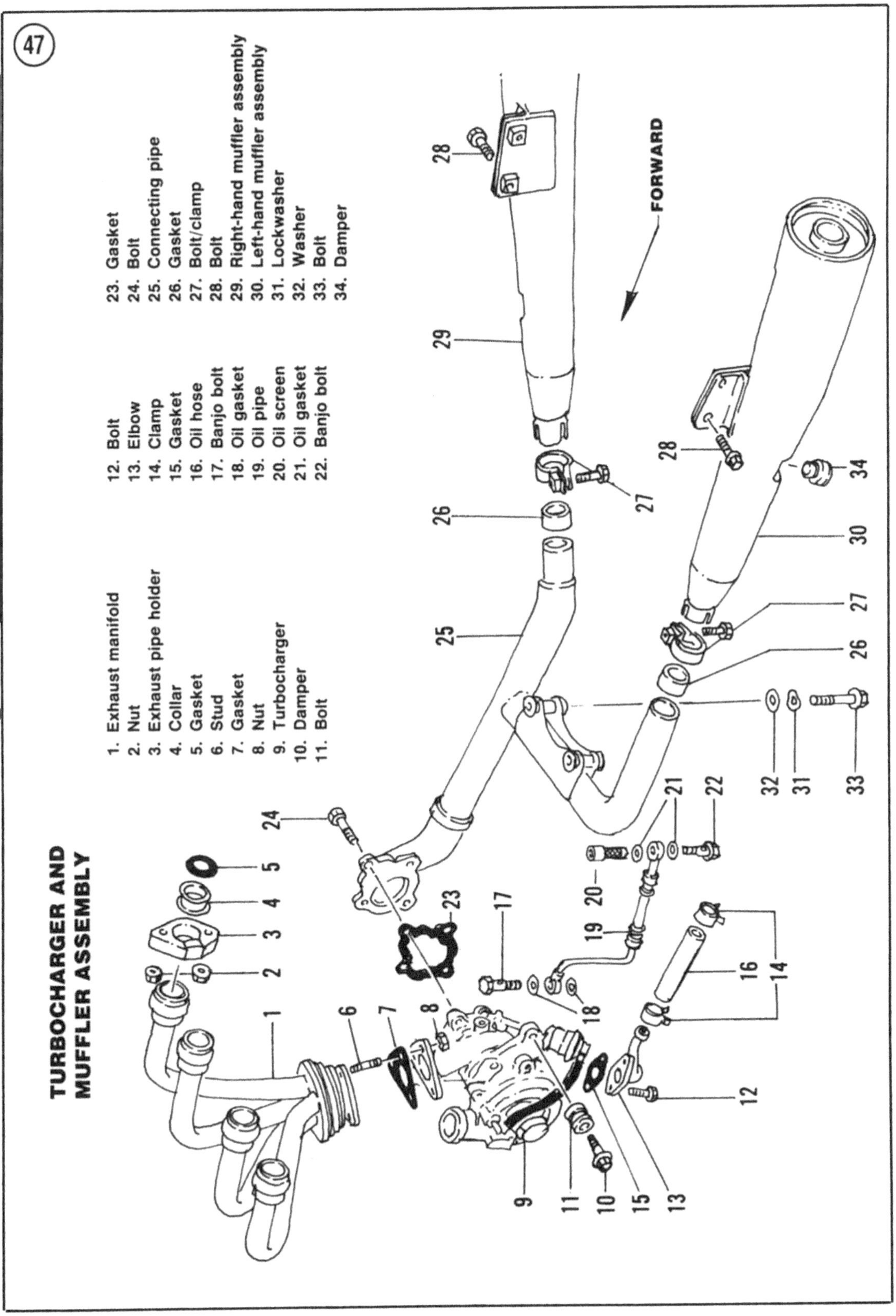

(47)

TURBOCHARGER AND MUFFLER ASSEMBLY

FORWARD

1. Exhaust manifold
2. Nut
3. Exhaust pipe holder
4. Collar
5. Gasket
6. Stud
7. Gasket
8. Nut
9. Turbocharger
10. Damper
11. Bolt
12. Bolt
13. Elbow
14. Clamp
15. Gasket
16. Oil hose
17. Banjo bolt
18. Oil gasket
19. Oil pipe
20. Oil screen
21. Oil gasket
22. Banjo bolt
23. Gasket
24. Bolt
25. Connecting pipe
26. Gasket
27. Bolt/clamp
28. Bolt
29. Right-hand muffler assembly
30. Left-hand muffler assembly
31. Lockwasher
32. Washer
33. Bolt
34. Damper

11

c. Install the turbocharger onto the crankcase.

d. Install the banjo bolt on the oil pan.

e. Tighten the connecting pipe and the turbocharger.

f. Refill the engine oil.

Engine Temperature Sensor Testing

1. Turn the ignition switch off.

2. Disconnect the electrical lead from the engine temperature sensor lead.

3. Connect one ohmmeter lead to the sensor terminal and the other lead to a good engine ground.

4. Set the ohmmeter at R×1K. The correct reading at 68° F is 2.0-3.0K ohms. If this reading is incorrect, replace the engine temperature sensor.

5. The engine temperature sensor is installed in the cylinder on the intake side (**Figure 48**). To remove it, disconnect the electrical connector at the sensor and unscrew it with a wrench or socket. Install it and tighten to 1.3 mkg (9.5 ft.-lb.).

Main/Pump Relay Testing

The relay is comprised of 2 parts: the main and fuel pump relays. This procedure describes separate test procedures for both relays. The following equipment will be required to perform the test procedures:

a. 12-volt battery.

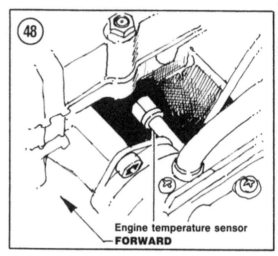

Engine temperature sensor
FORWARD

MAIN RELAY INSPECTION (49)

Test	Battery connections Positive lead	Negative lead	Test results*
1	#86c	#85	0 ohms
2	#85	#86c	Infinity

* Check continuity between terminals #88z and #88a when performing Tests 1 and 2.

FUEL PUMP RELAY INSPECTION

Test	Battery connections Positive lead	Negative lead	Test results**
1	#86	#85	0 ohms
2	#85	#86	Infinity
3	#86a	#85	0 ohms
4	#85	#86a	Infinity
5	#86b	#85	0 ohms

** Check continuity between terminals #88y and #88d when performing Tests 1-5.

RELAY TERMINALS

88d   86c   86   88b

88z   88a

86a   88y   88c

85   86b

b. 2 test leads.

c. Test light (12 volt, 3-3.5 watt light).

d. Ohmmeter.

Remove the relay from the left-hand outer frame tube behind the surge tank.

Main relay check

Referring to **Figure 49**, connect a 12-volt battery to the terminals specified and check the continuity between terminals 88a and 88z. If the readings are as shown in **Figure 49**, the main relay is in good condition. If not, replace the relay assembly.

Fuel pump relay test 1

Referring to **Figure 49**, connect a 12-volt battery to the terminals specified and check the continuity between terminals 88d and 88y. If the readings are as shown in **Figure 49**, the fuel pump relay is in good condition. If not, replace the relay assembly.

Fuel pump relay test 2

Refer to **Figure 49**.

1. Connect the battery positive (+) cable to the relay No. 85 terminal.

2. Connect the test light between the battery negative (-) cable and the No. 86b relay terminal.

CAUTION
Failure to install the test light in the test circuit will damage the relay resistor.

3. Check the continuity between the No. 88d and No. 88y terminals as in Test 1. Interpret results as follows:

   a. If the test light goes on and the relay does not show continuity between the relay terminals, the fuel pump relay is good.

   b. If the test light does not go on and/or the relay teminals show continuity, replace the relay assembly.

FUEL SYSTEM TESTING

Leak Check

Referring to **Figure 50**, check all connections for fuel leakage.

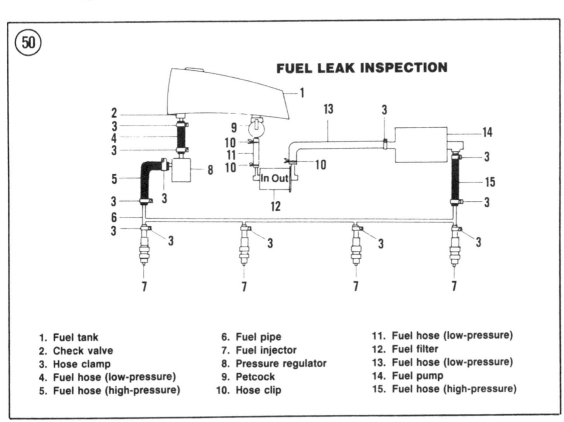

FUEL LEAK INSPECTION

1. Fuel tank
2. Check valve
3. Hose clamp
4. Fuel hose (low-pressure)
5. Fuel hose (high-pressure)
6. Fuel pipe
7. Fuel injector
8. Pressure regulator
9. Petcock
10. Hose clip
11. Fuel hose (low-pressure)
12. Fuel filter
13. Fuel hose (low-pressure)
14. Fuel pump
15. Fuel hose (high-pressure)

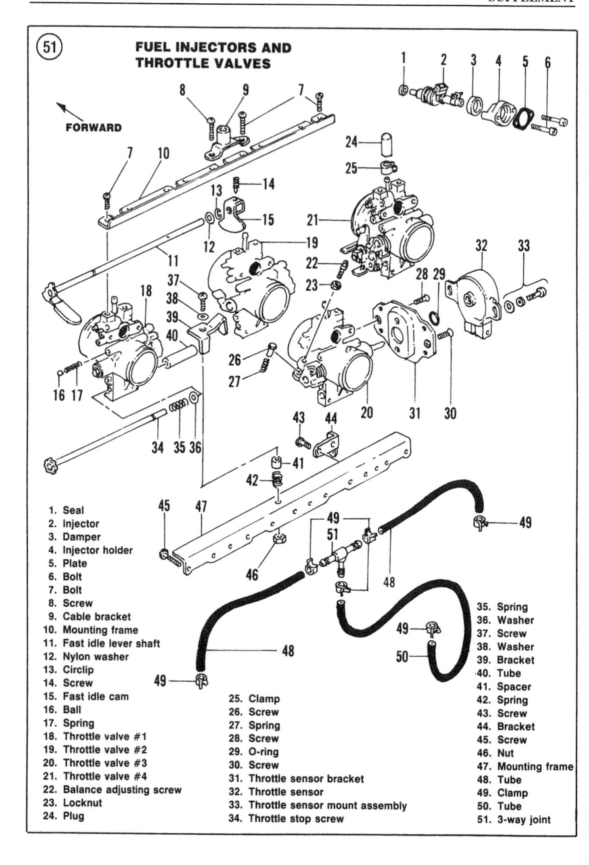

(51) **FUEL INJECTORS AND THROTTLE VALVES**

FORWARD

1. Seal
2. Injector
3. Damper
4. Injector holder
5. Plate
6. Bolt
7. Bolt
8. Screw
9. Cable bracket
10. Mounting frame
11. Fast idle lever shaft
12. Nylon washer
13. Circlip
14. Screw
15. Fast idle cam
16. Ball
17. Spring
18. Throttle valve #1
19. Throttle valve #2
20. Throttle valve #3
21. Throttle valve #4
22. Balance adjusting screw
23. Locknut
24. Plug

25. Clamp
26. Screw
27. Spring
28. Screw
29. O-ring
30. Screw
31. Throttle sensor bracket
32. Throttle sensor
33. Throttle sensor mount assembly
34. Throttle stop screw

35. Spring
36. Washer
37. Screw
38. Washer
39. Bracket
40. Tube
41. Spacer
42. Spring
43. Screw
44. Bracket
45. Screw
46. Nut
47. Mounting frame
48. Tube
49. Clamp
50. Tube
51. 3-way joint

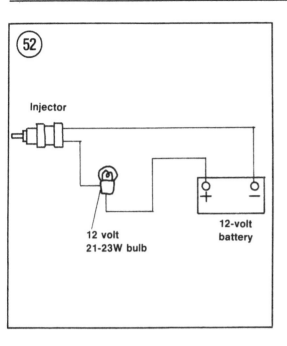

Injector

12 volt
21-23W bulb

12-volt
battery

Fuel Pump Inspection

1. Disconnect the electrical connector at the fuel pump.
2. Connect a 12-volt battery to the connector as follows:
   a. Positive (+): Orange/black lead.
   b. Negative (-): Black/yellow lead.
3. The fuel pump should operate when the leads are connected. If not, replace the fuel pump as described in this chapter.

Injector Inspection

This procedure must be performed with the engine cold. Refer to **Figure 51**.
1. Turn the ignition switch off.
2. Remove the fuel tank.
3. Disconnect the electrical connector from the injector to be tested.
4. Set an ohmmeter on the ohms×1 scale. Attach the leads to the injector leads. The correct reading is 1.8-3.0 ohms.
5. If the meter reading is incorrect, replace the injector. If the reading was correct, proceed to Step 6.
6. Attach a test light and 12-volt battery to the injector as shown in **Figure 52**.

CAUTION
Failure to install the test light in the test circuit will damage the relay resistor.

7. The injector should click every time the negative battery lead contacts the battery. If the injector does not click, it must be replaced.

NOTE
Have the fuel injectors replaced by a Kawasaki dealer as special tools and procedures are required during reassembly.

Pressure Regulator Check

This procedure requires a special pressure gauge and adapter. Refer all service to a Kawasaki dealer.

Air Leak Check

Check the connections between the parts shown in **Figure 53** for air leaks. Replace all hoses and clamps or other worn or damaged parts as required.

Surge Tank Draining

Periodically remove the surge tank drain plugs (**Figure 54**) and drain the tank of water and oil build-up. At the same time, check the drain plugs for wear or damage and replace them if necessary. Reinstall the drain plugs and tighten securely.

CAUTION
A damaged or loose drain bolt will allow air to be drawn into the system and the fuel injection system will not operate properly.

EVAPORATIVE EMISSION CONTROL SYSTEM (1984-ON CALIFORNIA MODELS)

An evaporative emission control system is installed on all 1984 and later models sold in California. See **Figure 55** (turbo) or **Figure 56** (non-turbo). The evaporative emission control system routes fuel vapors collected from the fuel system into the engine during operation and stores the fuel vapors in a canister when the engine is stopped.

Service Notes

No periodic adjustments are required to maintain the evaporative emission control

11

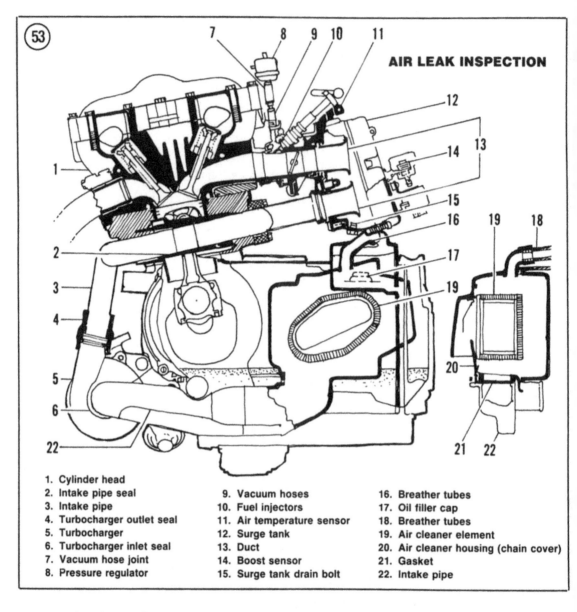

53   AIR LEAK INSPECTION

1. Cylinder head
2. Intake pipe seal
3. Intake pipe
4. Turbocharger outlet seal
5. Turbocharger
6. Turbocharger inlet seal
7. Vacuum hose joint
8. Pressure regulator

9. Vacuum hoses
10. Fuel injectors
11. Air temperature sensor
12. Surge tank
13. Duct
14. Boost sensor
15. Surge tank drain bolt

16. Breather tubes
17. Oil filler cap
18. Breather tubes
19. Air cleaner element
20. Air cleaner housing (chain cover)
21. Gasket
22. Intake pipe

system other than periodic inspection and the replacement of worn or damaged parts. When checking the system, observe the following.

WARNING
When servicing the evaporative emission control system, fuel vapors will be present. Always perform service work with the ignition switch turned OFF and in a well ventilated area free of any source of flame or sparks.

1. If the canister becomes contaminated with gasoline, water, solvent or any other liquid, the canister must be replaced.

2. Always reconnect hoses to the part from which they were removed. Refer to **Figure 55** or **Figure 56**.

3. After completing a repair job, check the evaporative system hoses to make sure they have not been pinched or cut.

4. Always hold the liquid/vapor separator in its installed position to prevent fuel from flowing into or out of the separator housing.

5. Inspect the canister and the liquid/separator canister for cracks or other damage; replace if necessary.

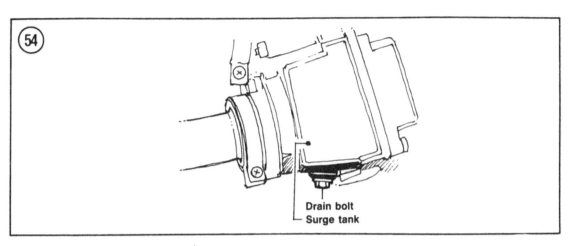

(54)

Drain bolt
Surge tank

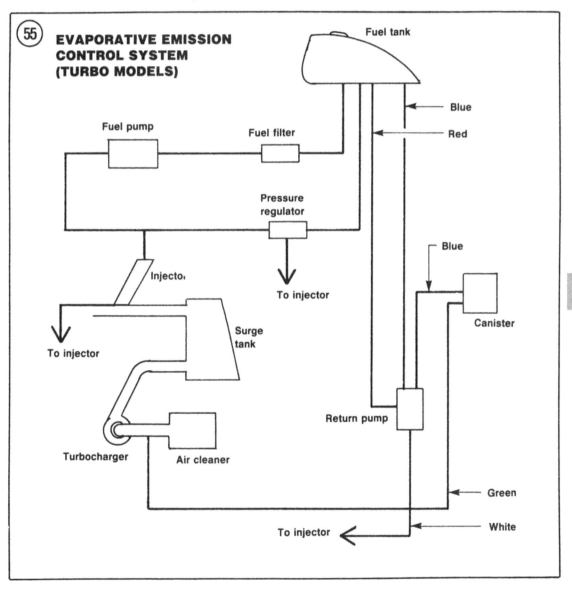

(55) **EVAPORATIVE EMISSION
CONTROL SYSTEM
(TURBO MODELS)**

Fuel tank

Blue

Red

Fuel pump

Fuel filter

Pressure
regulator

Blue

Injector

To injector

Canister

To injector

Surge
tank

Return pump

Turbocharger

Air cleaner

Green

To injector

White

11

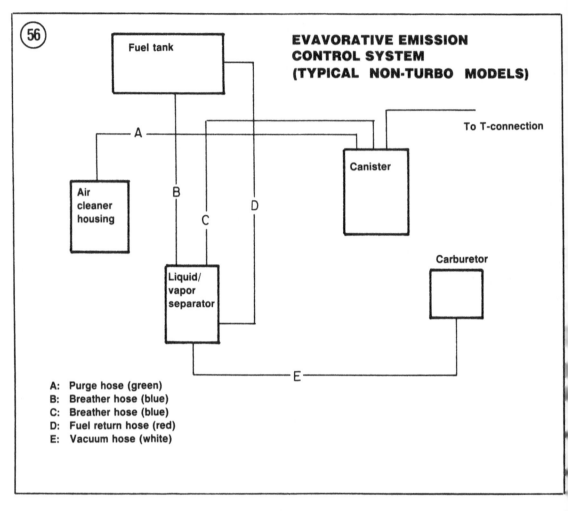

EVAVORATIVE EMISSION CONTROL SYSTEM (TYPICAL NON-TURBO MODELS)

A: Purge hose (green)
B: Breather hose (blue)
C: Breather hose (blue)
D: Fuel return hose (red)
E: Vacuum hose (white)

Table 18 MIKUNI CARBURETOR SPECIFICATIONS

Carburetor type	BS34
Main jet	110
Needle jet	Y-9
Jet needle	
Non-U.S. models	
1982 KZ750R	4BE3
1983 KZ/Z750L	4BE3
1983 ZX750A	4BC6
U.S. models	
1982 KZ750R	4BE04
1983 KZ/Z750L	4BE04
1983 ZX750A	4BC7
Jet needle clip position	
Non-U.S. models	3rd groove from the top
U.S. models	Fixed
Pilot jet	37.5
Pilot air screw	
U.S. models	Non-adjustable
Non-U.S. models	2 turns out
Float level	16.6-20.6 mm
	(0.65-0.81 in.)

CHAPTER EIGHT

ELECTRICAL SYSTEM

FUSES

Fuse Replacement (ZX750)

1. Remove the left-hand side cover and remove the screws (**Figure 57**) securing the junction box cover. The circuits controlled by each fuse are shown on the label attached to the junction box cover.

2. Refer to **Figure 58** and carefully remove the desired fuse. Refer to **Figure 59** to determine if the fuse is blown.
3. Replace any blown fuses. Press the new fuse fully into the terminals to ensure a good connection.
4. Always investigate the reason for a blown fuse and correct any short circuit or defective electrical component in the motorcycle wiring.
5. Install the junction box cover and motorcycle side cover.

BATTERY

Removal/Installation
(1982 KZ750R; 1983 ZX750A)

1. Remove the battery cover (**Figure 60**).

11

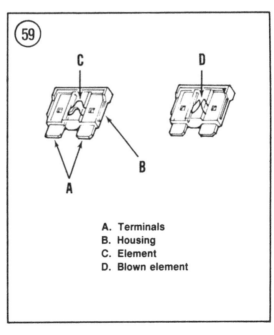

A. Terminals
B. Housing
C. Element
D. Blown element

2. Disconnect the battery electrolyte level sensor wire (**Figure 61**).

3. Complete the battery removal and installation as outlined under *Battery* in Chapter Eight of the main book.

IGNITION SYSTEM
(1983-ON ZX750 MODELS)

The ignition system on 1983-on ZX750 models functions essentially the same as other models. However, the mechanical advance mechanism is replaced by an electronic advance circuit within the IC igniter unit. Certain functions of the ignition circuit are routed through the main electrical junction box as shown in **Figure 62**.

Operation and Troubleshooting

Several circuits interact within the ignition system. If starting troubles or apparent ignition malfunctions are encountered, it is necessary to carefully and systematically check the associated circuits and switches. The following conditions must be present before the motorcycle will start:

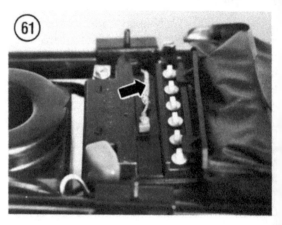

   a. The ignition switch must be in the ON position.

   b. The engine stop switch must be in the RUN position.

   c. The gear selector must be in NEUTRAL and/or the clutch lever must be pulled in.

If the engine turns over but will not start and the above switch positions are correct, refer to **Figure 62** and perform the following procedure. If the engine will not turn over, check the starting circuit as outlined in this supplement.

1. Turn off the ignition and disconnect the 10-pin connector from the IC igniter unit.

2. Connect the positive (+) lead of a voltmeter to the red lead in the male half (motorcycle side) of the igniter unit wiring harness. Connect the voltmeter negative (-) lead to the black/yellow wire in the male half of the connector.

3. Turn on the ignition switch. The voltmeter must indicate battery voltage. If battery voltage is present, check the following items:

   a. Carefully check all connectors. All connections must be clean and tight.

   b. Sidestand switch.

   c. Starter lockout switch (clutch lever).

   d. Neutral switch.

   e. Starter circuit diodes in the junction box.

   f. Have the IC igniter unit checked by dealer.

4. If battery voltage is not present in Step 2, check the following items:

   a. Check battery connections and associated wiring for shorts or open connections. Make sure all contacts are clean and all connections are tight.

   b. Engine stop switch.

   c. Ignition switch.

   d. Main fuse (30 amp).

5. Check the continuity of the associated switches as outlined under *Switch Continuity Checks* in this supplement.

6. Reconnect the IC igniter 10-pin connector.

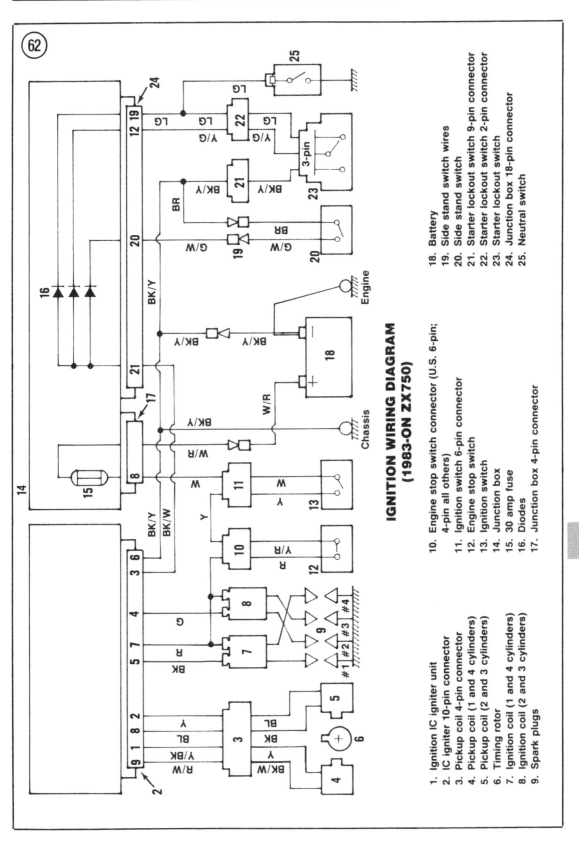

IGNITION WIRING DIAGRAM (1983-ON ZX750)

1. Ignition IC igniter unit
2. IC igniter 10-pin connector
3. Pickup coil 4-pin connector
4. Pickup coil (1 and 4 cylinders)
5. Pickup coil (2 and 3 cylinders)
6. Timing rotor
7. Ignition coil (1 and 4 cylinders)
8. Ignition coil (2 and 3 cylinders)
9. Spark plugs
10. Engine stop switch connector (U.S. 6-pin; 4-pin all others)
11. Ignition switch 6-pin connector
12. Engine stop switch
13. Ignition switch
14. Junction box
15. 30 amp fuse
16. Diodes
17. Junction box 4-pin connector
18. Battery
19. Side stand switch wires
20. Side stand switch
21. Starter lockout switch 9-pin connector
22. Starter lockout switch 2-pin connector
23. Starter lockout switch
24. Junction box 18-pin connector
25. Neutral switch

11

Ignition Advance Check

To check the operation of the electronic advance circuit within the IC igniter unit perform the following procedure.

1. Remove the screws securing the ignition cover and remove the cover to gain access to the ignition pickup coils and rotor. Take care not to damage the cover gasket.

2. Connect a stroboscopic timing light to No. 1 or No. 4 spark plug according to the manufacturer's instructions.

3. Start the engine and run at an idle. Aim the timing light at the ignition rotor. The "F" mark should align with the timing mark on the engine (**Figure 63**).

4. Increase the engine above the following rpm range:

    a. 1983-on ZX750A: 3,900 rpm.

    b. 1984-on ZX750E: 3,600 rpm.

    c. 1984-on KZ/Z750L: 3,800 rpm.

With the engine speed above the specified rpm range, the advance timing mark must be aligned with the timing mark on the engine. If the ignition timing does not advance correctly, replace the IC igniter unit.

5. Install the ignition cover.

IC Igniter Check

A very accurate ohmmeter is necessary to check out the IC igniter. Refer the IC igniter to your Kawasaki dealer or a qualified specialist for this testing.

Pickup Coil Removal/Installation

1. Remove the screws securing the ignition cover on the right-hand side of the engine. Carefully remove the cover so the gasket is not damaged.

2. Remove the right-hand side cover and disconnect the pickup coil connector (**Figure 64**).

3. Remove the screws securing the pickup coil assembly to the engine and remove the assembly (**Figure 65**).

4. Installation is the reverse of these steps.

SWITCH CONTINUITY CHECKS

Refer to **Figure 62** for the following procedures.

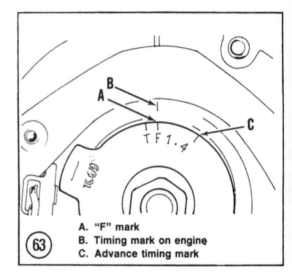

A. "F" mark
B. Timing mark on engine
C. Advance timing mark

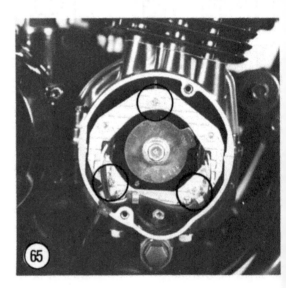

Engine Stop Switch

1. Disconnect the 4-or 6-pin connector from the engine stop switch.
2. Connect the leads of an ohmmeter or continuity device between the red and yellow/red wires in the switch half of the connector.
3. Continuity must be present when the switch is in the RUN position. No continuity must be present when the switch is OFF. Replace the switch assembly if defective.
4. Reconnect the engine stop switch connector.

Sidestand Switch

1. Disconnect the wires from the sidestand switch (**Figure 66**).
2. Connect the leads of an ohmmeter or continuity device to the switch wires.
3. Continuity must be present when the stand is up and not present when the stand is down. Replace the switch if defective.
4. Reconnect the sidestand switch wires.

Clutch Lever and Neutral Switch

1. Disconnect the 9-pin and 2-pin connectors from the left handlebar switch assembly.

2. Connect one lead of an ohmmeter or continuity device to the black/yellow wire in the switch half of the 9-pin connector. Connect the other meter lead to the yellow/green wire in the switch half of the 2-pin connector. Continuity must be present when the clutch lever is pulled in and not present when the lever is released. Replace the switch if defective.
3. Connect the leads of the ohmmeter to the yellow/green and light green wires in the switch half of the 2-pin connector. Continuity must be present when the clutch lever is released. Replace the switch if defective.
4. Connect the leads of the ohmmeter between the light green wire on the motorcycle half of the 2-pin connector and a good chassis ground. Continuity must be present when the gear shift lever is in NEUTRAL and not present when any gear is selected. Replace the neutral switch if defective.
5. Reconnect the 2-pin and 9-pin connectors.

Ignition Switch

1. Disconnect the 6-pin connecter from the ignition switch wiring harness.
2. Connect an ohmmeter or continuity device between the following pairs of wires. Continuity must be present when the switch is ON and not present when the switch is OFF. Replace the switch if defective:
   a. Brown to white.
   b. Brown to yellow.
   c. White to yellow.
   d. Black to red.
   e. White/black to orange/green (U.S. models only).
3. Connect an ohmmeter or continuity device between the following pairs of wires. Continuity must be present when the switch is in PARK and not present when the switch is OFF. Replace the switch if defective:
   a. White to red.
   b. White/black to orange/green (U.S. models only).

CHARGING SYSTEM

Refer to **Figure 67** to remove and install the rectifier/regulator on 1983 ZX750A models.

11

ALTERNATOR ROTOR

A 12 mm bolt is used to secure the alternator rotor on 1983 models. Torque the bolt to 13.0 mkg (94 ft.-lb.).

STARTING SYSTEM
(1983-ON ZX750 MODELS)

The starting system on 1983-on ZX750 models operates essentially the same as other models. Additional associated circuits are routed through the junction box and other switches that may affect the starting system operation (**Figure 68**).

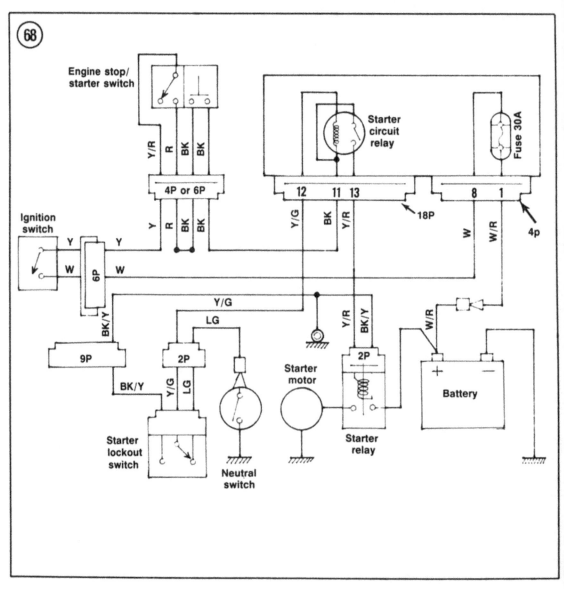

Operation and Troubleshooting

Several circuits interact within the starting system. If starting malfunctions are encountered, it is necessary to carefully and systematically check the associated circuits and switches. The following conditions must be present before the motorcycle will start:

a. The ignition switch must be ON.

b. The engine stop switch must be in the RUN position.

c. The gear selector must be in NEUTRAL and/or the clutch lever must be pulled in.

If the engine will not turn over and the above switch positions are correct, check the starting circuit relay as outlined under *Junction Box* in this supplement. If the starting circuit relay is good, refer to **Figure 68** and perform the following procedure to determine if battery power is reaching the starter relay (solenoid).

1. Remove the left-hand side cover and disconnect the 2-pin connecter from the starter relay (**Figure 69**).

2. Connect the positive (+) lead of a voltmeter to the yellow/red wire in the female half (motorcycle side) of the starter circuit wiring harness. Connect the negative (-) lead to the black/yellow wire in the female half of the connector.

3. Turn on the ignition switch. Place the gear shift lever in NEUTRAL and/or pull in the clutch lever. The voltmeter must indicate battery voltage. If battery voltage is present, check the starter relay as outlined in this section of the supplement.

4. If battery voltage is not present, check the following items:

a. Check battery connections and associated wiring for shorts or open connections. Make sure all contacts are clean and all connections are tight.

b. Main fuse.

c. Starting circuit relay. Refer to *Junction Box* in this supplement.

d. Perform continuity checks of the neutral switch, starter lockout switch, ignition switch and engine stop switch as described in this section of the supplement.

Starter Relay Check

Refer to **Figure 68** for this procedure.

1. Remove the left-hand side cover and disconnect the 2-pin connecter from the starter relay (**Figure 69**).

2. Connect one end of a jumper wire to the battery positive (+) terminal.

3. Momentarily connect the other end of the jumper wire to the yellow/red wire in the male terminal on the starter relay. Check the female half of the connector to make sure of the location of the yellow/red wire terminal. When the jumper wire is touched to the yellow/red terminal the starter relay should "clack" loudly and the engine should turn over.

4. If a loud "clack" can be heard from the relay but the engine does not turn over, remove the starter motor. Replace the relay if no sound is heard when the terminal is jumped to battery voltage.

5. Reconnect the 2-pin starter relay connector and install the motorcycle side cover.

STARTER MOTOR

A new starter motor with permanent magnet field coils is fitted to all 1982 and later models (**Figure 70**). Operation and service of this starter model is essentially the same as the unit installed on earlier models. Refer to *Starting System* in Chapter Eight of the main book.

REAR BRAKE LIGHT SWITCH REPLACEMENT

Refer to **Figure 71** for this procedure.

1. Remove the right-hand motorcycle side cover.

2. Disconnect the wires from the brake light switch.

3. Unhook the spring connecting the light switch plunger to the brake pedal.

4. Loosen the locknuts securing the switch to the frame and remove the switch.

5. Installation of the new switch is the reverse of these steps. Adjust the switch position in the frame so that the brake light goes on when the brake pedal is pressed.

11

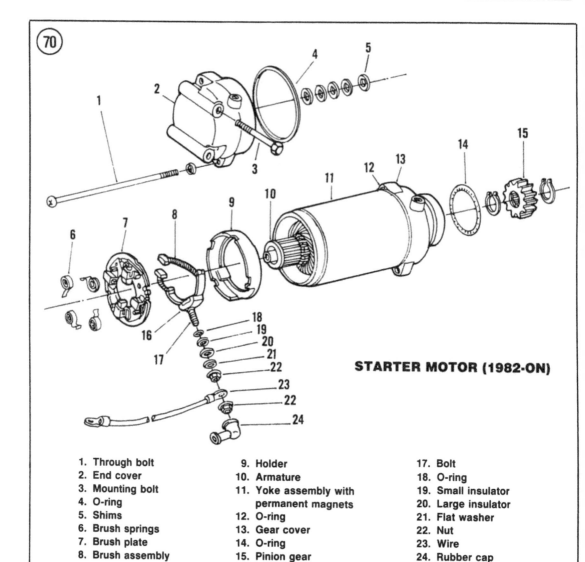

STARTER MOTOR (1982-ON)

1. Through bolt
2. End cover
3. Mounting bolt
4. O-ring
5. Shims
6. Brush springs
7. Brush plate
8. Brush assembly
9. Holder
10. Armature
11. Yoke assembly with permanent magnets
12. O-ring
13. Gear cover
14. O-ring
15. Pinion gear
16. Insulator
17. Bolt
18. O-ring
19. Small insulator
20. Large insulator
21. Flat washer
22. Nut
23. Wire
24. Rubber cap

FRONT BRAKE LIGHT SWITCH REPLACEMENT

Refer to **Figure 72** for this procedure.
1. Disconnect the switch wires from the switch unit.
2. Remove the screw securing the switch unit to the master cylinder assembly and remove the switch.
3. Installation of the new switch is the reverse of these steps.

JUNCTION BOX (1983-ON ZX750 MODELS)

The junction box contains the fuses, 3 relays, an accessory 2-pin connector and 2 diode assemblies. See **Figure 73** for a wiring diagram of the junction box. The following procedures check the 3 relays and diode assemblies. Refer to the applicable procedure in this supplement to check the individual circuitry associated with the ignition, starting

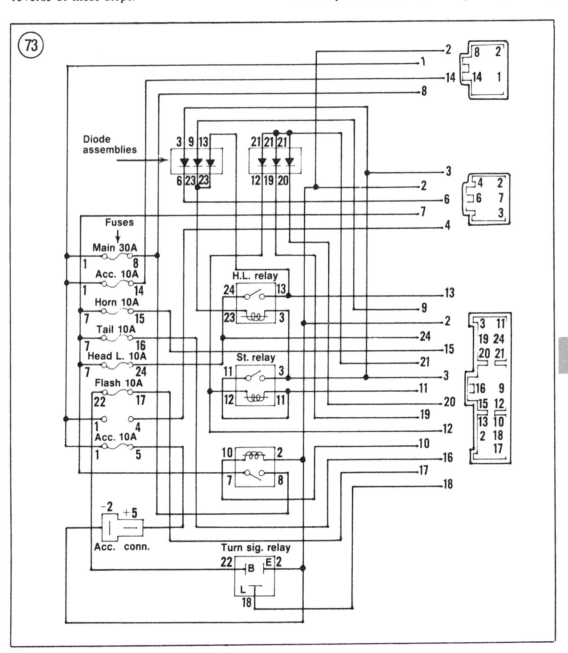

11

and headlight systems. Refer to *Fuses* in this supplement to check and replace the fuses.

Diode Assembly Check/Replacement

1. Remove the left-hand side cover. Remove the screws securing the junction box cover and remove the cover (**Figure 74**).
2. Refer to **Figure 75** and carefully unplug one diode assembly from the junction box panel.
3. Connect the leads of an ohmmeter to each diode one at a time, then reverse the leads and check each diode one at a time. The meter should indicate very low resistance (continuity) in one direction and very high resistance with the meter leads connected in the opposite direction.
4. If any diode shows low or high resistance in both directions, replace the diode assembly.
5. Install the diode assembly into the junction box panel.
6. Install the junction box cover and the side cover.

Relay Check/Replacement

This procedure applies to the main, starting circuit and headlight circuit relays. All relays are the same and interchangeable.
1. Remove the left-hand side cover. Remove the screws securing the junction box cover and remove the cover (**Figure 74**).
2. Refer to **Figure 75** and carefully unplug one relay assembly from the junction box panel.
3. Connect an ohmmeter or continuity device to terminals 3 and 4 on the relay as shown in **Figure 76**.
4. Connect a 12 volt power supply (jumper wires from the battery) to terminals 1 and 2 as shown in **Figure 76**.

> *CAUTION*
> *Take care when connecting the ohmmeter to the relay terminals. Ensure that the ohmmeter is not connected to the 12 volt power supply or the meter may be damaged.*

5. When voltage is applied to the relay terminals, the relay must energize and the ohmmeter must indicate continuity. If

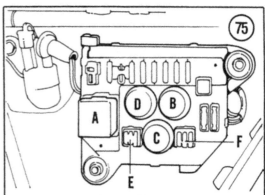

JUNCTION BOX
(1983-ON ZX750 MODELS)

A. Turn signal relay
B. Main relay
C. Starter circuit relay
D. Headlight circuit relay
E. Headlight relay diode assembly
F. Starter circuit relay diode assembly

continuity is not present, the relay is defective.
6. Carefully install the relay in the junction box panel. Install the junction box cover and side cover.

Junction Box Circuitry Check/Removal/Installation

1. To check the circuitry within the junction box, remove the bolts securing the junction box and carefully pull the box out far enough to gain access to the connectors. Take care not to lose the grommets and spacers on the junction box mounting bolts.

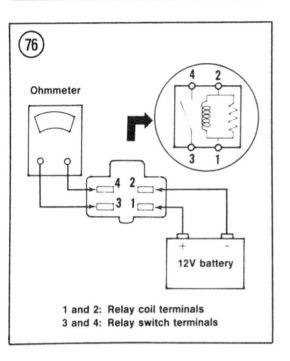

Ohmmeter

4 2
3 1

12V battery

1 and 2: Relay coil terminals
3 and 4: Relay switch terminals

2. Carefully squeeze the locking tabs on the connectors and disconnect the 3 connectors from the back of the junction box.
3. Refer to **Figure 73** and perform continuity checks of the circuitry within the junction box. Continuity must be present on all terminals with the same numbers and not present on terminals with different numbers. Replace the junction box if any open or shorted circuits are present in the circuitry.
4. Make sure all terminals in the junction box and mating connectors are clean and tight. Make sure the connectors fit securely when connected to the back of the junction box. Install the junction box and the cover. Install the side cover.

LIGHTING SYSTEM
(1983-ON U.S. AND CANADA ZX750)

The 1983-on ZX750 models delivered in the U.S. and Canada contain additional circuitry and components in the junction box (**Figure 77**). These additional circuits allow the headlight to stay off when the ignition is first turned ON until the engine starter switch is engaged. The headlight then stays on until the ignition switch is turned OFF. The headlight will also go out whenever the starter

switch is pushed to start the engine if the engine is stalled.

Lighting Relay Check

Check the lighting relay as outlined under *Junction Box* in this section of the supplement.

Reserve Lighting Unit Replacement

Remove the fairing to gain access to the reserve lighting unit (**Figure 78**).

Signal Light Flasher Replacement

1. Remove the left-hand side cover.
2. Remove the screws securing the junction box cover and remove the cover (**Figure 74**).
3. Remove the flasher unit from the junction box panel (**Figure 79**).
4. Install the new flasher. Install the junction box cover and the side cover.

TACHOMETER/VOLTMETER

The combination tachometer/voltmeter uses 2 separate circuits with one common pointer. When the TACHO/VOLT switch is pressed, the meter indicates battery voltage. The instrument normally indicates engine revolutions. The engine speed is picked up from the primary windings in one ignition coil.

Visual Inspection

1. Remove the fuel tank as outlined in Chapter Seven of the main book to gain access to the white 9-pin connector under the tank. Make sure the connection is clean and tight.
2. Remove the fairing and the back cover from the instrument panel.
3. Check that all wire connections on the instruments are tight.
4. Inspect the 4 rubber dampers in the meter mounting bracket; replace them if they are not in good condition.
5. Ensure that all mounting fasteners are tight.

Meter Test

The test assumes that the battery and electrical system are in good operating

11

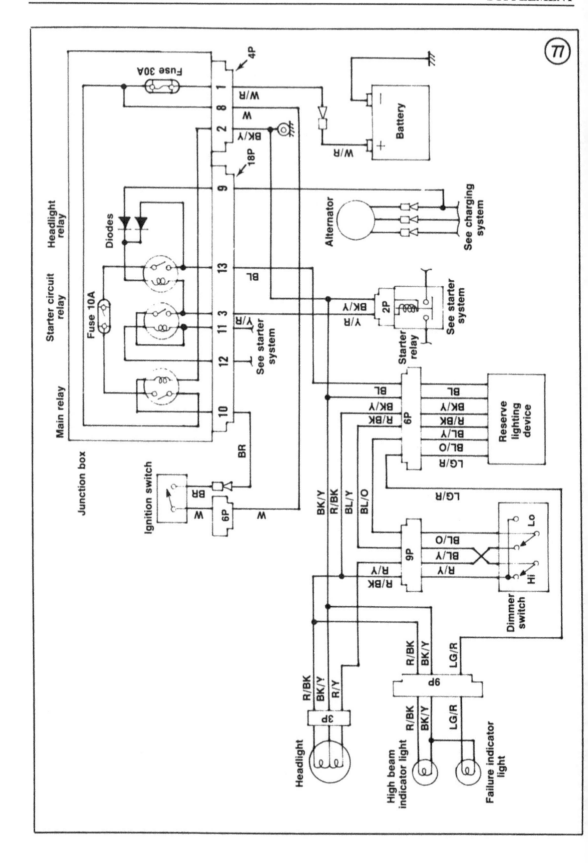

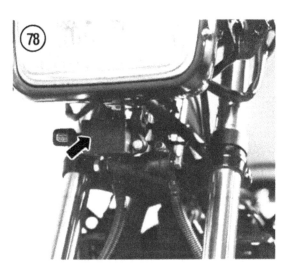

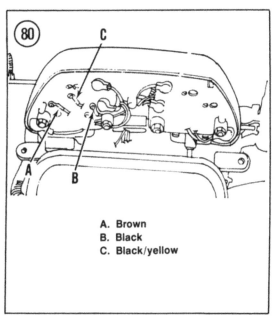

A. Brown
B. Black
C. Black/yellow

condition and that a malfunction exists only in the tachometer/voltmeter circuit.

1. Perform the *Visual Check* to ensure that the problem is not a loose connection or such.

2. Turn on the ignition switch.

NOTE
An electric tachometer without a voltmeter is installed in 1983 KZ/Z750L models. To test the operation of the tachometer on these models, proceed to Step 4.

3. Refer to **Figure 80** and connect the (+) lead of a voltmeter to the brown wire on the back of the tachometer/voltmeter. Connect the (-) lead of the voltmeter to the black/yellow wire on the instrument. The voltmeter must indicate battery voltage. If battery voltage is not present at the instrument, the problem is a broken wire or bad connection in the wiring harness. If battery voltage is present at the instrument, proceed to the next step.

4. Connect the voltmeter (+) lead to the brown wire and the (-) lead to the black wire (green on 1983 ZX750A models). See **Figure 81**. The voltmeter must indicate 0 volts when the engine is stopped. Start the engine and check the voltage at the instrument. When the engine is running the voltmeter indication should be 2-4 volts.

5. If the instrument malfunctions in the tachometer or voltmeter mode, it is defective and must be replaced.

11

GAUGE AND WARNING LIGHT SYSTEM (1982 KZ750R AND 1983-ON ZX750)

The gauge and warning light system monitors critical operating circuits on the motorcycle. The system consists of a liquid crystal display (LCD) combined with a red light emitting diode (LED) warning light to monitor the engine oil level, fuel level, battery electrolyte level and position status of the sidestand.

Though the visual display and terminology for the 1982-on models differ slightly, the operation is essentially the same.

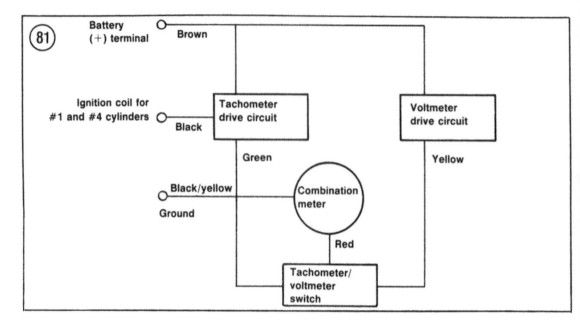

System Operation

The ignition switch must be ON before any warning lights will come on.

1. *Warning light*—This red LED flashes any time one of the LCD warning indicators is on. This LED is part of the LCD panel on 1982 models and is in the instrument panel of 1983 models.

2. *Sidestand warning*—This 2-segment red LCD flashes whenever the sidestand is down. On 1982 models it also flashes when the engine switch is OFF.

3. *Oil level*—This 2-segment red LCD flashes whenever the engine oil level is below a predetermined level.

4. *Battery electrolyte level*—This 2-segment red LCD flashes whenever the electrolyte level in the battery is lower than a predetermined level.

5. *Fuel gauge and low fuel*—This 9-segment black LCD indicates a full fuel tank when all 9 segments are lighted. Segments go out as fuel is consumed until at a predetermined level the last segment flashes to indicate low fuel level. A built-in time delay circuit helps stabilize the display. Each segment takes approximately 3-12 seconds to appear or disappear. The bottom segment takes approximately 3-7 seconds to begin or stop flashing.

6. *ZX750E: Digital Fuel Injection*—This 2-segment red LCD flashes whenever the DFI system is operating incorrectly.

The system is initiated each time the ignition is turned ON. The built-in computer cycles through an LCD and LED display unit check. If all displays are correct, the computer begins to monitor the motorcycle functions. When the ignition is turned ON the self-test is as follows:

a. Nine fuel gauge segments appear one at a time and stay on.

b. All LCD warning indicators come on one at a time and stay on.

c. The red LED warning light comes on for approximately one second to indicate the end of the test.

Gauge and Warning System Test

The following test assumes the battery is fully charged. Refer to **Figure 82** for 1982 models and **Figure 83** for 1983-on models.

1A. On 1982 models, perform the following to gain access to the warning system connector:

a. Remove the fairing.

b. Remove the screws securing the headlight assembly to the housing and carefully remove the light assembly.

c. Disconnect the red 6-pin connector.

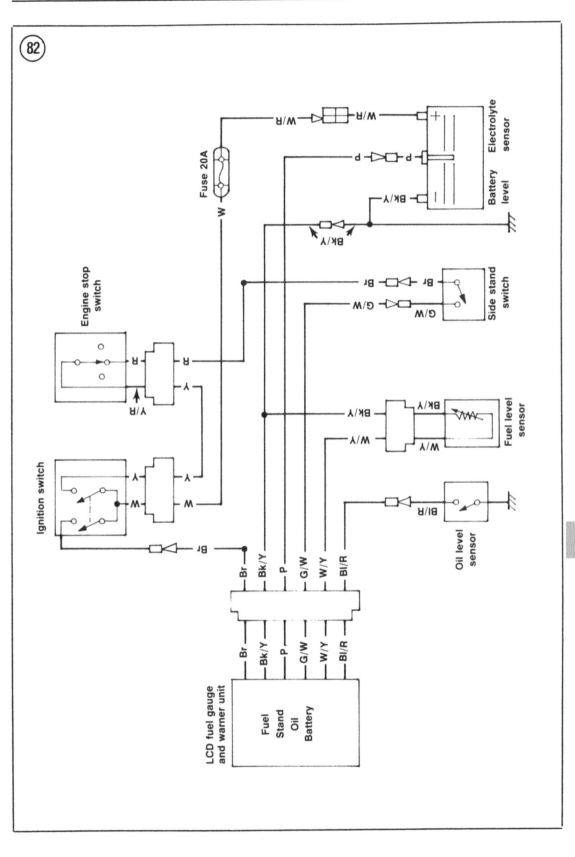

1B. On 1983 and later models, remove the fuel tank to gain access to the warning system connector under the fuel tank.

2. Connect a voltmeter to the motorcycle half of the warning system wiring harness as follows:

  a. Voltmeter positive (+) lead to the brown/white wire (brown wire on 1982 models).

  b. Voltmeter negative (-) lead to the black/yellow wire.

3. Turn on the ignition switch. The voltmeter must indicate battery voltage. If battery voltage is not present in the warning system wiring harness, check for a bad connection or a broken wire. If battery voltage is present at the connector, continue this procedure.

CAUTION
Take care during the next test steps not to short any wires or connector terminals together or to ground. The warning system gauge unit may be destroyed.

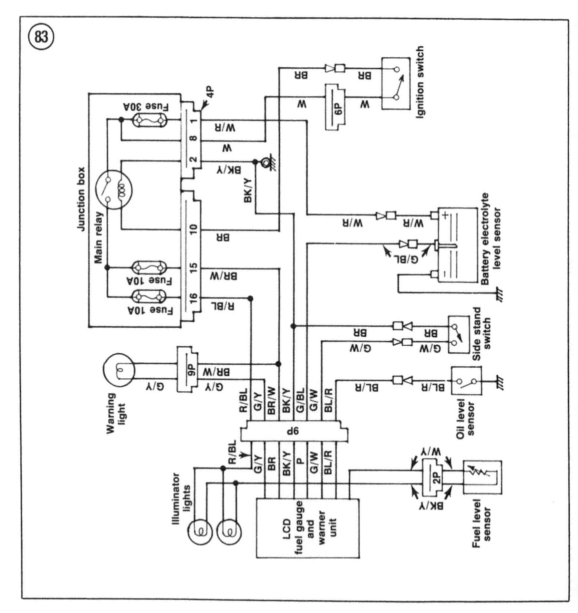

4. Prepare jumper wires to connect battery power and to simulate the warning sensors. The following number of jumper wires are required:

    a. 1982 models: 6 jumper wires.

    b. 1983-on non-turbo models: 7 jumper wires.

    c. 1984-on turbo models: 8 jumper wires.

5. Refer to **Figure 82** or **Figure 83** and connect jumper wires as outlined in **Table 19** to simulate warning circuit connections.

6. Use 2 jumper wires to connect power to the warning unit as follows:

    a. On 1982 models, connect jumper wires between the brown (+) wires in each half of the warning unit connector. Then connect the black/yellow (-) wires in each half of the connector.

    b. On 1983 and later non-turbo models, connect jumper wires between the brown/white (+) wires of each half of the warning unit connector. Then

connect the black/yellow (-) wires in each half of the connector.

7. On 1984 and later turbo models, connect a jumper wire between the green/yellow wires of each half of the warning unit connector.

8. When battery power is connected to the warning unit, the self-test procedure should start as described in *System Operation*. When the self-test is complete the warning unit display should appear as shown in section A in **Figures 84-86**. If the self-test did not start or if any malfunction exists in the display, the warning unit is defective and must be replaced.

9. One-by-one disconnect the jumper wires simulating the warning circuits. The LCD warning light should flash and the warning unit display should flash for each simulated malfunction. See section B in **Figures 84-86**.

10. If the warning and gauge unit tests good but a malfunction still exists when the unit is connected to the motorcycle wiring harness, a

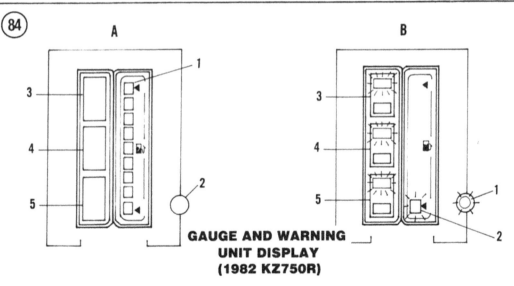

GAUGE AND WARNING UNIT DISPLAY (1982 KZ750R)

A
1. Fuel gauge—9 segments lighted.
2. Warning light—not lighted.
3. Side stand warning—not lighted.
4. Oil level warning—not lighted.
5. Battery electrolye level warning—not lighted.

*A time delay is included in the fuel gauge circuit. It takes approximately 3-12 seconds for each segment to appear or disappear and 3-7 seconds for the bottom segment to begin or stop flashing.

B
1. Warning light—Flashes when any simulating jumper wire is disconnected.
2. Fuel warning—Flashes when white/yellow jumper is disconnected.*
3. Side stand warning—Flashes when green/white jumper is disconnected.
4. Oil level warning—Flashes when blue/red jumper is disconnected.
5. Battery electrolyte level warning—Flashes when pink jumper is disconnected.

11

fault may exist in the wiring or connector. Perform *Connector and Wiring Test*.

Connector and Wiring Test

NOTE
The following test requires the use of a VOM (volt-ohm-milliammeter) as described in Chapter One of the main book.

1. Perform *Gauge and Warning System Test*. If the warning unit tests good, the following test will check the main motorcycle harness and sensor wiring as well as all the related connections. If any indication is not as specified, check for faulty or corroded connections as well as broken or shorted wires. If the wiring and connectors appear to

be good, perform the applicable warning sensor test to determine if a particular sensor is defective.

NOTE
All of the following test steps connections are made on the motorcycle half of the gauge and warning system connector.

2A. On 1982 models, make sure the engine stop switch is in the RUN position. Connect the positive (+) voltmeter lead to the green/white wire and the negative (-) voltmeter lead to the black/yellow wire. The voltmeter must indicate battery voltage when the sidestand is up and 0 volts when the stand is down.

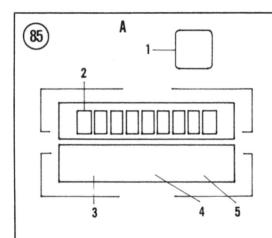

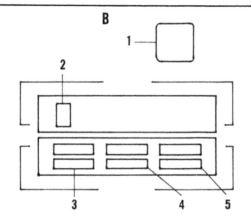

GAUGE AND WARNING UNIT DISPLAY (1983-ON ZX750)

A
1. Fuel gauge—9 segments lighted.
2. Warning light—not lighted.
3. Side stand warning—not lighted.
4. Oil level warning—not lighted.
5. Battery electrolye level warning—not lighted.

B
1. Warning light—Flashes when any simulating jumper wire is disconnected.
2. Fuel warning—Flashes when white/yellow jumper is disconnected.*
3. Side stand warning—Flashes when green/white jumper is disconnected.
4. Oil level warning—Flashes when blue/red jumper is disconnected.
5. Battery electrolyte level warning—Flashes when pink jumper is disconnected.

*A time delay is included in the fuel gauge circuit. It takes approximately 3-12 seconds for each segment to appear or disappear and 3-7 seconds for the bottom segment to begin or stop flashing.

2B. On 1983 models, connect the leads of an ohmmeter between the green/white wire and the black/yellow wires. The ohmmeter must indicate 0 ohms when the sidestand is up and continuity when the stand is down.

3. Connect the ohmmeter leads between the blue/red and black/yellow wires. The ohmmeter must indicate less than 0.5 ohm when the engine oil level is higher than the "low level" line on the oil level window gauge. The ohmmeter must indicate continuity when the oil level is much lower than the "low level" line on the engine oil level window gauge.

4. On 1983 models, disconnect the fuel level sensor connector.

5. Connect the ohmmeter leads between the white/yellow and black/yellow wires. The ohmmeter must indicate 1-117 ohms, depending on the level of fuel in the tank (117 ohms for an empty tank).

6. Connect the positive (+) lead of the voltmeter to the pink wire and the negative (-) lead of the voltmeter to the black/yellow wire. The voltmeter must indicate more than 6 volts when the battery electrolyte is above the "low level" line and 0 volts when the electrolyte is below the "low level" line.

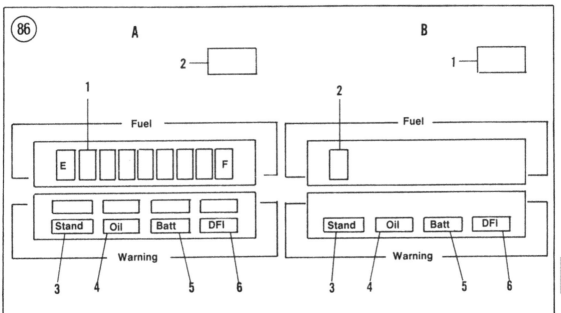

GAUGE AND WARNING UNIT DISPLAY (1984-ON TURBO)

A
1. Fuel gage—9 segments lighted.
2. Warning light—not lighted.
3. Side stand warning—not lighted.
4. Oil pressure warning—not lighted.
5. Battery electrolyte level—not lighted.
6. DFI warning—not lighted.

B
1. Warning light—Flashes when any simulating jumper wire is disconnected.
2. Fuel warning—Flashes when white/yellow jumper is disconnected.*
3. Side stand warning—Flashes when green/white jumper is disconnected.
4. Oil level warning—Flashes when blue/red jumper is disconnected.
5. Battery electrolyte level warning—Flashes when pink jumper is disconnected.
6. DFI warning—Flashes when green/red jumper is connected to the negative battery terminal.

* A time delay is included in the fuel gauge circuit. It takes approximately 3-12 seconds for each segment to appear or disappear and 3-7 seconds for the bottom segment to begin or stop flashing.

11

Sidestand Sensor Test

1. Turn off the motorcycle ignition switch and disconnect the switch leads from the sidestand switch.

2. Connect an ohmmeter between the sidestand switch wires (green/white and brown or red). The ohmmeter must indicate as follows:

   a. On models with a red switch wire, there must be 0 ohms when the sidestand is fully loaded or fully up. The meter must indicate continuity when the motorcycle weight is not on the stand but the stand is not kicked into the up position.

   b. On models with a brown switch wire, the ohmmeter must indicate 0 ohms when the stand is up and continuity when the stand is down.

3. Replace the sidestand switch if defective.

Oil Level Sensor Test

1. Refer to Chapter Three in the main book and drain the engine oil.

2. Remove the bolts securing the oil level sensor to the crankcase and remove the sensor (**Figure 87**).

3. Connect an ohmmeter to the oil level sensor leads. The ohmmeter must indicate continuity when the sensor is held upright and 0 ohms when the sensor is turned over. Replace the sensor if defective.

4. Install the sensor in the crankcase. Add engine oil as outlined in Chapter Three of the main book.

Battery Electrolyte Sensor Test

1. Remove the seat and remove the cover from the battery.

2. Disconnect the pink sensor wire at the connector. Check for the following:

   a. Make sure that the sensor is fully installed into the battery and the arrow mark on the sensor points toward the terminal side of the battery.

   b. Ensure that the battery electrolyte level is above the "low level" line and that the battery is fully charged.

3. Connect the positive (+) lead of a voltmeter to the pink sensor wire and the negative (-) lead to a good ground on the

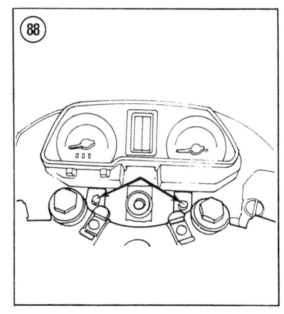

motorcycle frame. The voltmeter must indicate more than 6 volts. If the meter indication is incorrect, replace the sensor.

Fuel Level Sensor Test

1. Remove the fuel tank as outlined in Chapter Seven of the main book and drain the tank into a suitable container.

2. Remove the fuel level sensor unit from the tank as outlined in this supplement.

3. Check that the float on the sensor moves smoothly under its own weight without binding or tight spots.

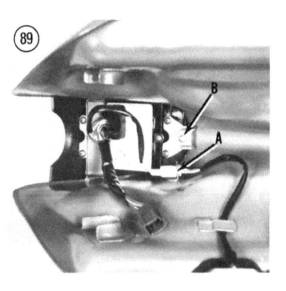

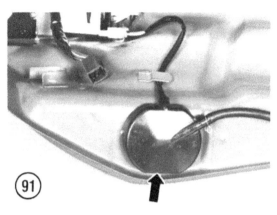

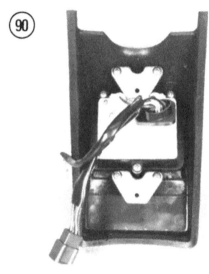

4. Connect an ohmmeter to the sensor wires. The ohmmeter must indicate 1-5 ohms with the float in the full (highest) position and 103-117 ohms with the float in the empty (lowest) position. If the ohmmeter indication is erratic or the values are not as specified, replace the sensor.

Gauge and Warning Unit
Removal/Installation (1982)

1. Remove the fairing.
2. Remove the screws securing the headlight assembly to the housing and carefully remove the light assembly.

3. Disconnect the red 6-pin connector.
4. Remove the screws securing the bottom cover on the meter assembly and remove the cover.
5. Remove the mounting bolts and nuts securing the instrument assembly to the motorcycle (**Figure 88**).
6. Remove the gauge and warning unit from the instrument assembly.
7. Installation is the reverse of these steps.

Gauge and Warning Unit
Removal/Installation (1983)

1. Remove the fuel tank. Drain the tank enough so the tank can be turned on its side.
2. Disconnect the fuel level sending unit connector (A, **Figure 89**).
3. Remove the screws securing the gauge panel to the fuel tank and carefully remove the gauge panel (B, **Figure 89**).
4. Remove the screws securing the gauge unit to the panel and remove the gauge unit (**Figure 90**).
5. Installation is the reverse of these steps.

Fuel Level Sensor
Removal/Installation

1. Remove the fuel tank and completely drain the fuel into a suitable container.
2. Carefully unsnap the plastic cover/drain from the sensor unit (**Figure 91**).
3. Remove the bolts securing the sensor unit to the tank and remove the sensor unit. Take care not to damage the gasket.

11

4. Installation is the reverse of these steps. Keep the following points in mind:

   a. Ensure that the sensor unit gasket is in good condition. Replace the gasket if not perfect or fuel leaks may occur.

   b. On 1982-on non-turbo models, install the sensor unit with the arrow pointing toward the front of the tank (**Figure 92**).

   c. On 1984-on turbo models, install the sensor unit so that the float is positioned directly forward of the sensor body.

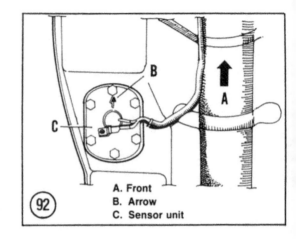

A. Front
B. Arrow
C. Sensor unit

Table 19 WARNING UNIT SIMULATED CIRCUIT TEST CONNECTIONS

Connection	Warning circuit
Green/white to battery (–) (1983 models)	Side stand
Green/white to battery (+) (1982 models)	Side stand
Blue/red to battery (–)	Oil level
Pink to battery (+)	Battery electrolyte
White/yellow to battery (–)	Fuel gauge

CHAPTER NINE

WHEELS, TIRES, AND BRAKES

REAR WHEEL REMOVAL/INSTALLATION (UNI-TRAK MODELS)

Remove the rear wheel as outlined in Chapter Nine of the main book. Loosen the clamp bolt (A, **Figure 93**) and rotate the cam-type chain adjusters (B, **Figure 93**) forward until maximum slack exists in the drive chain. Tighten the rear wheel components as specified in **Table 7**.

BRAKES (1983-ON ZX750)

The front and rear brake calipers on 1983 and later ZX750 models are changed slightly from the components used on other models. However, the service and operation of the new calipers are essentially the same. See **Figure 94** and **Figure 95**.

Pad Replacement

Pad replacement is nearly identical for front and rear calipers.
1. Remove the bolts securing the front caliper to the fork leg (**Figure 96**) or the rear caliper to the mounting bracket (**Figure 97**).
2. Carefully pull the caliper away from the brake disc and slide out the inside brake pad (**Figure 98**).

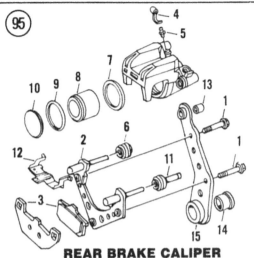

REAR BRAKE CALIPER (1983-ON ZX750)

1. Mounting bolts
2. Caliper holder
3. Brake pads
4. Rubber dust cap
5. Bleeder valve
6. Rubber boot
7. Fluid seal
8. Piston
9. Dust seal
10. Insulator
11. Friction boot
12. Anti-rattle spring
13. Small spacer
14. Large spacer
15. Caliper bracket

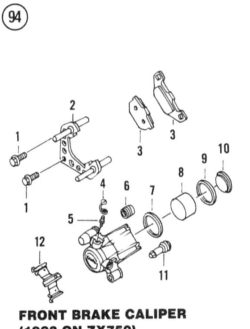

FRONT BRAKE CALIPER (1983-ON ZX750)

1. Mounting bolts
2. Caliper holder
3. Brake pads
4. Rubber dust cap
5. Bleeder valve
6. Rubber boot
7. Fluid seal
8. Piston
9. Dust seal
10. Insulator
11. Friction boot
12. Anti-rattle spring

11

3. Disengage the outside pad from the ends of the caliper holder and remove the pad (**Figure 99**).

4. Use a clean rag and carefully wipe the pad sliding surfaces on the caliper holder and anti-rattle spring (A, **Figure 100**).

5. Remove the cover from the front master cylinder reservoir. Wrap a rag around the reservoir to catch any brake fluid spills.

6. Carefully push the caliper piston back into the caliper body as far as it will go (B, **Figure 100**).

> *NOTE*
> *If the reservoir is full, brake fluid may overflow slightly when the caliper piston is pushed completely back into the caliper.*

> *CAUTION*
> *Do not allow any brake fluid to spill on painted surfaces or the paint will be damaged.*

7. Install new inner and outer brake pads. Ensure that each pad fits properly in the caliper holder.

8. Install the caliper mounting bolts and torque the bolts to 3.3 mkg (24 ft.-lb.).

9. Top up the master cylinder reservoir with approved brake fluid, if necessary, and install the reservoir cap.

10. Spin the front or rear wheel and apply the brake a few times to ensure that the brakes operate properly and the pads adjust correctly.

BRAKE BLEEDING

On models equipped with anti-dive front suspension, it is necessary to bleed the

anti-dive mechanism and junction block as well as the caliper. See **Figure 101** and **Figure 102** for the location of the bleed valves.

Use the techniques for bleeding the brakes as outlined in Chapter Nine of the main book. Bleed the anti-dive mechanism and brake calipers in the following manner:

    a. Left caliper bleed valve.
    b. Left anti-dive bleed valve.
    c. Left junction block valve.
    d. Right caliper bleed valve.
    e. Right anti-dive bleed valve.
    f. Right junction block valve.

CHAPTER TEN

CHASSIS

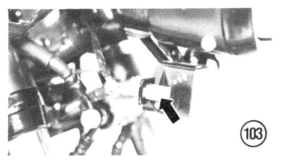

FAIRING REMOVAL/INSTALLATION

1. Disconnect the electrical connector from under the right side of the fairing (**Figure 103**, fairing removed for clarity).
2. Remove the acorn nuts securing each side of the fairing to the mounting bracket.
3. Carefully pull the fairing forward and clear of the motorcycle.
4. Installation is the reverse of these steps. Do not overtighten the acorn nuts securing the fairing. Make sure the electrical connection is tight.

HANDLEBAR ASSEMBLY (1982 KZ750R AND 1983-ON ZX750A, KZ/Z750-L)

Handlebar Removal/Installation

Refer to **Figure 104** for this procedure.
1. On the left handlebar perform the following:
    a. Loosen the bolt securing the clutch lever.
    b. Remove the screws securing the left switch housing and open the housing.

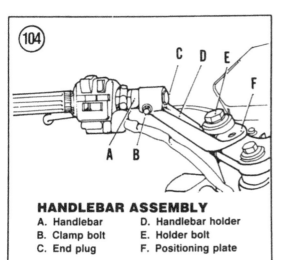

HANDLEBAR ASSEMBLY

A. Handlebar
B. Clamp bolt
C. End plug
D. Handlebar holder
E. Holder bolt
F. Positioning plate

c. Remove the end plug (A, **Figure 104**)
and clamp bolt (B, **Figure 104**) and pull
the handlebar out of the holder.

2. On the right handlebar perform the
following:

   a. Loosen the bolts securing the front
master cylinder.

   b. Remove the screws securing the right
switch housing and slide off the throttle
grip assembly.

   c. Remove the end plug (A, **Figure 104**)
and clamp bolt (B, **Figure 104**) and pull
the handlebar out of the holder.

3. Installation is the reverse of these steps.
Keep the following points in mind:

   a. On 1983 models, align the mark on the
handlebar with the split opening in the
holder.

   b. Torque the handlebar clamp bolt to 3.0
mkg (22 ft.-lb.).

   c. Install both left and right switch
housings so that the small tip on the
housing fits into the hole as shown in
Figure 105.

   d. Tighten the upper master cylinder
clamp bolt to 0.9 mkg (6.5 ft.-lb.). Then
tighten the lower clamp bolt to the same
torque value. The gap in between the
mounting clamps must be on the
bottom.

   e. Ensure that all control cables are free
and operate smoothly. Make sure the
mirrors are properly secured.

Handlebar Holder Removal/Installation

Refer to **Figure 104** for this procedure.

1. Remove the screws securing the cover to
the steering stem and remove the cover. Pull
off the rubber cap.

2. Remove the handlebar holder bolt (E,
Figure 104).

3. Remove the Allen bolt securing the
positioning plate and remove the plate. Lift
off the handlebar holder.

4. Installation is the reverse of these steps.
Keep the following points in mind.

5. If the fork tubes were removed, install the
fork tubes so that the top of the tube is just
slightly below the surface of the steering stem
head as shown in **Figure 106**.

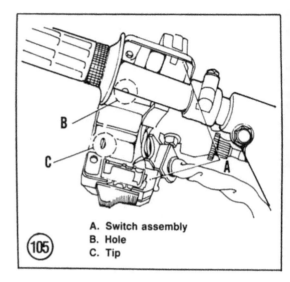

A. Switch assembly
B. Hole
C. Tip

(105)

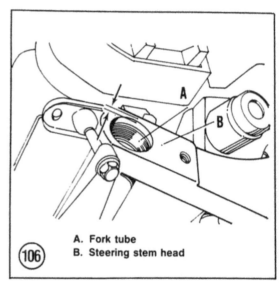

A. Fork tube
B. Steering stem head

(106)

6. Tighten the upper and lower clamp bolts
securing the fork tubes finger-tight.

7. Install the holder on the steering stem head
and secure it in place with the holder bolt and
positioning plate.

8A. On 1982 KZ750R models, install the
positioning plate so that the arrow points to
the rear as shown in **Figure 107**.

8B. On 1983 ZX750A models, install the
positioning plate so that the arrow points to
the front of the motorcycle.

8C. On 1984 and later KZ/Z750-L models,
install the positioning plate so that the
triangular mark faces to the front of the
motorcycle. Then install the handlebars so

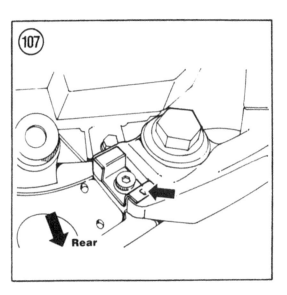

that the mark on the handlebar aligns with the slit on the handlebar holder.

9. Tighten the holder bolt and positioning plate Allen bolt finger-tight.

10. Tighten the fasteners in the following order:

    a. Holder bolt—10.0 mkg (72 ft.-lb.).

    b. Upper fork clamp bolt—2.0 mkg (14.5 ft.-lb.).

    c. Lower fork clamp bolt— 3.8 mkg (27 ft.-lb.).

    d. Positioning plate Allen bolt—snug.

HANDLEBARS (1983 KZ/Z750L)

Install the handlebar clamps so that the arrow on each clamp points to the front of the motorcycle.

FRONT FORKS

The front forks on 1982 KZ750R and 1983 ZX750A models are changed slightly from other models. Both models are equipped with a connecting hose assembly which acts as an equalizer so the air pressure in each fork tube is exactly the same. Removal and disassembly of the forks on these models is essentially the same as outlined under *Front Forks* in Chapter Ten of the main book. Refer to **Figure 108** for 1982 KZ750R models and **Figure 109** for 1983-on ZX750 models.

Refer to **Table 9** for fork oil specifications for all 1982 and later models. Fill the forks

and set the air pressure as outlined in Chapter Three of this supplement and the main book. When setting the fork oil level on models equipped with the connecting hose assembly, be sure that both fork tubes are compressed at the same time, or the fork oil will be forced from one fork tube to the other through the connecting hose.

FRONT FORK ANTI-DIVE ASSEMBLY (1983-ON ZX750 MODELS)

The anti-dive assembly consists of 2 basic components: the brake plunger (A, **Figure 110**) and the anti-dive valve (B, **Figure 110**). Neither component is repairable. If a component is defective and the anti-dive function does not operate correctly, the faulty component must be replaced.

Brake Plunger
Test/Removal/Installation

1. Remove the bolts securing the brake plunger assembly to the anti-dive valve (A, **Figure 110**). Carefully remove the plunger assembly. Take care not to damage the O-ring between the plunger and the anti-dive valve.

2. Loosen one of the fittings on the metal brake line and shift the plunger assembly around until the plunger rod inside the assembly is accessible. Retighten the brake line fitting enough to keep the fluid from running out.

> *CAUTION*
> *Do not attempt to move the plunger assembly without first loosening one of the fittings on the metal brake line or the brake line will be bent and distorted.*

> *CAUTION*
> *Take care not to spill brake fluid on the painted surfaces of the front forks or the paint may be damaged.*

3. Hold your finger over the plunger rod and apply the front brake lightly. Check that the plunger rod extends approximately 2 mm (0.08 in.) as the front brake is applied. Release the brake lever and push the plunger rod back into the plunger body. The plunger rod must move smoothly in both directions. If the rod

11

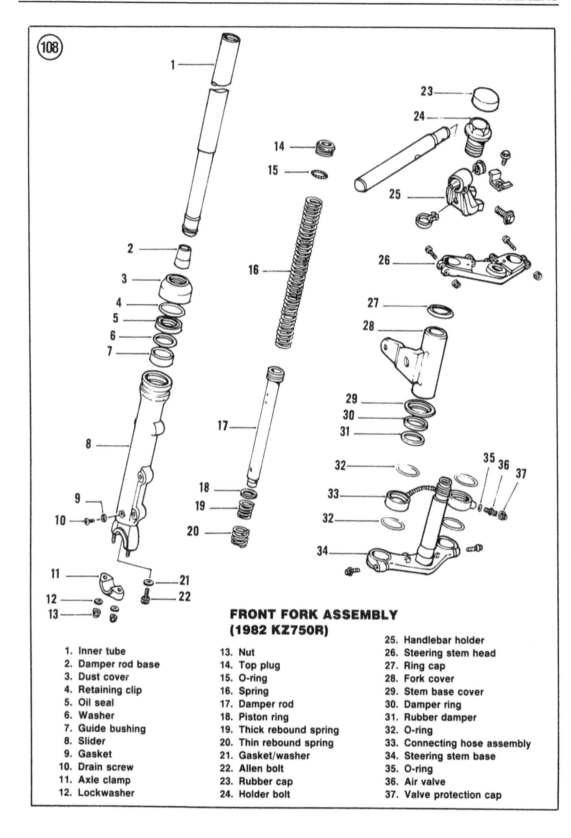

**FRONT FORK ASSEMBLY
(1982 KZ750R)**

1. Inner tube
2. Damper rod base
3. Dust cover
4. Retaining clip
5. Oil seal
6. Washer
7. Guide bushing
8. Slider
9. Gasket
10. Drain screw
11. Axle clamp
12. Lockwasher
13. Nut
14. Top plug
15. O-ring
16. Spring
17. Damper rod
18. Piston ring
19. Thick rebound spring
20. Thin rebound spring
21. Gasket/washer
22. Allen bolt
23. Rubber cap
24. Holder bolt
25. Handlebar holder
26. Steering stem head
27. Ring cap
28. Fork cover
29. Stem base cover
30. Damper ring
31. Rubber damper
32. O-ring
33. Connecting hose assembly
34. Steering stem base
35. O-ring
36. Air valve
37. Valve protection cap

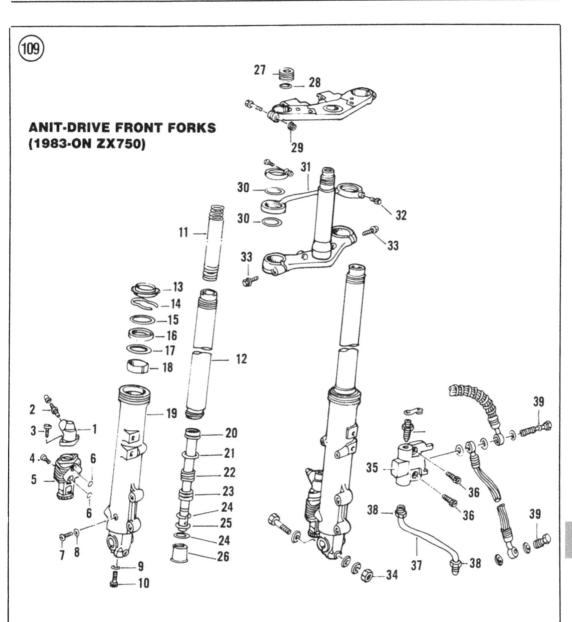

ANIT-DRIVE FRONT FORKS
(1983-ON ZX750)

1. Anti-dive brake plunger assembly
2. Bleeder valve
3. Bolt
4. Mounting bolt
5. Anti-dive valve assembly
6. O-ring
7. Drain screw
8. Gasket
9. Gasket
10. Bottom bolt
11. Spring
12. Inner fork tube
13. Dust seal
14. Retaining clip
15. Washer
16. Oil seal
17. Washer
18. Guide bushing
19. Slider
20. Damping rod
21. Piston ring
22. Thicker spring
23. Thinner spring
24. Spring
25. Washer
26. Damper rod base
27. Top plug
28. O-ring
29. Clamp bolt nut
30. O-ring
31. Connecting pipe assembly
32. Air valve
33. Clamp bolt
34. Nut
35. Junction block
36. Mounting bolt
37. Metal brake pipe
38. Fittings
39. Banjo bolt

11

is stuck or does not move smoothly, replace the brake plunger assembly.

4. Install the brake plunger assembly. Keep the following points in mind:
   a. Make sure the O-ring is in good condition. Replace it if necessary.
   b. Retighten the metal brake line fitting and bleed the brakes as outlined in this supplement.

Anti-dive Valve
Test/Removal/Installation

1. Remove the brake plunger assembly from the anti-dive valve as described in this section of this supplement.

2. Remove the fork tube from the motorcycle. Remove the top bolt and remove the main fork spring.

3. Tape over the air pressure equalizer hole in the fork leg to keep the fork oil from running out.

4. Hold the fork leg vertically and compress the fork tube. Alternately press and release the plunger rod in the anti-dive assembly as the fork tube is compressed. The fork compression should be noticably stiffer when the plunger rod is pressed than when released. The extension (rebound) stroke of the fork tube should be the same whether or not the plunger is pressed. Replace the anti-dive valve if it does not function smoothly and correctly.

5. Installation is the reverse of the removal steps. Keep the following points in mind:
   a. If the anti-dive valve was removed, ensure that O-rings between the valve assembly and the fork tube are in good condition.
   b. Install the fork spring and fork top bolt and install the fork on the motorcycle.
   c. Install the brake plunger assembly as described in this section of the supplement. Be sure to bleed the brakes.

STEERING (1982 KZ750R AND 1983-ON ZX750 MODELS)

The steering head assembly on these models is equipped with tapered roller bearings on the top and bottom. Replacement

of the bearings and races requires special tools; refer these tasks to a Kawasaki dealer. If the bearings are kept properly lubricated and adjusted, they should last the life of the motorcycle.

Adjustment

The adjustment procedure is very similiar to earlier models as outlined under *Steering* in Chapter Ten of the main book. Refer to Chapter Ten and keep the following steps in mind.

1. Loosen the lower fork clamp bolts. It is not necessary to loosen the upper fork clamp bolts.

2. There are 2 notched steering stem adjusters on the steering stem (**Figure 111**). Make sure the upper adjuster is backed off enough to allow the lower adjuster to be moved.

3. Tighten the lower adjuster until the steering just becomes hard to turn. Do not overtighten. Back off the adjuster very

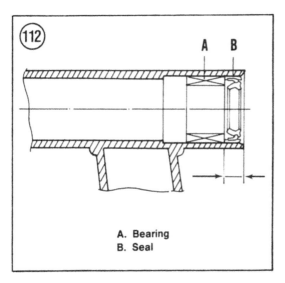

A. Bearing
B. Seal

REAR SHOCK ABSORBER
(1983-ON ZX750)

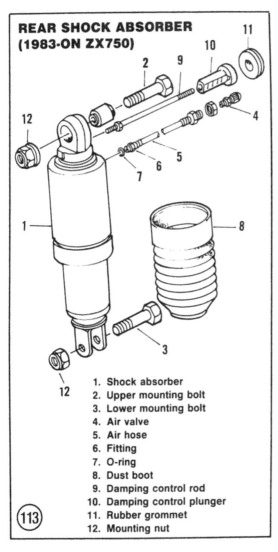

1. Shock absorber
2. Upper mounting bolt
3. Lower mounting bolt
4. Air valve
5. Air hose
6. Fitting
7. O-ring
8. Dust boot
9. Damping control rod
10. Damping control plunger
11. Rubber grommet
12. Mounting nut

slightly, not over 1/8 turn. Hold the lower adjuster and tighten the upper adjuster against it to hold it in place.

4. Turn the handlebars from lock to lock several times to make sure the adjustment is correct. Readjust if necessary.

5. Tighten the lower fork clamp bolts and the steering stem head bolt as specified in **Table 6, 7,** or **8.**

SWING ARM
(DUAL-SHOCK MODELS)

Removal/Lubrication/Installation

The swing arm on all 1982 and later dual-shock models is equipped with 2 needle bearings instead of 4 as on earlier models. Remove the swing arm as outlined in Chapter Ten of the main book. If the bearings must be replaced, have the job performed by a Kawasaki dealer, as special tools are required. New grease seals should also be installed whenever bearings are replaced. Ensure that the new bearings are installed 5 mm from the edge of the swing arm to allow room for the grease seals (**Figure 112**). Lubricate the bearings with a good grade of chassis grease before installing the swing arm in the motorcycle.

UNI-TRAK REAR SUSPENSION
(1983-ON ZX750)

Shock Absorber Removal/Installation

Refer to **Figure 113** for this procedure.

1. Remove the rear wheel as outlined in this section of the supplement.

2. Remove the seat and side covers.

3. Remove screws securing the forward portion of the rear fender and remove the fender (**Figure 114**).

4. Remove the bolts securing the chain guard to the swing arm and remove the chain guard.

5. Loosen the nut securing the air valve to the frame mounting tab and slide the valve out of the frame (**Figure 115**).

6. Loosen the locknut securing the damping adjuster rod and unscrew the rod from the shock absorber (**Figure 116**).

7. Remove the upper and lower bolts securing the shock absorber (**Figure 117** and **Figure 118**).

8. Remove the spacer from the lower shock mounting hole in the rocker arm (**Figure 119**).

9. Support the swing arm with a wire or Bungee cord and lift the shock absorber out of the frame (**Figure 120**).

10. Perform *Shock Absorber Inspection*.

11. Installation is the reverse of these steps. Keep the following points in mind:

    a. Grease the needle bearing in the lower shock mounting hole in the rocker arm (**Figure 121**). Inspect the seals installed on the outer edges of the needle bearing (**Figure 122**) and replace them if they are worn or damaged.

    b. Install the spacer into the lower shock mounting hole (**Figure 119**).

    c. Torque the shock absorber mounting bolts as specified in **Table 7** or **Table 8**.

Shock Absorber Inspection

1. Remove the rubber dust boot and check the body of the shock absorber for damage.

2. Check for loss of oil around the joint between the upper and lower portions of the shock body (**Figure 123**). Any more than a light film on the lower shock body indicates a bad internal seal; the shock absorber must be replaced.

3. Examine the upper shock mount for damage (**Figure 124**) and replace if necessary. If the upper mount is damaged or worn in any way, excessive play will be present in the action of the rear suspension.

4. Inspect the air hose and replace it if damaged. Make sure a new O-ring is fitted to the hose connection at the shock absorber (**Figure 125**).

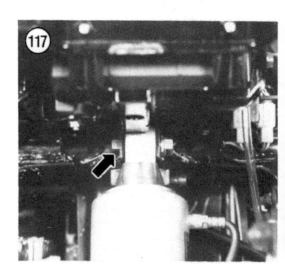

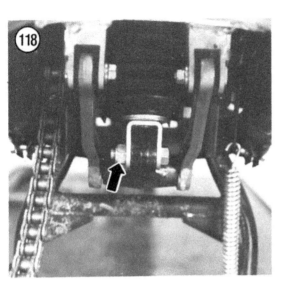

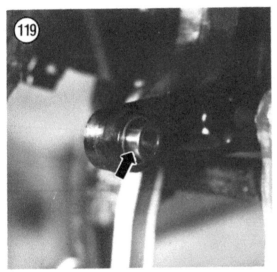

11

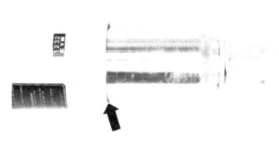

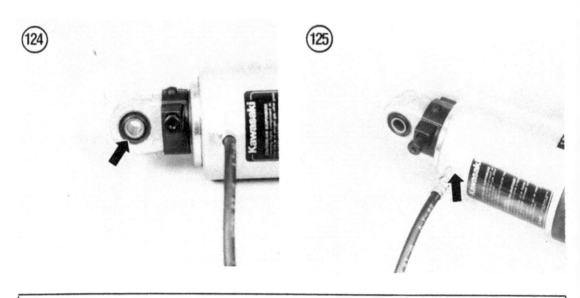

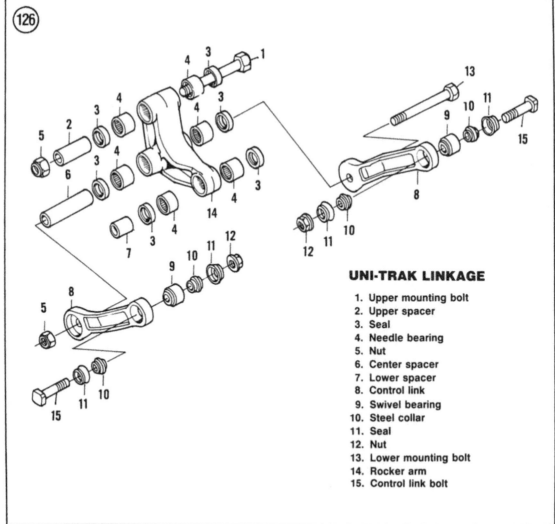

UNI-TRAK LINKAGE

1. Upper mounting bolt
2. Upper spacer
3. Seal
4. Needle bearing
5. Nut
6. Center spacer
7. Lower spacer
8. Control link
9. Swivel bearing
10. Steel collar
11. Seal
12. Nut
13. Lower mounting bolt
14. Rocker arm
15. Control link bolt

Swing Arm Removal/Installation

Refer to **Figure 126** and **Figure 127** for this procedure.

1. Remove the shock absorber as described in this supplement.

2. Remove the bolts securing the lower control links to the swing arm (**Figure 128**). Note the dust seals and metal rings on each side of each control link.

3. Disconnect the rear brake hose from the plastic clips on the swing arm.

4. Remove the nut securing the swing arm pivot bolt and carefully tap the pivot bolt out of the frame. Carefully slide the swing arm out of the frame.

5. Installation is the reverse of these steps. Keep the following points in mind.

6. Carefully check the condition of the grease seals on the outside of each swing arm bearing. Lightly grease the swing arm needle bearings and install the swing arm into the frame.

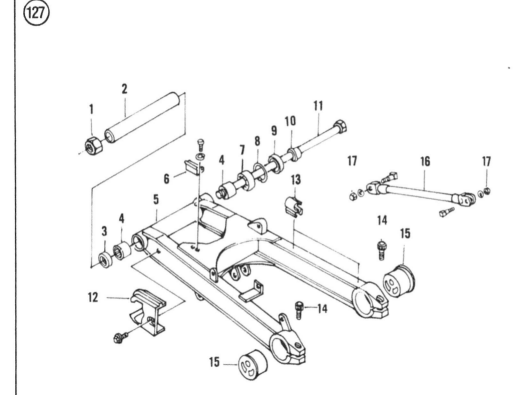

SWING ARM (UNI-TRAK)

1. Pivot shaft nut
2. Center spacer
3. Seal
4. Needle bearing
5. Swing arm
6. Chain guard clip
7. Ball bearing
8. Circlip
9. Seal
10. Collar
11. Pivot bolt
12. Chain protector
13. Brake hose clip
14. Adjuster clamp bolt
15. Chain adjuster
16. Torque link
17. Nut

11

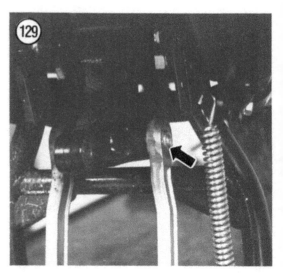

7. Loosen the forward mounting bolt securing both control links (**Figure 129**). It is much easier to install one link at a time.

8. Inspect the control link grease seals and metal rings. Replace the components if worn or damaged.

9. Lightly grease the swivel joint on each control link (**Figure 130**). Install the metal ring into the hole on each side of the swivel joint.

10. Install the grease seals over each side of one swivel joint and slide the joint into the swing arm mounting point. Repeat for the other swivel joint.

11. Install the shock absorber and tighten all the rear suspension mounting bolts as specified in **Table 7** and **Table 8**.

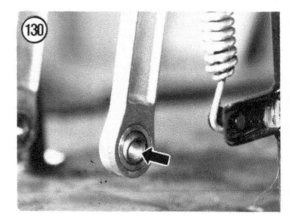

INDEX

12

12

1980 KZ750E—U.S. and Canada

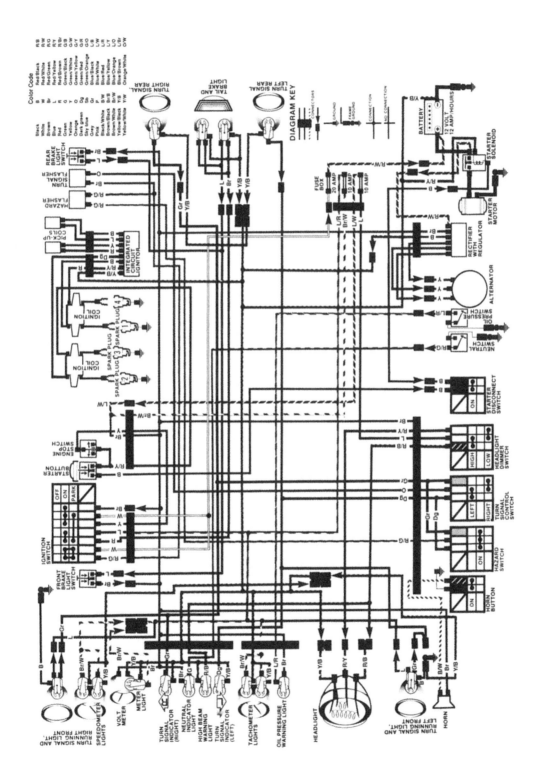

1980 KZ750E—Europe

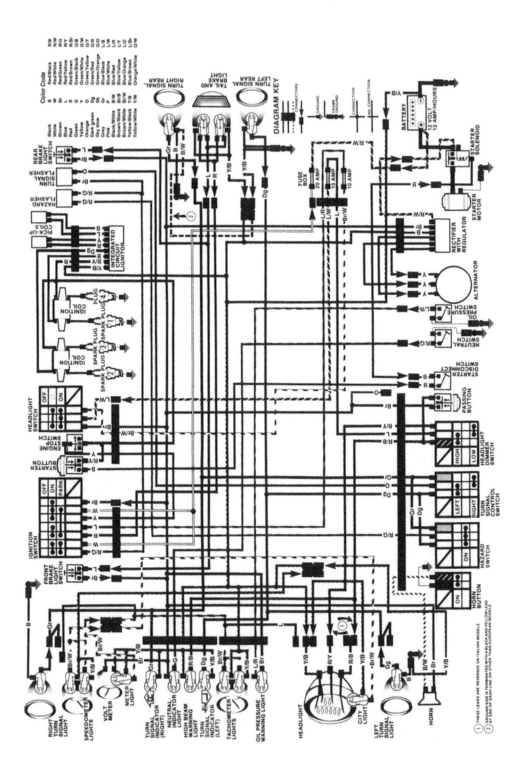

1980 KZ750H—U.S. and Canada

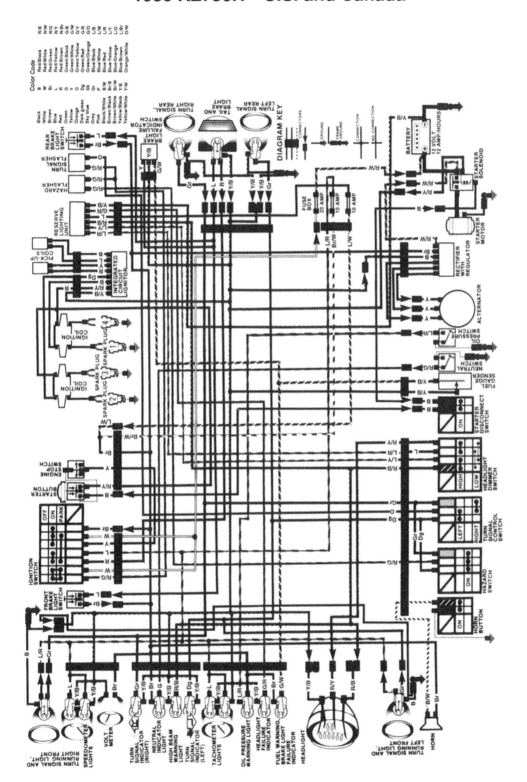

1980 KZ750H—Europe

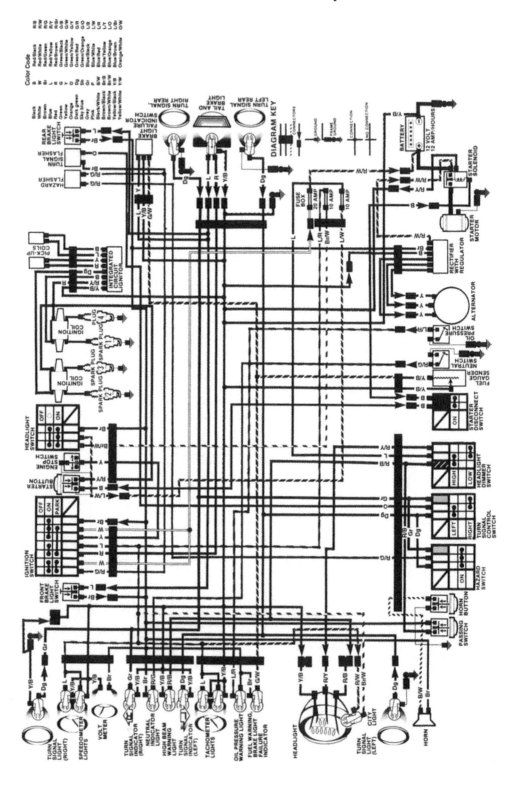

1981 KZ750E—U.S. and Canada

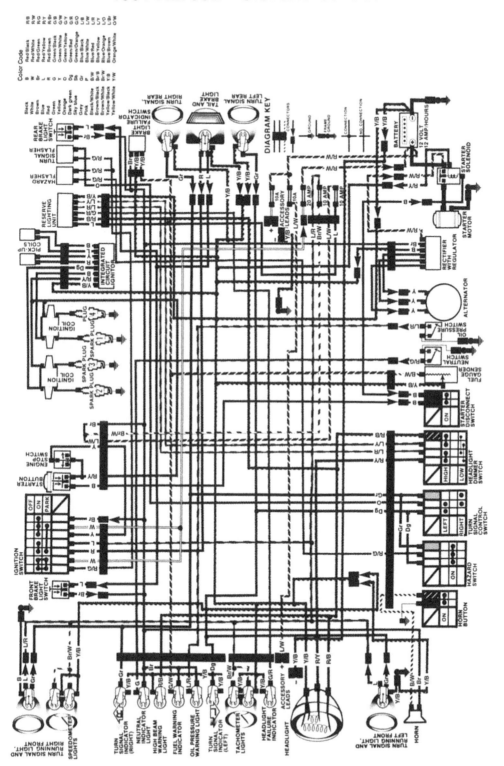

1981 KZ750H—U.S. and Canada

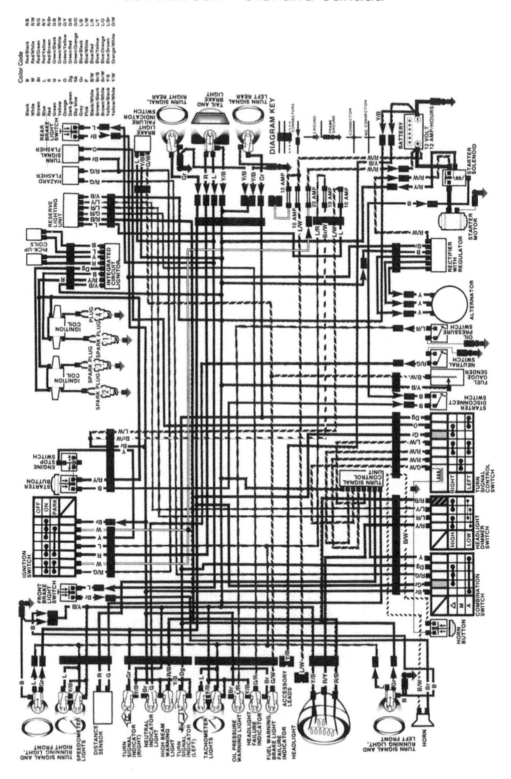

1981 KZ750H—Europe

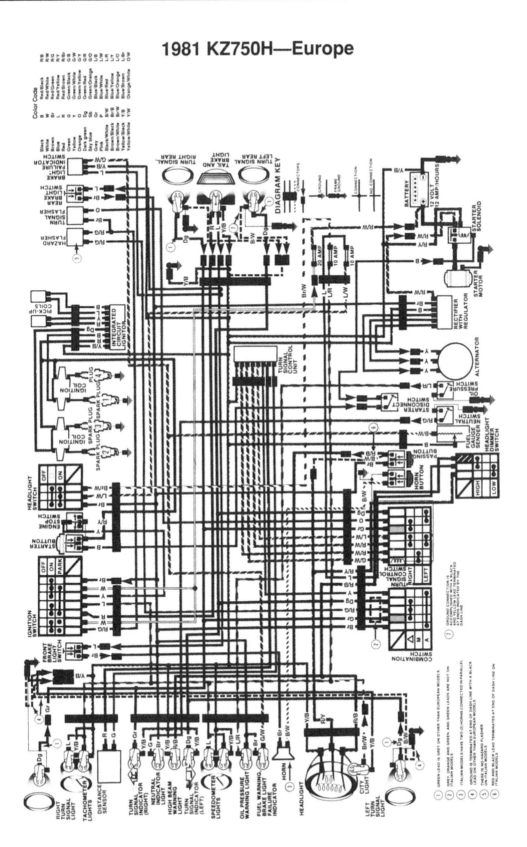

13

1981 KZ750L—Europe

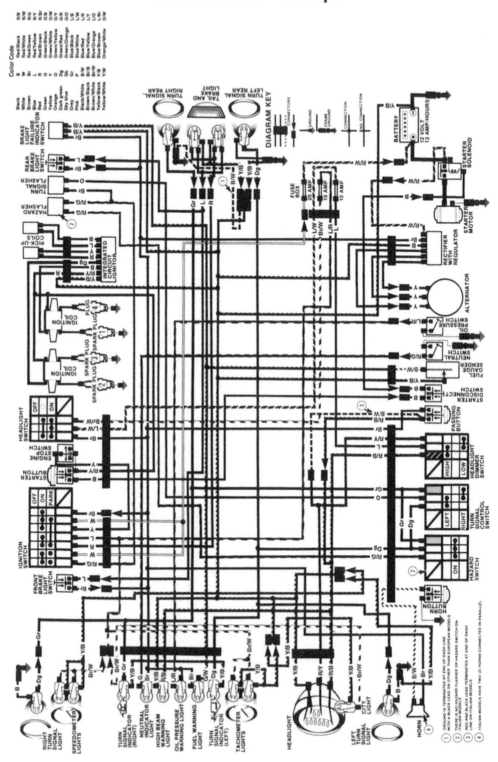

1982 KZ750E3—U.S. and Canada

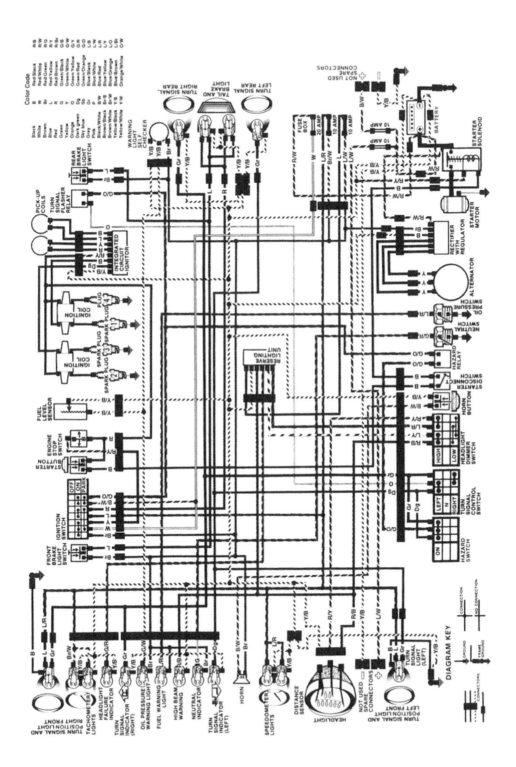

1982 KZ750H3—U.S. and Canada

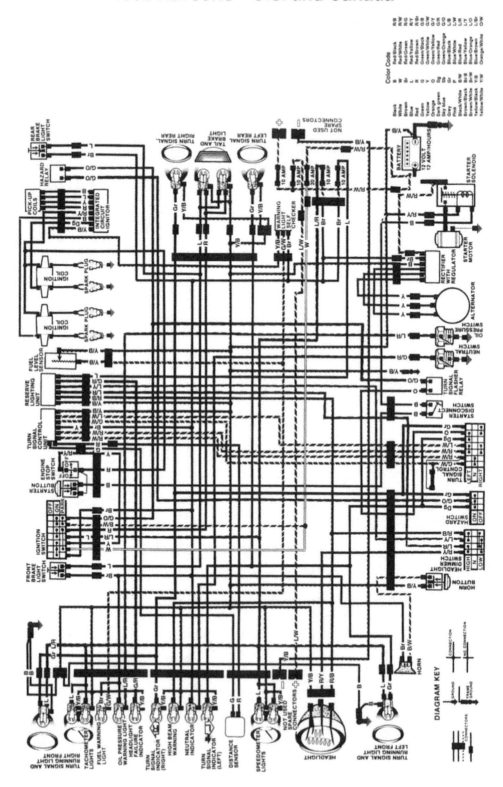

1982 KZ750R1—U.S. and Canada

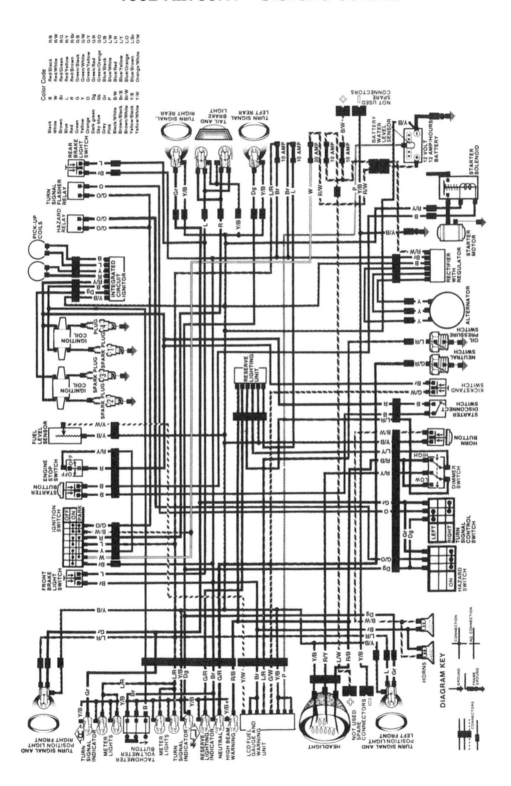

13

1982 Z750H3—Europe and General Export

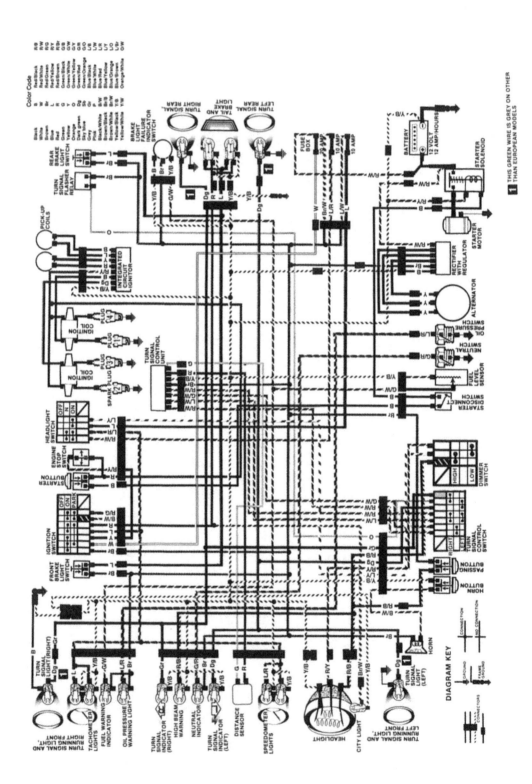

1982 Z750L2—Europe and General Export

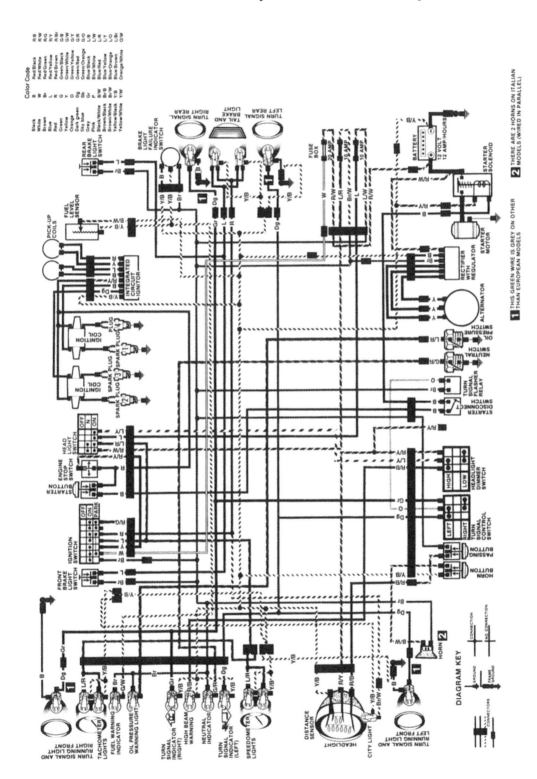

1982 Z750R1—Europe and General Export

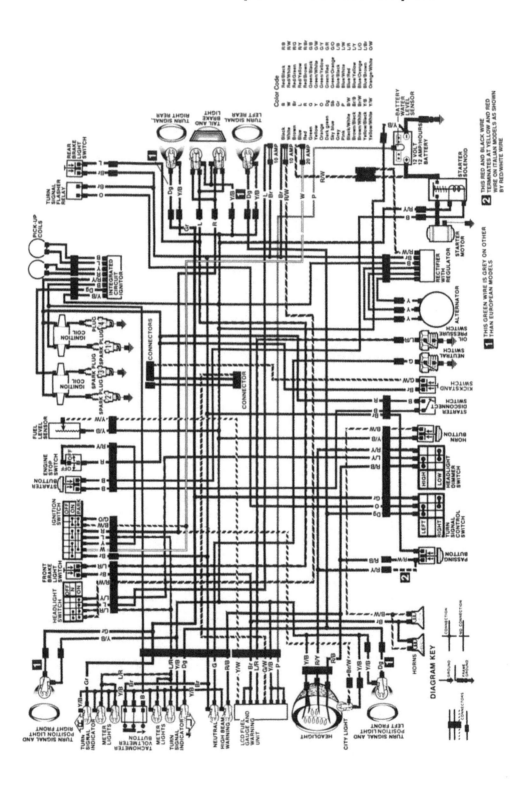

1983 KZ750H4

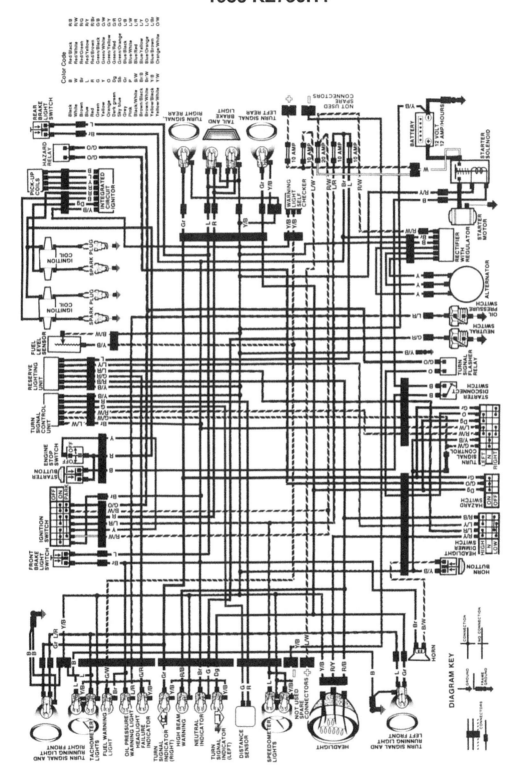

1983 KZ750L3—U.S. and Canada

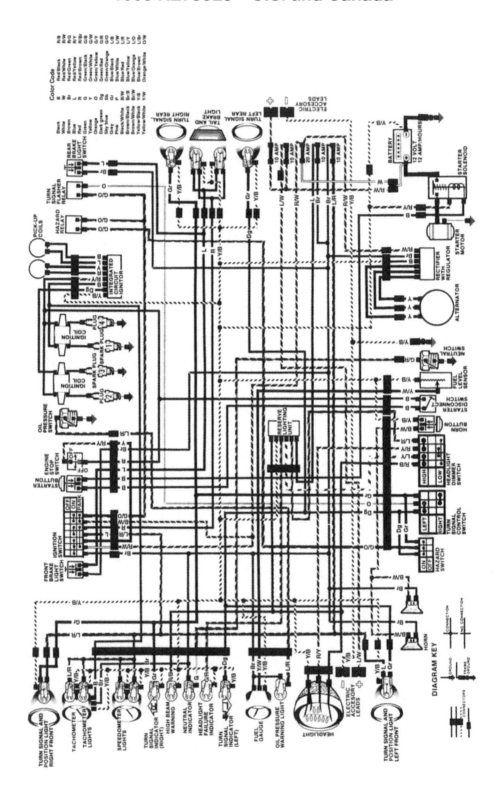

1983 Z750L3—Europe and General Export

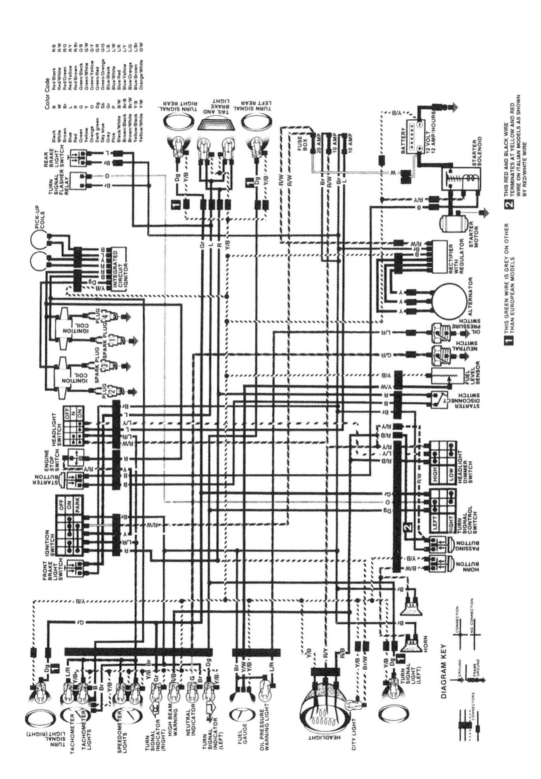

1983 ZX750A1—U.S. and Canada

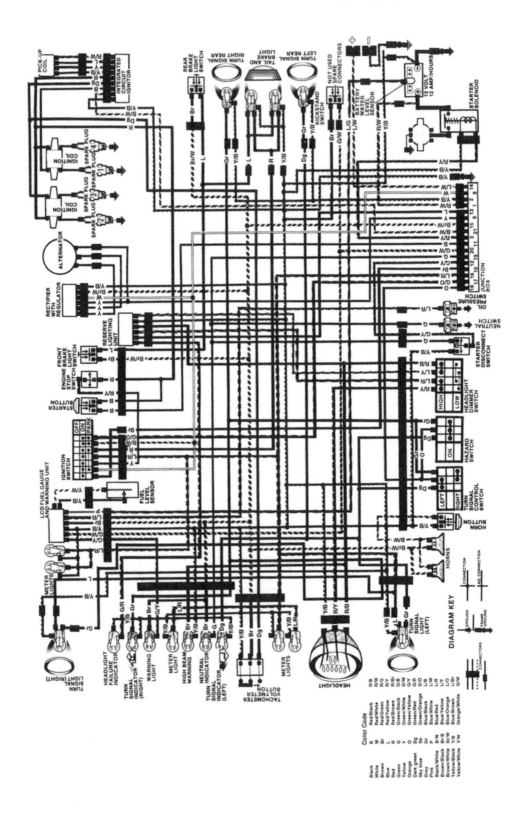

1983 ZX750A1—Europe and General Export

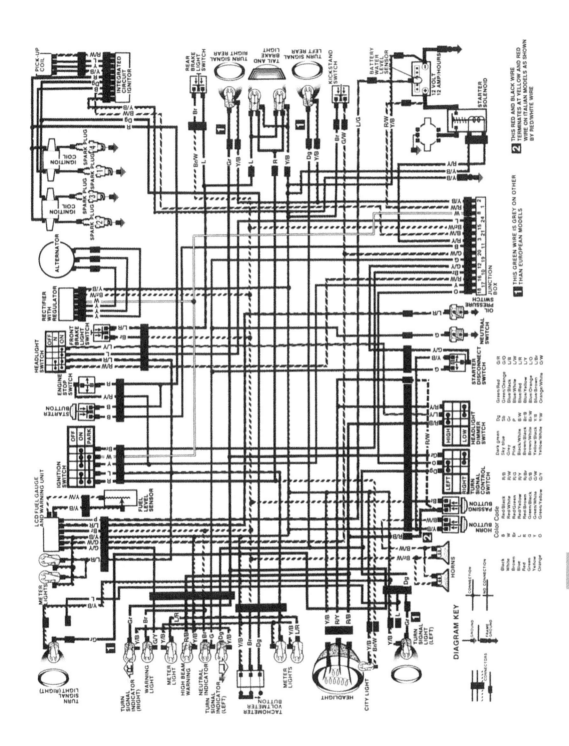

13

1984-1985 KZ750L—U.S. and Canada

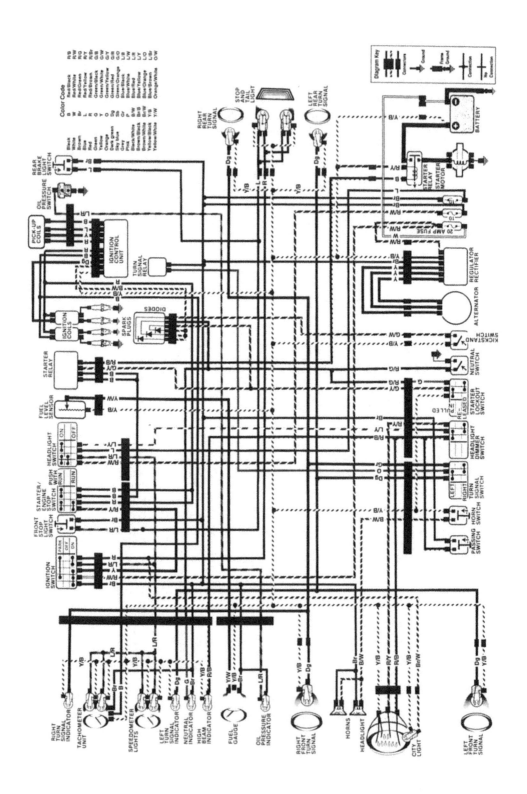

1984-1985 Z750L—Except U.S. and Canada

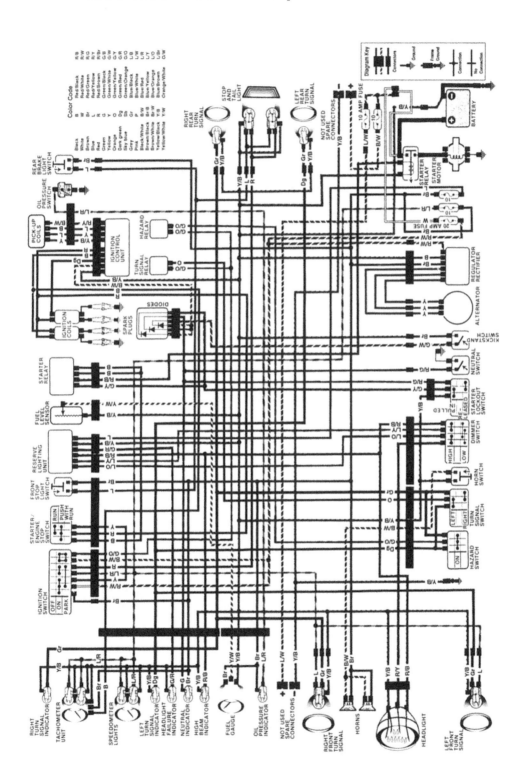

13

1984-1985 ZX750E-1—U.S. and Canada

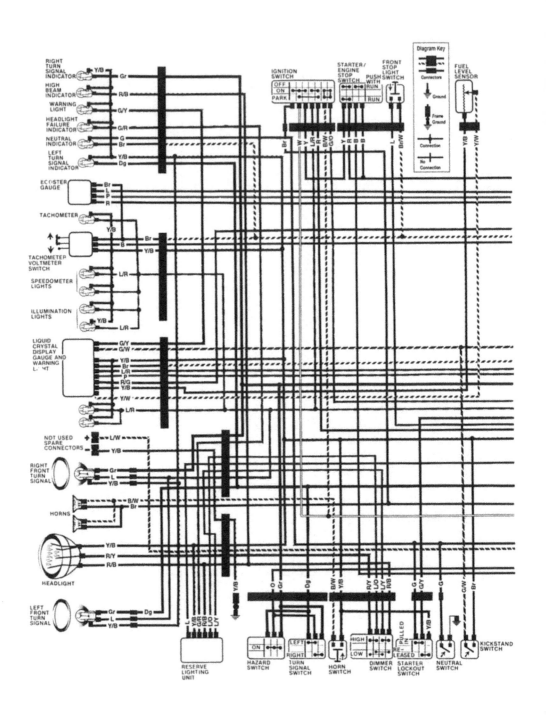

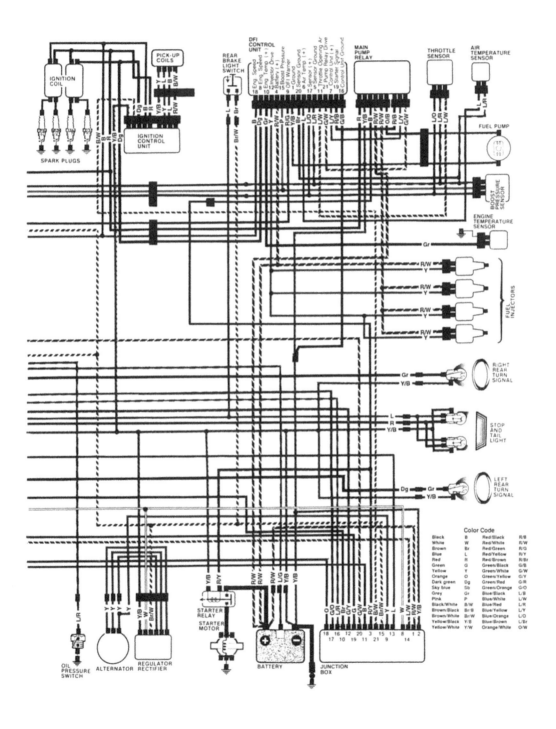

1984-1985 ZX750E-1—Except U.S. and Canada

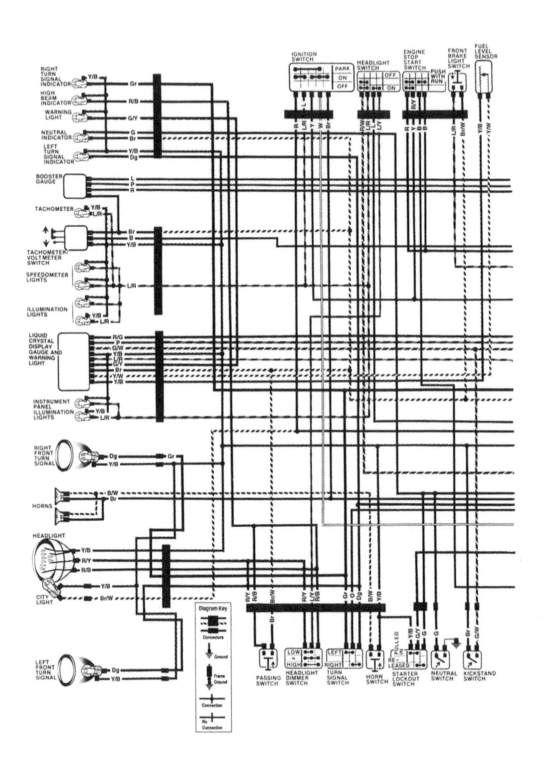

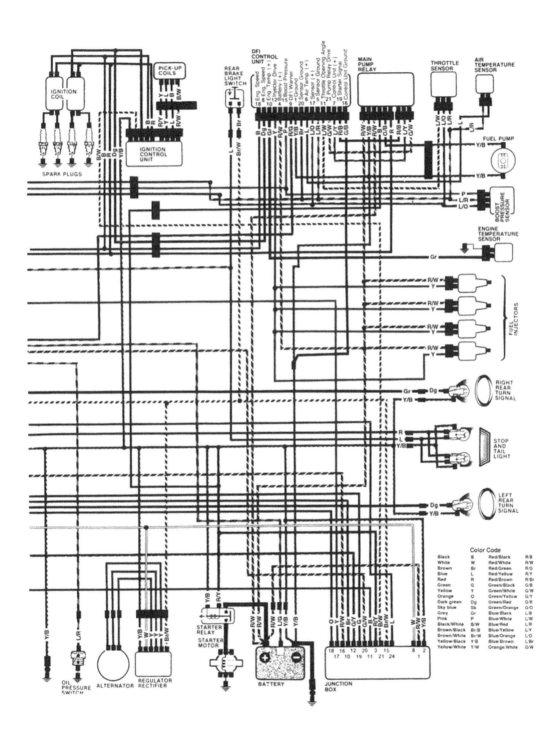

Color Code

Black	B	Red/Black	R/B
White	W	Red/White	R/W
Brown	Br	Red/Green	R/G
Blue	L	Red/Yellow	R/Y
Red	R	Red/Brown	R/Br
Green	G	Green/Black	G/B
Yellow	Y	Green/White	G/W
Orange	O	Green/Yellow	G/Y
Dark green	Dg	Green/Red	G/R
Sky blue	Sb	Green/Orange	G/O
Grey	Gr	Blue/Black	L/B
Pink	P	Blue/White	L/W
Black/White	B/W	Blue/Red	L/R
Brown/Black	Br/B	Blue/Yellow	L/Y
Brown/White	Br/W	Blue/Orange	L/O
Yellow/Black	Y/B	Blue/Brown	L/Br
Yellow/White	Y/W	Orange/White	O/W

13

MAINTENANCE LOG

Date	Miles	Type of Service